LONDON MATHEMATICAL SOCIETY LECTURE NOTE SERIES

Managing Editor: Professor M. Reid, Mathematics Institute,
University of Warwick, Coventry CV4 7AL, United Kingdom

The titles below are available from booksellers, or from Cambridge University Press at
http://www.cambridge.org/mathematics

287 Topics on Riemann surfaces and Fuchsian groups, E. BUJALANCE, A.F. COSTA & E. MARTÍNEZ (eds)
288 Surveys in combinatorics, 2001, J.W.P. HIRSCHFELD (ed)
289 Aspects of Sobolev-type inequalities, L. SALOFF-COSTE
290 Quantum groups and Lie theory, A. PRESSLEY (ed)
291 Tits buildings and the model theory of groups, K. TENT (ed)
292 A quantum groups primer, S. MAJID
293 Second order partial differential equations in Hilbert spaces, G. DA PRATO & J. ZABCZYK
294 Introduction to operator space theory, G. PISIER
295 Geometry and integrability, L. MASON & Y. NUTKU (eds)
296 Lectures on invariant theory, I. DOLGACHEV
297 The homotopy category of simply connected 4-manifolds, H.-J. BAUES
298 Higher operads, higher categories, T. LEINSTER (ed)
299 Kleinian groups and hyperbolic 3-manifolds, Y. KOMORI, V. MARKOVIC & C. SERIES (eds)
300 Introduction to Möbius differential geometry, U. HERTRICH-JEROMIN
301 Stable modules and the D(2)-problem, F.E.A. JOHNSON
302 Discrete and continuous nonlinear Schrödinger systems, M.J. ABLOWITZ, B. PRINARI & A.D. TRUBATCH
303 Number theory and algebraic geometry, M. REID & A. SKOROBOGATOV (eds)
304 Groups St Andrews 2001 in Oxford I, C.M. CAMPBELL, E.F. ROBERTSON & G.C. SMITH (eds)
305 Groups St Andrews 2001 in Oxford II, C.M. CAMPBELL, E.F. ROBERTSON & G.C. SMITH (eds)
306 Geometric mechanics and symmetry, J. MONTALDI & T. RATIU (eds)
307 Surveys in combinatorics 2003, C.D. WENSLEY (ed.)
308 Topology, geometry and quantum field theory, U.L. TILLMANN (ed)
309 Corings and comodules, T. BRZEZINSKI & R. WISBAUER
310 Topics in dynamics and ergodic theory, S. BEZUGLYI & S. KOLYADA (eds)
311 Groups: topological, combinatorial and arithmetic aspects, T.W. MÜLLER (ed)
312 Foundations of computational mathematics, Minneapolis 2002, F. CUCKER *et al* (eds)
313 Transcendental aspects of algebraic cycles, S. MÜLLER-STACH & C. PETERS (eds)
314 Spectral generalizations of line graphs, D. CVETKOVIĆ, P. ROWLINSON & S. SIMIĆ
315 Structured ring spectra, A. BAKER & B. RICHTER (eds)
316 Linear logic in computer science, T. EHRHARD, P. RUET, J. Y. GIRARD & P. SCOTT (eds)
317 Advances in elliptic curve cryptography, I.F. BLAKE, G. SEROUSSI & N.P. SMART (eds)
318 Perturbation of the boundary in boundary-value problems of partial differential equations, D. HENRY
319 Double affine Hecke algebras, I. CHEREDNIK
320 L-functions and Galois representations, D. BURNS, K. BUZZARD & J. NEKOVÁŘ (eds)
321 Surveys in modern mathematics, V. PRASOLOV & Y. ILYASHENKO (eds)
322 Recent perspectives in random matrix theory and number theory, F. MEZZADRI & N.C. SNAITH (eds)
323 Poisson geometry, deformation quantisation and group representations, S. GUTT *et al* (eds)
324 Singularities and computer algebra, C. LOSSEN & G. PFISTER (eds)
325 Lectures on the Ricci flow, P. TOPPING
326 Modular representations of finite groups of Lie type, J.E. HUMPHREYS
327 Surveys in combinatorics 2005, B.S. WEBB (ed)
328 Fundamentals of hyperbolic manifolds, R. CANARY, D. EPSTEIN & A. MARDEN (eds)
329 Spaces of Kleinian groups, Y. MINSKY, M. SAKUMA & C. SERIES (eds)
330 Noncommutative localization in algebra and topology, A. RANICKI (ed)
331 Foundations of computational mathematics, Santander 2005, L.M. PARDO, A. PINKUS, E. SÜLI & M.J. TODD (eds)
332 Handbook of tilting theory, L. ANGELERI HÜGEL, D. HAPPEL & H. KRAUSE (eds)
333 Synthetic differential geometry (2nd Edition), A. KOCK
334 The Navier–Stokes equations, N. RILEY & P. DRAZIN
335 Lectures on the combinatorics of free probability, A. NICA & R. SPEICHER
336 Integral closure of ideals, rings, and modules, I. SWANSON & C. HUNEKE
337 Methods in Banach space theory, J.M.F. CASTILLO & W.B. JOHNSON (eds)
338 Surveys in geometry and number theory, N. YOUNG (ed)
339 Groups St Andrews 2005 I, C.M. CAMPBELL, M.R. QUICK, E.F. ROBERTSON & G.C. SMITH (eds)
340 Groups St Andrews 2005 II, C.M. CAMPBELL, M.R. QUICK, E.F. ROBERTSON & G.C. SMITH (eds)
341 Ranks of elliptic curves and random matrix theory, J.B. CONREY, D.W. FARMER, F. MEZZADRI & N.C. SNAITH (eds)
342 Elliptic cohomology, H.R. MILLER & D.C. RAVENEL (eds)
343 Algebraic cycles and motives I, J. NAGEL & C. PETERS (eds)
344 Algebraic cycles and motives II, J. NAGEL & C. PETERS (eds)

345 Algebraic and analytic geometry, A. NEEMAN
346 Surveys in combinatorics 2007, A. HILTON & J. TALBOT (eds)
347 Surveys in contemporary mathematics, N. YOUNG & Y. CHOI (eds)
348 Transcendental dynamics and complex analysis, P.J. RIPPON & G.M. STALLARD (eds)
349 Model theory with applications to algebra and analysis I, Z. CHATZIDAKIS, D. MACPHERSON, A. PILLAY & A. WILKIE (eds)
350 Model theory with applications to algebra and analysis II, Z. CHATZIDAKIS, D. MACPHERSON, A. PILLAY & A. WILKIE (eds)
351 Finite von Neumann algebras and masas, A.M. SINCLAIR & R.R. SMITH
352 Number theory and polynomials, J. MCKEE & C. SMYTH (eds)
353 Trends in stochastic analysis, J. BLATH, P. MÖRTERS & M. SCHEUTZOW (eds)
354 Groups and analysis, K. TENT (ed)
355 Non-equilibrium statistical mechanics and turbulence, J. CARDY, G. FALKOVICH & K. GAWEDZKI
356 Elliptic curves and big Galois representations, D. DELBOURGO
357 Algebraic theory of differential equations, M.A.H. MACCALLUM & A.V. MIKHAILOV (eds)
358 Geometric and cohomological methods in group theory, M.R. BRIDSON, P.H. KROPHOLLER & I.J. LEARY (eds)
359 Moduli spaces and vector bundles, L. BRAMBILA-PAZ, S.B. BRADLOW, O. GARCÍA-PRADA & S. RAMANAN (eds)
360 Zariski geometries, B. ZILBER
361 Words: Notes on verbal width in groups, D. SEGAL
362 Differential tensor algebras and their module categories, R. BAUTISTA, L. SALMERÓN & R. ZUAZUA
363 Foundations of computational mathematics, Hong Kong 2008, F. CUCKER, A. PINKUS & M.J. TODD (eds)
364 Partial differential equations and fluid mechanics, J.C. ROBINSON & J.L. RODRIGO (eds)
365 Surveys in combinatorics 2009, S. HUCZYNSKA, J.D. MITCHELL & C.M. RONEY-DOUGAL (eds)
366 Highly oscillatory problems, B. ENGQUIST, A. FOKAS, E. HAIRER & A. ISERLES (eds)
367 Random matrices: High dimensional phenomena, G. BLOWER
368 Geometry of Riemann surfaces, F.P. GARDINER, G. GONZÁLEZ-DIEZ & C. KOUROUNIOTIS (eds)
369 Epidemics and rumours in complex networks, M. DRAIEF & L. MASSOULIÉ
370 Theory of p-adic distributions, S. ALBEVERIO, A.YU. KHRENNIKOV & V.M. SHELKOVICH
371 Conformal fractals, F. PRZYTYCKI & M. URBAŃSKI
372 Moonshine: The first quarter century and beyond, J. LEPOWSKY, J. MCKAY & M.P. TUITE (eds)
373 Smoothness, regularity and complete intersection, J. MAJADAS & A. G. RODICIO
374 Geometric analysis of hyperbolic differential equations: An introduction, S. ALINHAC
375 Triangulated categories, T. HOLM, P. JØRGENSEN & R. ROUQUIER (eds)
376 Permutation patterns, S. LINTON, N. RUŠKUC & V. VATTER (eds)
377 An introduction to Galois cohomology and its applications, G. BERHUY
378 Probability and mathematical genetics, N. H. BINGHAM & C. M. GOLDIE (eds)
379 Finite and algorithmic model theory, J. ESPARZA, C. MICHAUX & C. STEINHORN (eds)
380 Real and complex singularities, M. MANOEL, M.C. ROMERO FUSTER & C.T.C WALL (eds)
381 Symmetries and integrability of difference equations, D. LEVI, P. OLVER, Z. THOMOVA & P. WINTERNITZ (eds)
382 Forcing with random variables and proof complexity, J. KRAJÍČEK
383 Motivic integration and its interactions with model theory and non-Archimedean geometry I, R. CLUCKERS, J. NICAISE & J. SEBAG (eds)
384 Motivic integration and its interactions with model theory and non-Archimedean geometry II, R. CLUCKERS, J. NICAISE & J. SEBAG (eds)
385 Entropy of hidden Markov processes and connections to dynamical systems, B. MARCUS, K. PETERSEN & T. WEISSMAN (eds)
386 Independence-friendly logic, A.L. MANN, G. SANDU & M. SEVENSTER
387 Groups St Andrews 2009 in Bath I, C.M. CAMPBELL *et al* (eds)
388 Groups St Andrews 2009 in Bath II, C.M. CAMPBELL *et al* (eds)
389 Random fields on the sphere, D. MARINUCCI & G. PECCATI
390 Localization in periodic potentials, D.E. PELINOVSKY
391 Fusion systems in algebra and topology M. ASCHBACHER, R. KESSAR & B. OLIVER
392 Surveys in combinatorics 2011, R. CHAPMAN (ed)
393 Non-abelian fundamental groups and Iwasawa theory, J. COATES *et al* (eds)
394 Variational Problems in Differential Geometry, R. BIELAWSKI, K. HOUSTON & M. SPEIGHT (eds)
395 How groups grow, A. MANN
396 Arithmetic Differential Operators over the p-adic Integers, C.C. RALPH & S.R. SIMANCA
397 Hyperbolic geometry and applications in quantum Chaos and cosmology, J. BOLTE & F. STEINER (eds)
398 Mathematical models in contact mechanics, M. SOFONEA & A. MATEI
399 Circuit double cover of graphs, C.-Q. ZHANG

London Mathematical Society Lecture Note Series: 398

Mathematical Models in Contact Mechanics

M. SOFONEA
University of Perpignan

A. MATEI
University of Craiova

CAMBRIDGE
UNIVERSITY PRESS

CAMBRIDGE
UNIVERSITY PRESS

University Printing House, Cambridge CB2 8BS, United Kingdom

One Liberty Plaza, 20th Floor, New York, NY 10006, USA

477 Williamstown Road, Port Melbourne, VIC 3207, Australia

314-321, 3rd Floor, Plot 3, Splendor Forum, Jasola District Centre, New Delhi - 110025, India

103 Penang Road, #05-06/07, Visioncrest Commercial, Singapore 238467

Cambridge University Press is part of the University of Cambridge.

It furthers the University's mission by disseminating knowledge in the pursuit of education, learning and research at the highest international levels of excellence.

www.cambridge.org
Information on this title: www.cambridge.org/9781107606654

First published 2012

A catalogue record for this publication is available from the British Library

Library of Congress Cataloging in Publication data
Sofonea, Mircea.
Mathematical models in contact mechanics / M. Sofonea, A. Matei.
p. cm. – (London mathematical society lecture note series ; 398)
Includes bibliographical references and index.
ISBN 978-1-107-60665-4 (pbk.)
1. Contact mechanics – Mathematical models.
I. Matei, Andaluzia. II. Title.
TA353.S55 2004
620.1´05 – dc23 2012023155

ISBN 978-1-107-60665-4 Paperback

To the memory of my parents
(MIRCEA SOFONEA)

To my family
(ANDALUZIA MATEI)

Contents

Preface *page* **xi**

I Introduction to variational inequalities **1**

1 Preliminaries on functional analysis **3**
1.1 Normed spaces 3
1.1.1 Basic definitions 3
1.1.2 Linear continuous operators 6
1.1.3 Fixed point theorems 8
1.2 Hilbert spaces 11
1.2.1 Projection operators 11
1.2.2 Orthogonality 15
1.2.3 Duality and weak convergence 17
1.3 Elements of nonlinear analysis 19
1.3.1 Monotone operators 19
1.3.2 Convex lower semicontinuous functions 24
1.3.3 Minimization problems 29

2 Elliptic variational inequalities **33**
2.1 Variational inequalities of the first kind 33
2.1.1 Existence and uniqueness 34
2.1.2 Penalization 35

2.2 Variational inequalities of the second kind 40
2.2.1 Existence and uniqueness 40
2.2.2 A convergence result 42
2.2.3 Regularization 43
2.3 Quasivariational inequalities 49
2.3.1 The Banach fixed point argument 49
2.3.2 The Schauder fixed point argument 51
2.3.3 A convergence result 54

3 History-dependent variational inequalities 57
3.1 Nonlinear equations with history-dependent operators 57
3.1.1 Spaces of vector-valued functions 58
3.1.2 Two examples 61
3.1.3 The general case 65
3.2 History-dependent quasivariational inequalities 67
3.2.1 A basic existence and uniqueness result 67
3.2.2 A convergence result 73
3.3 Evolutionary variational inequalities 75
3.3.1 Existence and uniqueness 75
3.3.2 Convergence results 78

II Modelling and analysis of contact problems 81

4 Modelling of contact problems 83
4.1 Function spaces in contact mechanics 84
4.1.1 Preliminaries 84
4.1.2 Spaces for the displacement field 85
4.1.3 Spaces for the stress field 88
4.1.4 Spaces for piezoelectric contact problems 89
4.2 Physical setting and constitutive laws 91
4.2.1 Physical setting 91
4.2.2 Elastic constitutive laws 92
4.2.3 Viscoelastic constitutive laws 95
4.2.4 Viscoplastic constitutive laws 98
4.2.5 The von Mises convex 100
4.3 Modelling of elastic contact problems 103
4.3.1 Preliminaries 104
4.3.2 Contact conditions 104
4.3.3 Friction laws 107
4.4 Modelling of elastic-viscoplastic contact problems 111
4.4.1 Preliminaries 111
4.4.2 Contact conditions and friction laws 112

4.5 Modelling of piezoelectric contact problems 114
4.5.1 Physical setting and preliminaries 114
4.5.2 Constitutive laws 117
4.5.3 Contact conditions 119

5 Analysis of elastic contact problems 123
5.1 The Signorini contact problem 123
5.1.1 Problem statement 123
5.1.2 Existence and uniqueness 126
5.1.3 Penalization 128
5.1.4 Dual variational formulation 131
5.1.5 Minimization 137
5.1.6 One-dimensional example 139
5.2 Frictional contact problems 143
5.2.1 Statement of the problems 144
5.2.2 Existence and uniqueness 147
5.2.3 A convergence result 148
5.2.4 Regularization 149
5.2.5 Dual variational formulation 155
5.2.6 Minimization 160
5.3 A frictional contact problem with normal compliance 162
5.3.1 Problem statement 162
5.3.2 The Banach fixed point argument 164
5.3.3 The Schauder fixed point argument 166
5.3.4 Convergence results 167

6 Analysis of elastic-visco plastic contact problems 173
6.1 Bilateral frictionless contact problems 173
6.1.1 Contact of materials with short memory 174
6.1.2 Contact of materials with long memory 176
6.2 Viscoelastic contact problems with long memory 178
6.2.1 Frictionless contact with unilateral constraint 178
6.2.2 Frictional contact with normal compliance 181
6.2.3 A convergence result 183
6.3 Viscoelastic contact problems with short memory 185
6.3.1 Contact with normal compliance 186
6.3.2 Contact with normal damped response 189
6.3.3 Other frictional contact problems 192
6.3.4 Convergence results 196
6.4 Viscoplastic frictionless contact problems 200
6.4.1 Contact with normal compliance 200
6.4.2 Contact with unilateral constraint 205
6.4.3 A convergence result 208

7 Analysis of piezoelectric contact problems **217**
7.1 An electro-elastic frictional contact problem 217
7.1.1 Problem statement 218
7.1.2 Existence and uniqueness 220
7.1.3 Dual variational formulation 223
7.2 An electro-viscoelastic frictional contact problem 227
7.2.1 Problem statement 227
7.2.2 Existence and uniqueness 231
7.3 An electro-viscoplastic frictionless contact problem 237
7.3.1 Problem statement 237
7.3.2 Existence and uniqueness 241

Bibliographical notes **251**

List of symbols **257**

References **262**

Index **275**

Preface

Contact processes between deformable bodies abound in industry and everyday life and, for this reason, considerable efforts have been made in their modelling and analysis. Owing to their inherent complexity, contact phenomena lead to new and interesting mathematical models. Here and everywhere in this book by a mathematical model we mean a system of partial differential equations, associated with boundary conditions and initial conditions, eventually, which describes a specific contact process.

The purpose of this book is to introduce the reader to some representative mathematical models which arise in Contact Mechanics. Our aim is twofold: first, to present a sound and rigorous description of the way in which the mathematical models are constructed; second, to present the mathematical analysis of such models which includes the variational formulation, existence, uniqueness and convergence results. To this end, we use results on various classes of variational inequalities in Hilbert spaces, that we present in an abstract functional framework. Also, we use various functional methods, including monotonicity, compactness, penalization, regularization and duality methods. Moreover, we pay particular attention to the mechanical interpretation of our results and, in this way, we illustrate the cross fertilization between modelling and applications on the one hand, and nonlinear analysis on the other hand.

This book is intended as a unified and readily accessible source for graduate students, as well as mathematicians, engineers and scientists. Its reading requires only basic knowledge of linear algebra, general topology, functional analysis and mechanics of continua.

The book is divided into two parts and seven chapters.

Part I contains Chapters 1–3 and represents a brief introduction to the study of variational inequalities. The material presented here was selected with emphasis on useful mathematical tools needed in the study of contact problems.

Part II contains Chapters 4–7 and represents the main part of the book. It is devoted to the modelling and analysis of representative frictionless or frictional contact problems and includes original results.

A brief description of the chapters of the book follows.

Chapter 1 is devoted to preliminaries on functional analysis which are fundamental to the developments later in this book. The material presented in this chapter is standard and can be found in many textbooks and monographs. For this reason we provide proofs only for the results which are frequently used in the rest of the book, i.e. the Banach fixed point theorem, the projection lemma, the Riesz representation theorem and the Weierstrass theorem, among others.

Chapter 2 is devoted to the study of several classes of elliptic variational inequalities with strongly monotone Lipschitz continuous operators. For each class of inequalities, we establish existence and uniqueness results for the solution by using arguments of monotonicity, fixed point or compactness. We also prove existence results by using penalization and regularization arguments. And, sometimes, we state and prove convergence results.

Chapter 3 deals with the study of a special class of variational inequalities, the so-called history-dependent variational inequalities, and is based on our original research. We present existence, uniqueness and convergence results, then we particularize them in the study of various evolutionary variational inequalities. Our aim in this chapter is to introduce a general framework in which a large number of quasistatic contact problems can be cast, and to construct mathematical tools which are useful in the study of these problems.

Chapter 4 introduces preliminary material needed in the study of contact problems. Here we introduce the spaces of functions used in Contact Mechanics, together with their main properties. We also present basic notions on modelling of contact problems, including the constitutive laws, the contact conditions and the friction laws used in the rest of the book. Finally, we pay particular attention to the modelling of contact processes with piezoelectric materials.

Chapter 5 is devoted to the study of static frictionless or frictional contact problems with nonlinearly elastic materials. The contact is either bilateral or is modelled with normal compliance or with the Signorini condition. Friction is modelled with versions of Coulomb's law, in which the friction bound is either given or depends on the normal stress. For each of the problems studied in this chapter we provide a variational formulation, which is in the form of an elliptic variational or quasivariational inequality for the displacement field. Then we use the abstract results of Chapter 2 in order to establish existence, uniqueness and convergence results.

Chapter 6 deals with quasistatic frictionless and frictional contact problems. We model the material's behavior with viscoelastic or viscoplastic constitutive laws and, in the case of viscoelastic constitutive laws, we consider the cases of both short and long memory. The contact is either bilateral or is modelled with normal compliance with or without unilateral constraint, or with the so-called normal damped response condition. Friction is modelled with versions of Coulomb's law, in which the friction bound is either given or depends on the normal stress. For each of the problems studied in this chapter we provide a variational formulation, which is in the form of a history-dependent or evolutionary variational inequality for the displacement or the velocity field. Then we use the abstract results of Chapter 3 in order to establish existence, uniqueness and convergence results.

Chapter 7 deals with static and quasistatic frictionless and frictional contact problems for piezoelectric materials. We model the material's behavior with electro-elastic, electro-viscoelastic or electro-viscoplastic constitutive laws. The contact is either bilateral or is modelled with normal compliance. The foundation is assumed to be either electrically conductive or an insulator. For each of the problems studied in this chapter we provide a variational formulation, which is in the form of a system of variational inequalities for the displacement and the electric potential field. Then we use the abstract results of Chapters 2 and 3 in order to establish existence and uniqueness results.

Most of the results presented in this book can be extended to more general cases; however, since our aim is to provide an accessible presentation of the mathematical models which arise in the study of contact problems, we restrict ourselves to some introductory problems and we avoid giving the results in the most general abstract form, so that it is easier for the reader to understand more clearly the essential ideas involved. For instance, in the study of the variational inequalities presented in Part I of the book we restrict ourselves to the Hilbertian case, convex functions and strongly monotone Lipschitz continuous operators; also, in the study of history-dependent inequalities we avoid using the Bochner–Sobolev spaces $W^{k,p}(0,T;X)$ and we restrict ourselves to using spaces of continuous and continuously differentiable functions defined on the time interval $[0,T]$ with values in X. Nevertheless, we refer the reader to the section of bibliographical notes at the end of the book, where we provide references in which more information on the topics related to the body of the text can be found.

The present book is a result of cooperation between the authors during the past several years and was supported by the CNRS (France) and the Romanian Academy, under the *LEA Math-Mode* program, as well as the Romanian National Authority for Scientific Research (CNCS – UEFiSCDI) within the project PN-II-RU-TE-2011-3-0223. Its contents are based on our recent research as well as on various post-graduate courses of Applied Mathematics and Contact Mechanics we delivered in several universities in France, Poland, Romania and Spain. In writing this book we have benefited from many individuals to whom we express our thanks. We especially thank Professors Mikaël Barboteu, Weimin Han, Jiří Jarušek, Constantin Niculescu, Stanisław Migórski, Anna Ochal, Vicenţiu Rădulescu, Meir Shillor, Juan M. Viaño and Barbara Wohlmuth for their beneficial cooperation and for their constant support. We extend our thanks to Joëlle Sulian who prepared the figures of this book. Finally, we express our gratitude to the staff of Cambridge University Press for their help in producing the book in your hand.

Perpignan, France — *Mircea Sofonea*
Craiova, Romania — *Andaluzia Matei*

Part I

Introduction to variational inequalities

1

Preliminaries on functional analysis

This chapter presents preliminary material from functional analysis which will be used in subsequent chapters. Some of the results are stated without proofs, since they are standard and can be found in many references. Nevertheless, we pay particular attention to the results which are repeatedly used in the following chapters of the book, for which we present the details in proofs. They include the Banach fixed point theorem, the projection lemma, the Riesz representation theorem and the Weiersterass theorem, among others. All the linear spaces considered in this book including abstract normed spaces, Banach spaces, Hilbert spaces and various function spaces are assumed to be real linear spaces.

1.1 Normed spaces

We start this section with basic definitions, notation and results concerning the normed spaces. Then we recall two main fixed point results: the Banach fixed point theorem and the Schauder fixed point theorem.

1.1.1 Basic definitions

Given a linear space X, we recall that a *norm* $\|\cdot\|_X$ is a function from X to $\mathbb{R}$ with the following properties.

(1) $\|u\|_X \geq 0 \ \ \forall\, u \in X$, and $\|u\|_X = 0$ iff $u = 0_X$.

(2) $\|\alpha u\|_X = |\alpha| \, \|u\|_X \quad \forall u \in X, \forall \alpha \in \mathbb{R}.$

(3) $\|u + v\|_X \leq \|u\|_X + \|v\|_X \quad \forall u, v \in X.$

The pair $(X, \|\cdot\|_X)$ is called a *normed space*. Here and everywhere in this book 0_X will denote the zero element of X. Also, we will simply say X is a normed space when the definition of the norm is understood from the context.

On a linear space various norms can be defined. Sometimes it is desirable to know if two norms are related. Let $\|\cdot\|^{(1)}$ and $\|\cdot\|^{(2)}$ be two norms over a linear space X. The two norms are said to be *equivalent* if there exist two constants $c_1, c_2 > 0$ such that

$$c_1 \|u\|^{(1)} \leq \|u\|^{(2)} \leq c_2 \|u\|^{(1)} \quad \forall u \in X. \tag{1.1}$$

We recall that a sequence $\{u_n\} \subset X$ is said to *converge* (strongly) to $u \in X$ if

$$\|u_n - u\|_X \to 0 \quad \text{as } n \to \infty. \tag{1.2}$$

In this case u is called the (strong) *limit* of the sequence $\{u_n\}$ and we write

$$u = \lim_{n\to\infty} u_n \quad \text{or} \quad u_n \to u \quad \text{in } X.$$

It is straightforward to verify that the limit of a sequence, if it exists, is unique. The adjective "strong" is introduced in the previous definition to distinguish this convergence from the weak convergence which will be introduced on page 6.

A sequence $\{u_n\} \subset X$ is said to be *bounded* if there exists $M > 0$ such that

$$\|u_n\|_X \leq M \qquad \forall n \in \mathbb{N} \tag{1.3}$$

or, equivalently, if

$$\sup_n \|u_n\|_X < \infty.$$

To test the convergence of a sequence without knowing its limit, it is usually convenient to refer to the notion of a Cauchy sequence. Let X be a normed space. A sequence $\{u_n\} \subset X$ is called a *Cauchy sequence* if

$$\|u_m - u_n\|_X \to 0 \quad \text{as } m, n \to \infty.$$

Obviously, a convergent sequence is a Cauchy sequence, but in a general infinite dimensional space, a Cauchy sequence may fail to converge. This justifies the following definition.

Definition 1.1 A normed space is said to be *complete* if every Cauchy sequence from the space converges to an element in the space. A complete normed space is called a *Banach space*.

From this definition it follows that $(X, \|\cdot\|_X)$ is a Banach space iff for every Cauchy sequence $\{u_n\} \subset X$ there exists an element $u \in X$ such that $u_n \to u \quad$ in X. Moreover, using (1.1) and (1.2) it is easy to see that, for two equivalent norms, convergence in one norm implies convergence in the other norm. As a consequence, if $\|\cdot\|^{(1)}$ and $\|\cdot\|^{(2)}$ are equivalent norms on the linear space X then $(X, \|\cdot\|^{(1)})$ is a Banach space iff $(X, \|\cdot\|^{(2)})$ is a Banach space.

We introduce in what follows a particular type of normed space, in which the norm is defined in a special way. Given a linear space X we recall that an *inner product* $(\cdot,\cdot)_X$ is a function from $X \times X$ to $\mathbb{R}$ with the following properties.

(1) $(u,u)_X \geq 0 \quad \forall\, u \in X$, and $(u,u)_X = 0$ iff $u = 0_X$.

(2) $(u,v)_X = (v,u)_X \quad \forall\, u,v \in X$.

(3) $(\alpha\, u + \beta\, v, w)_X = \alpha\,(u,w)_X + \beta\,(v,w)_X \qquad \forall\, u,v,w \in X,\ \forall\, \alpha,\beta \in \mathbb{R}$.

The pair $(X, (\cdot,\cdot)_X)$ is called an *inner product space*. When the definition of the inner product $(\cdot,\cdot)_X$ is clear from the context, we simply say X is an inner product space.

Next, it is well known that an inner product $(\cdot,\cdot)_X$ induces a norm through the formula

$$\|u\|_X = \sqrt{(u,u)_X} \qquad \forall\, u \in X \tag{1.4}$$

and we note that everywhere in this book the norm in an inner product space is the one induced by the inner product through the above formula. For an inner product space we have the *Cauchy–Schwarz inequality*:

$$|(u,v)_X| \leq \|u\|_X \|v\|_X \qquad \forall\, u,v \in X, \tag{1.5}$$

with the equality holding iff u and v are linearly dependent. Moreover, the following identity holds:

$$\|u+v\|_X^2 + \|u-v\|_X^2 = 2\left(\|u\|_X^2 + \|v\|_X^2\right) \qquad \forall\, u,v \in X. \tag{1.6}$$

Identity (1.6) is called the *parallelogram identity* or the *parallelogram law*.

Among the inner product spaces, of particular importance are the Hilbert spaces.

Definition 1.2 A complete inner product space is called a *Hilbert space*.

From the definition, we see that an inner product space X is a Hilbert space if X is a Banach space under the norm induced by the inner product.

1.1.2 Linear continuous operators

Let $(X, \|\cdot\|_X)$ and $(Y, \|\cdot\|_Y)$ be two normed spaces and let $L : X \to Y$ be an operator. We recall that $L : X \to Y$ is *linear* if

$$L(\alpha_1 v_1 + \alpha_2 v_2) = \alpha_1 L(v_1) + \alpha_2 L(v_2) \quad \forall v_1, v_2 \in X, \ \forall \alpha_1, \alpha_2 \in \mathbb{R}.$$

The operator L is said to be *continuous* if

$$u_n \to u \ \text{ in } X \implies L(u_n) \to L(u) \ \text{ in } Y.$$

It can be proved that, if L is linear, then L is continuous if and only if it is *bounded*, i.e., there exists $M > 0$ such that

$$\|L(v)\|_Y \leq M\|v\|_X \quad \forall v \in X.$$

We will use the notation $\mathcal{L}(X, Y)$ for the set of all linear continuous operators from X to Y. For $L \in \mathcal{L}(X, Y)$, the quantity

$$\|L\|_{\mathcal{L}(X,Y)} = \sup_{0 \neq v \in X} \frac{\|Lv\|_Y}{\|v\|_X} \tag{1.7}$$

is called the *operator norm* of L, and $L \mapsto \|L\|_{\mathcal{L}(X,Y)}$ defines a norm on the space $\mathcal{L}(X, Y)$. Moreover, if Y is a Banach space then $\mathcal{L}(X, Y)$ is also a Banach space. For a linear operator L, we usually write $L(v)$ as Lv, but sometimes we also write Lv even when L is not linear.

For a normed space X, the space $\mathcal{L}(X, \mathbb{R})$ is called the *dual* space of X and is denoted by X'. The elements of X' are linear continuous functionals on X. Recall that a linear functional $\ell : X \to \mathbb{R}$ belongs to X' iff it is *continuous*, i.e.

$$u_n \to u \ \text{ in } X \implies \ell(u_n) \to \ell(u) \ \text{ in } \mathbb{R}$$

or, equivalently, if it is *bounded*, i.e. there exists $M > 0$ such that

$$|\ell(v)| \leq M\|v\|_X \quad \forall v \in X.$$

The duality pairing between X' and X is usually denoted by $\ell(v)$ or $\langle v', v\rangle$ or $\langle v', v\rangle_{X' \times X}$ for $\ell, v' \in X'$ and $v \in X$. It follows from (1.7) that a norm on X' is

$$\|\ell\|_{X'} = \sup_{0 \neq v \in X} \frac{|\ell(v)|}{\|v\|_X}. \tag{1.8}$$

Moreover, $(X', \|\cdot\|_{X'})$ is always a Banach space.

We can now introduce another kind of convergence in a normed space. A sequence $\{u_n\} \subset X$ is said to *converge weakly* to $u \in X$ if for each $\ell \in X'$,

$$\ell(u_n) \to \ell(u) \quad \text{as } n \to \infty.$$

In this case u is called the *weak limit* of $\{u_n\}$ and we write

$$u_n \rightharpoonup u \quad \text{in } X.$$

It follows from the Hahn–Banach theorem that the weak limit of a sequence, if it exists, is unique. Moreover, it is easy to see that strong convergence implies weak convergence, i.e., if $u_n \to u$ in X, then $u_n \rightharpoonup u$ in X. The converse of this property is not true in general.

Assume now that $(X, (\cdot,\cdot)_X)$ is an inner product space and note that in this case $u \mapsto (u,v)_X$ is a linear continuous functional on X, for all $v \in X$. Therefore, it follows from the definition of the weak convergence that

$$u_n \rightharpoonup u \quad \text{in } X \implies (u_n, v)_X \to (u,v)_X \quad \text{as } n \to \infty, \quad \forall\, v \in X. \tag{1.9}$$

We shall see in Section 1.2 that, in the case when $(X, (\cdot,\cdot)_X)$ is a Hilbert space, the converse of (1.9) is true.

The convergence of sequences is used to define closed subsets in a normed space.

Definition 1.3 Let X be a normed space. A subset $K \subset X$ is called:

(i) *(strongly) closed* if the limit of each convergent sequence of elements of K belongs to K, that is

$$\{u_n\} \subset K, \quad u_n \to u \quad \text{in } X \implies u \in K;$$

(ii) *weakly closed* if the limit of each weakly convergent sequence of elements of K belongs to K, that is

$$\{u_n\} \subset K, \quad u_n \rightharpoonup u \quad \text{in } X \implies u \in K.$$

Evidently, every weakly closed subset of X is (strongly) closed, but the converse is not true, in general. An exception is provided by the class of convex subsets of a Banach space, as shown in the following result.

Theorem 1.4 (Mazur's theorem) *A convex subset of a Banach space is (strongly) closed if and only if it is weakly closed.*

Here, for the convenience of the reader, we recall that a subset K of a linear space is said to be *convex* if it has the property

$$u, v \in K \implies (1-t)\,u + t\,v \in K \quad \forall\, t \in [0,1].$$

Let X and Y be linear spaces. A mapping $a : X \times Y \to \mathbb{R}$ is called a *bilinear form* if it is linear in each argument, that is for any u_1, u_2, $u \in X$, v_1, v_2, $v \in Y$, and α_1, $\alpha_2 \in \mathbb{R}$,

$$a(\alpha_1 u_1 + \alpha_2 u_2, v) = \alpha_1\, a(u_1, v) + \alpha_2\, a(u_2, v),$$
$$a(u, \alpha_1 v_1 + \alpha_2 v_2) = \alpha_1\, a(u, v_1) + \alpha_2\, a(u, v_2).$$

For the case $X = Y$ we say that a bilinear form is *symmetric* if

$$a(u,v) = a(v,u) \qquad \forall\, u\,, v \in X.$$

Let now $(X, \|\cdot\|_X)$ and $(Y, \|\cdot\|_Y)$ be two normed spaces. A bilinear form $a : X \times Y \to \mathbb{R}$ is said to be *continuous* if there exists a constant $M > 0$ such that

$$a(u,v) \leq M\, \|u\|_X \|v\|_Y \quad \forall\, u \in X,\ \forall\, v \in Y.$$

For the case $X = Y$ we say that a bilinear form is *positive* if

$$a(u,u) \geq 0 \qquad \forall\, u \in X$$

and X-*elliptic* if there exists a constant $m > 0$ such that

$$a(u,u) \geq m\, \|u\|_X^2 \qquad \forall\, u \in X.$$

It is easy to see that if the bilinear form a is X-elliptic then it is positive. The converse of this property is not true, in general.

1.1.3 Fixed point theorems

Let X be a Banach space with the norm $\|\cdot\|_X$, and K a subset of X. Also, let $\Lambda : K \to X$ be an operator defined on K. We are interested in the existence of a solution $u \in K$ of the operator equation

$$\Lambda u = u. \tag{1.10}$$

An element $u \in K$ which satisfies (1.10) is called a *fixed point* of the operator Λ.

We introduce in what follows two main theorems which state the existence of the fixed points of nonlinear operators: the Banach fixed point theorem and the Schauder fixed point theorem. Both of them represent major results of functional analysis.

Theorem 1.5 (The Banach fixed point theorem) *Let K be a nonempty closed subset of a Banach space $(X, \|\cdot\|_X)$. Assume that $\Lambda : K \to K$ is a* contraction, *i.e. there exists a constant $\alpha \in [0,1)$ such that*

$$\|\Lambda u - \Lambda v\|_X \leq \alpha\, \|u - v\|_X \quad \forall\, u, v \in K. \tag{1.11}$$

Then there exists a unique $u \in K$ such that $\Lambda u = u$.

Proof Let $u_0 \in K$ be an arbitrary element of K and let $\{u_n\}$ be the sequence defined by

$$u_{n+1} = \Lambda u_n \qquad \forall\, n = 0, 1, 2, \ldots$$

Since $\Lambda : K \to K$, the sequence $\{u_n\}$ is well-defined. Let us first prove that $\{u_n\}$ is a Cauchy sequence. Using the contractivity of the operator Λ, we have

$$\|u_{n+1} - u_n\|_X \leq \alpha \|u_n - u_{n-1}\|_X \leq \cdots \leq \alpha^n \|u_1 - u_0\|_X.$$

Then for any $m > n \geq 1$,

$$\begin{aligned} \|u_m - u_n\|_X &\leq \sum_{j=0}^{m-n-1} \|u_{n+j+1} - u_{n+j}\|_X \\ &\leq \sum_{j=0}^{m-n-1} \alpha^{n+j} \|u_1 - u_0\|_X \\ &\leq \frac{\alpha^n}{1-\alpha} \|u_1 - u_0\|_X. \end{aligned}$$

Since $\alpha \in [0, 1)$, it follows from the previous inequalities that

$$\|u_m - u_n\|_X \to 0 \quad \text{as} \quad m, n \to \infty.$$

Thus, $\{u_n\}$ is a Cauchy sequence and has a limit $u \in K$, since K is a closed subset of the Banach space X. Moreover, since $u_n \to u$ in X, it follows from (1.11) that $\Lambda u_n \to \Lambda u$ in X. Therefore, taking the limit in $u_{n+1} = \Lambda u_n$ it follows that $u = \Lambda u$, which concludes the existence part of the theorem.

Suppose now that $u_1, u_2 \in K$ are fixed points of Λ. Then from $u_1 = \Lambda u_1$ and $u_2 = \Lambda u_2$ we obtain

$$u_1 - u_2 = \Lambda u_1 - \Lambda u_2.$$

Hence, (1.11) yields

$$\|u_1 - u_2\|_X = \|\Lambda u_1 - \Lambda u_2\|_X \leq \alpha \|u_1 - u_2\|_X,$$

which implies $\|u_1 - u_2\|_X = 0$, since $\alpha \in [0, 1)$. So, if Λ has a fixed point it follows that this fixed point is unique, which concludes the proof. ∎

We also need a version of the Banach fixed point theorem which we recall in what follows. To this end, for an operator Λ, we define its powers inductively by the formula $\Lambda^m = \Lambda(\Lambda^{m-1})$ for $m \geq 2$.

Theorem 1.6 *Assume that K is a nonempty closed subset of a Banach space X and let $\Lambda : K \to K$. Assume also that $\Lambda^m : K \to K$ is a contraction for some positive integer m. Then Λ has a unique fixed point.*

Proof By Theorem 1.5, the mapping Λ^m has a unique fixed point $u \in K$. From

$$\Lambda^m(u) = u,$$

we obtain

$$\Lambda^m(\Lambda(u)) = \Lambda(\Lambda^m(u)) = \Lambda u.$$

Thus $\Lambda u \in K$ is also a fixed point of Λ^m. Since Λ^m has a unique fixed point, we must have

$$\Lambda u = u,$$

i.e. u is a fixed point of Λ. The uniqueness of a fixed point of Λ follows easily from that of Λ^m. ■

We turn now to the Schauder fixed point theorem. To introduce it we need the following preliminaries.

Definition 1.7 Let X be a normed space. A subset $K \subset X$ is called:

(i) *bounded* if there exists $M > 0$ such that

$$\|u\|_X \leq M \qquad \forall u \in K;$$

(ii) *relatively sequentially compact* if each sequence in K has a convergent subsequence in X.

It follows from Definition 1.7 (i) that if $K \subset X$ is bounded, then every sequence $\{u_n\} \subset K$ satisfies (1.3) and, therefore, is bounded. Moreover, Definition 1.7 (ii) shows that a subset $K \subset X$ is relatively sequentially compact if for each $\{u_n\} \subset K$ there exists a subsequence $\{u_{n_k}\}$ and an element $u \in X$ such that $u_{n_k} \to u$ in X. Note also that, for simplicity, everywhere below we shall use the terminology *relatively compact* instead of *relatively sequentially compact.*

Definition 1.8 Let X and Y be normed spaces. The operator $\Lambda : K \subset X \to Y$ is called:

(i) *continuous at the point* $u \in K$ if for each sequence $\{u_n\} \subset K$ which converges in X to u, the sequence $\{\Lambda u_n\} \subset Y$ converges to Λu in Y, that is

$$\{u_n\} \subset K, \quad u_n \to u \quad \text{in } X \implies \Lambda u_n \to \Lambda u \quad \text{in } Y;$$

(ii) *continuous* if it is continuous at each point $u \in K$;

(iii) *compact* if it is continuous and maps bounded sets into relatively compact sets.

It follows from Definitions 1.8 (iii) and 1.7 (ii) that a continuous operator $\Lambda : K \subset X \to Y$ is compact if and only if for each bounded sequence $\{u_n\} \subset K$ there exists a subsequence $\{u_{n_k}\}$ such that the sequence $\{\Lambda u_{n_k}\}$ is convergent in Y.

We proceed with the following result.

Theorem 1.9 (The Schauder fixed point theorem) *Let K be a nonempty closed convex bounded subset of a Banach space X and let $\Lambda : K \to X$ be a compact operator such that $\Lambda(K) \subset K$. Then Λ has at least one fixed point.*

The proof of Theorem 1.9 requires a number of preliminary results and, therefore, we skip it. Nevertheless, we indicate that such a proof can be found in [155, p. 56]. Also, note that, unlike the Banach fixed point theorem, the Schauder fixed point theorem does not provide the uniqueness of the fixed point of the operator Λ. We shall use Theorem 1.9 in Section 2.3 in order to prove the solvability of a class of elliptic quasivariational inequalities.

1.2 Hilbert spaces

We introduce in what follows some useful results which are valid in Hilbert spaces. This concerns the projection operators, some properties related to orthogonality and the Riesz representation theorem, together with its consequences.

1.2.1 *Projection operators*

The projection operators represent an important class of nonlinear operators defined in a Hilbert space; to introduce them we need the following existence and uniqueness result.

Theorem 1.10 (The projection lemma) *Let K be a nonempty closed convex subset of a Hilbert space X. Then, for each $f \in X$ there exists a unique element $u \in K$ such that*

$$\|u - f\|_X = \min_{v \in K} \|v - f\|_X. \tag{1.12}$$

Proof Let $f \in X$, denote

$$d = \inf_{v \in K} \|v - f\|_X, \tag{1.13}$$

and let $\{u_n\}$ be a sequence of elements of K such that

$$\|u_n - f\|_X \to d \qquad \text{as} \quad n \to \infty. \tag{1.14}$$

Since K is a convex subset of X we deduce that $\dfrac{u_m + u_n}{2} \in K$ for all $m,\ n \in \mathbb{N}$ and, therefore, (1.13) implies that

$$\left\|\frac{u_m + u_n}{2} - f\right\|_X \ge d \qquad \forall\, m,\ n \in \mathbb{N}. \tag{1.15}$$

We use identity (1.6) with $u = u_m - f$ and $v = u_n - f$ to see that

$$4\left\|\frac{u_m + u_n}{2} - f\right\|_X^2 + \|u_m - u_n\|_X^2 = 2\left(\|u_m - f\|_X^2 + \|u_n - f\|_X^2\right)$$

for all $m,\ n \in \mathbb{N}$ and, therefore, (1.15) yields

$$\|u_m - u_n\|_X^2 \leq 2\left(\|u_m - f\|_X^2 + \|u_n - f\|_X^2\right) - 4d^2 \tag{1.16}$$

for all $m,\ n \in \mathbb{N}$. We combine now (1.14) and (1.16) to obtain

$$\|u_m - u_n\|_X \to 0 \qquad \text{as} \quad m,\ n \to \infty,$$

which shows that $\{u_n\}$ is a Cauchy sequence in X. Therefore, since X is a Hilbert space and K is a closed subset of X, there exists an element $u \in K$ such that

$$\|u_n - u\|_X \to 0 \quad \text{as} \quad n \to \infty. \tag{1.17}$$

We use (1.13) again to see that

$$d \leq \|u - f\|_X \leq \|u - u_n\|_X + \|u_n - f\|_X \quad \forall\, n \in \mathbb{N}$$

and, by (1.14) and (1.17), we obtain that

$$d \leq \|u - f\|_X \leq d.$$

We conclude that $\|u - f\|_X = d$ which proves the existence part of the theorem.

Assume now that u and v are two elements of K such that

$$\|u - f\|_X = \|v - f\|_X = d = \inf_{w \in K} \|w - f\|_X. \tag{1.18}$$

By the convexity of K we have $\dfrac{u+v}{2} \in K$ and, therefore, (1.18) yields

$$\begin{aligned} d \leq \left\|\frac{u+v}{2} - f\right\|_X &= \left\|\frac{1}{2}(u - f) + \frac{1}{2}(v - f)\right\|_X \\ &\leq \frac{1}{2}\|u - f\|_X + \frac{1}{2}\|v - f\|_X = d. \end{aligned}$$

It follows from this inequality that

$$\left\|\frac{u+v}{2} - f\right\|_X = d. \tag{1.19}$$

Using now the identity (1.6) with $u - f$ and $v - f$ instead of u and v, respectively, we have

$$\|u - v\|_X^2 = 2\left(\|u - f\|_X^2 + \|v - f\|_X^2\right) - 4\left\|\frac{u+v}{2} - f\right\|_X^2. \tag{1.20}$$

We now combine (1.18)–(1.20) to see that $\|u - v\|_X^2 = 0$, which implies that $u = v$ and concludes the proof of the theorem. ■

Theorem 1.10 allows us to introduce the following definition.

Definition 1.11 Let K be a nonempty closed convex subset of a Hilbert space X. Then, for each $f \in X$ the element u which satisfies (1.12) is called the *projection* of f on K and is usually denoted $\mathcal{P}_K f$. Moreover, the operator $\mathcal{P}_K : X \to K$ is called the *projection operator* on K.

It follows from Definition 1.11 that

$$f = \mathcal{P}_K f \iff f \in K. \tag{1.21}$$

We conclude from (1.21) that the element $f \in X$ is a fixed point of the projection operator $\mathcal{P}_K$ iff $f \in K$.

Next, we present the following characterization of the projection.

Proposition 1.12 *Let K be a nonempty closed convex subset of a Hilbert space X and let $f \in X$. Then $u = \mathcal{P}_K f$ if and only if*

$$u \in K, \quad (u, v-u)_X \geq (f, v-u)_X \quad \forall\, v \in K. \tag{1.22}$$

Proof Assume that $u = \mathcal{P}_K f$ and let $v \in K$. Then, by the definition of the projection it follows that $u \in K$ and, for all $t \in (0,1)$, we have

$$\begin{aligned}\|u-f\|_X^2 &\leq \|(1-t)u + tv - f\|_X^2 = \|(u-f) + t(v-u)\|_X^2 \\ &= \|u-f\|_X^2 + 2t\,(u-f, v-u)_X + t^2\|v-u\|_X^2.\end{aligned}$$

This inequality implies that

$$2\,(u-f, v-u)_X + t\,\|v-u\|_X^2 \geq 0 \qquad \forall\, t \in (0,1).$$

We pass to the limit in the previous inequality as $t \to 0$ to obtain (1.22).

Conversely, assume that u satisfies (1.22) and, again, let $v \in K$. Using (1.4) we have

$$\|v-f\|_X^2 - \|u-f\|_X^2 = \|v\|_X^2 - \|u\|_X^2 - 2\,(f, v-u)_X$$

and, using (1.22), it follows that

$$\|v-f\|_X^2 - \|u-f\|_X^2 \geq \|v\|_X^2 - \|u\|_X^2 - 2\,(u, v-u)_X = \|v-u\|_X^2 \geq 0.$$

The last inequality implies that

$$\|u-f\|_X \leq \|v-f\|_X$$

and, since $u \in K$ and v is an arbitrary element in K, we deduce that u satisfies (1.12). Therefore, by Definition 1.11 it follows that $u = \mathcal{P}_K f$, which concludes the proof. ∎

Note that, besides the characterization of the projection in terms of inequalities, Proposition 1.12 provides, implicitly, the existence of a unique solution to the inequality (1.22). Moreover, using this proposition it is easy to prove the following results.

Proposition 1.13 *Let K be a nonempty closed convex subset of a Hilbert space X. Then the projection operator $\mathcal{P}_K$ satisfies the following inequalities:*

$$(\mathcal{P}_K u - \mathcal{P}_K v, u - v)_X \geq 0 \quad \forall\, u, v \in X, \tag{1.23}$$
$$\|\mathcal{P}_K u - \mathcal{P}_K v\|_X \leq \|u - v\|_X \quad \forall\, u, v \in X. \tag{1.24}$$

Proof Let $u,\ v \in X$. We use (1.22) to obtain

$$(\mathcal{P}_K u, \mathcal{P}_K v - \mathcal{P}_K u)_X \geq (u, \mathcal{P}_K v - \mathcal{P}_K u)_X,$$
$$(\mathcal{P}_K v, \mathcal{P}_K u - \mathcal{P}_K v)_X \geq (v, \mathcal{P}_K u - \mathcal{P}_K v)_X.$$

We add these inequalities to see that

$$(\mathcal{P}_K u - \mathcal{P}_K v, \mathcal{P}_K v - \mathcal{P}_K u)_X \geq (u - v, \mathcal{P}_K v - \mathcal{P}_K u)_X$$

and, therefore,

$$(\mathcal{P}_K u - \mathcal{P}_K v, u - v)_X \geq \|\mathcal{P}_K u - \mathcal{P}_K v\|_X^2. \tag{1.25}$$

Inequality (1.23) follows from (1.25) and inequality (1.24) follows from (1.25) and the Cauchy–Schwarz inequality. ■

Proposition 1.14 *Let K be a nonempty closed convex subset of a Hilbert space X and let $G_K : X \to X$ be the operator defined by*

$$G_K u = u - \mathcal{P}_K u \qquad \forall\, u \in X. \tag{1.26}$$

Then, the following properties hold:

$$(G_K u - G_K v, u - v)_X \geq 0 \quad \forall\, u, v \in X, \tag{1.27}$$
$$\|G_K u - G_K v\|_X \leq 2\,\|u - v\|_X \quad \forall\, u,\ v \in X, \tag{1.28}$$
$$(G_K u, v - u)_X \leq 0 \quad \forall\, u \in X,\ v \in K, \tag{1.29}$$
$$G_K u = 0_X \quad \text{iff} \quad u \in K. \tag{1.30}$$

Proof Let $u,\ v \in X$. We use (1.26), the Cauchy–Schwarz inequality (1.5) and (1.24) to see that

$$\begin{aligned}(G_K u - G_K v, u - v)_X &= (u - \mathcal{P}_K u - v + \mathcal{P}_K v, u - v)_X \\ &= \|u - v\|_X^2 - (\mathcal{P}_K u - \mathcal{P}_K v, u - v)_X \\ &\geq \|u - v\|_X^2 - \|\mathcal{P}_K u - \mathcal{P}_K v\|_X \|u - v\|_X \geq 0,\end{aligned}$$

which shows that (1.27) holds. Also, since

$$\begin{aligned}\|G_K u - G_K v\|_X &= \|(u - v) + (\mathcal{P}_K v - \mathcal{P}_K u)\|_X \\ &\leq \|u - v\|_X + \|\mathcal{P}_K v - \mathcal{P}_K u\|_X,\end{aligned}$$

it follows from (1.24) that (1.28) holds, too.

Next, let $u \in X$ and $v \in K$. Then, from (1.26) and Proposition 1.12 we see that

$$\begin{aligned}(G_K u, v - u)_X &= (u - \mathcal{P}_K u, v - u)_X \\ &= (u - \mathcal{P}_K u, v - \mathcal{P}_K u)_X + (u - \mathcal{P}_K u, \mathcal{P}_K u - u)_X \\ &\leq (u - \mathcal{P}_K u, \mathcal{P}_K u - u)_X = -\|u - \mathcal{P}_K u\|_X^2 \leq 0,\end{aligned}$$

and, therefore, we deduce (1.29). And, finally, (1.30) is a direct consequence of (1.21), which concludes the proof. ■

1.2.2 Orthogonality

The inner product allows us to introduce the concept of orthogonality of elements in an inner product space.

Definition 1.15 Let X be an inner product space. Two elements u, $v \in X$ are said to be *orthogonal* if $(u, v)_X = 0$. The *orthogonal complement* of a subset $A \subset X$ is defined by

$$A^\perp = \{\, u \in X \ : \ (u, v)_X = 0 \ \text{ for all } \ v \in A \,\}. \tag{1.31}$$

If u, $v \in X$ are two orthogonal elements we usually write $u \perp v$. Moreover, it is known that if A is an arbitrary subset of an inner product space X, then its orthogonal complement $A^\perp$ is a closed linear subspace of X. Indeed, by using the previous definition and the properties of the inner product it is easy to see that if u_1, $u_2 \in A^\perp$ and λ_1, $\lambda_2 \in \mathbb{R}$ then $\lambda_1 u_1 + \lambda_2 u_2 \in A^\perp$ and, moreover, if $\{u_n\} \subset A^\perp$ is such that $u_n \to u$ in X then $u \in A^\perp$.

An important role in the theory of Hilbert spaces is played by the following result.

Theorem 1.16 *Let M be a closed linear subspace of a Hilbert space X and let $M^\perp$ denote its orthogonal complement. Then any element $u \in X$ can uniquely be decomposed as $u = m + m'$, where $m \in M$ and $m' \in M^\perp$. In this case we write $X = M \oplus M^\perp$ and we say that X is the* direct sum *of the spaces M and $M^\perp$.*

Proof Let $u \in X$ and let m, m' be the elements of X given by $m = \mathcal{P}_M u$, $m' = u - \mathcal{P}_M u$ where $\mathcal{P}_M : X \to M$ is the projection operator on the closed subspace M. It follows from this that $m \in M$ and, clearly, $u = m + m'$. Next, we use Proposition 1.12 to see that

$$(m, v - m)_X \geq (u, v - m)_X \quad \forall\, v \in M$$

and, therefore,

$$(m', v - m)_X \leq 0 \quad \forall\, v \in M.$$

We take in the previous inequality $v = m \pm w$, where w is an arbitrary element of M, to obtain

$$(m', w)_X = 0 \quad \forall\, w \in M.$$

It follows from this that $m' \in M^\perp$, which proves the existence of the decomposition.

To prove the uniqueness, we assume in what follows that $u = m + m' = p + p'$, where $m,\ p \in M$ and $m',\ p' \in M^\perp$. These equalities yield

$$m - p = p' - m'. \tag{1.32}$$

On the other hand, since M and $M^\perp$ are linear subspaces of X it follows that $m - p \in M$ and $p' - m' \in M^\perp$. We now use (1.31) to see that $(m - p, p' - m')_X = 0$ and, therefore, (1.32) yields $(m - p, m - p)_X = 0$ which implies that $m = p$. It follows now from (1.32) that $m' = p'$, which proves the uniqueness of the decomposition. ■

Theorem 1.16 allows us to prove the following result which will be useful later in this book.

Theorem 1.17 *Let M be a closed linear subspace of a Hilbert space X and let $M^{\perp\perp}$ be the orthogonal complement of the space $M^\perp$, that is $M^{\perp\perp} = (M^\perp)^\perp$. Then $M^{\perp\perp} = M$.*

Proof Let $u \in M^{\perp\perp}$ and let $u = m + m'$ be the decomposition of u provided by Theorem 1.16. We have

$$(u, m')_X = (m + m', m')_X = (m, m')_X + (m', m')_X = (m', m')_X \tag{1.33}$$

since $m \in M$ and $m' \in M^\perp$ and, therefore, $(m, m')_X = 0$. On the other hand, since $u \in M^{\perp\perp}$ and $m' \in M^\perp$ we deduce that $(u, m')_X = 0$ and, therefore, (1.33) implies that $(m', m')_X = 0$, i.e. $m' = 0_X$. It follows from this that $u = m$ which shows that $u \in M$. We conclude from above that

$$M^{\perp\perp} \subset M. \tag{1.34}$$

Next, assume in what follows that $u \in M$ and consider an arbitrary element $v \in M^\perp$. Definition 1.15 implies that $(v, u)_X = (u, v)_X = 0$ and, therefore, $u \in (M^\perp)^\perp$. We conclude that

$$M \subset M^{\perp\perp}. \tag{1.35}$$

Theorem 1.17 is now a consequence of the inclusions (1.34) and (1.35). ■

1.2.3 Duality and weak convergence

On Hilbert spaces, linear continuous functionals are limited in the forms they can take. The following theorem makes this more precise.

Theorem 1.18 (The Riesz representation theorem) *Let $(X, (\cdot,\cdot)_X)$ be a Hilbert space and let $\ell \in X'$. Then there exists a unique $u \in X$ such that*

$$\ell(v) = (u, v)_X \quad \forall v \in X. \tag{1.36}$$

Moreover,

$$\|\ell\|_{X'} = \|u\|_X. \tag{1.37}$$

Proof If $\ell = 0_{X'}$ it is easy to see that the element $u = 0_X$ is the unique element of X which satisfies (1.36) and (1.37).

Assume now that $\ell \neq 0_{X'}$ and denote by M the *kernel* of the functional ℓ, that is

$$M = \{\, v \in X \;:\; \ell(v) = 0 \,\}.$$

Since $\ell \in X'$ it follows that M is a closed subspace of X and, since $\ell \neq 0_{X'}$ it follows that $M \neq X$, i.e. there exists $w \in X$ such that $w \notin M$. Let $w = m + m'$ be the orthogonal decomposition of w provided by Theorem 1.16 and note that, since $w \notin M$, we have $m' \neq 0_X$.

Let v be an arbitrary element of X and consider the elements u and z of X defined by the equalities

$$u = \frac{\ell(m')}{\|m'\|_X^2}\, m', \tag{1.38}$$

$$z = \ell(v)m' - \ell(m')v. \tag{1.39}$$

It is easy to see that $\ell(z) = 0$ and, therefore, $z \in M$. Next, since $m' \in M^\perp$ we have

$$(m', z)_X = 0. \tag{1.40}$$

We now use (1.39) and (1.40) to deduce that

$$\ell(v)\|m'\|_X^2 = \ell(m')(m', v)_X.$$

This last equality implies that

$$\ell(v) = \frac{\ell(m')}{\|m'\|_X^2}(m', v)_X$$

and, using (1.38), we obtain (1.36).

Assume now that u_1 and u_2 are two elements of X such that (1.36) holds. We take $v = u_1 - u_2$ in (1.36) to obtain

$$\ell(u_1 - u_2) = (u_1, u_1 - u_2)_X = (u_2, u_1 - u_2)_X,$$

which implies that $\|u_1-u_2\|_X^2 = 0$. It follows that $u_1 = u_2$, which concludes the first part of the theorem.

Next, to prove (1.37) we note that (1.36) and the Cauchy–Schwarz inequality (1.5) yield

$$|\ell(v)| \leq \|u\|_X \|v\|_X \quad \forall v \in X$$

and, therefore, (1.8) shows that

$$\|\ell\|_{X'} \leq \|u\|_X. \tag{1.41}$$

Moreover, (1.8) implies that

$$|\ell(u)| \leq \|\ell\|_{X'} \|u\|_X$$

and, since by (1.36) we have $\ell(u) = \|u\|_X^2$, we deduce that

$$\|u\|_X \leq \|\ell\|_{X'}. \tag{1.42}$$

Equality (1.37) is now a consequence of inequalities (1.41) and (1.42). ■

We proceed with some important consequences of the Riesz representation theorem concerning the weak convergence in a Hilbert space. First, we recall that if $(X, (\cdot,\cdot)_X)$ is an inner product space then, as shown on page 7, (1.9) holds.

Assume now that $(X, (\cdot,\cdot)_X)$ is a Hilbert space. Then, it follows from Riesz's representation theorem that the converse of (1.9) is true, i.e.

$$(u_n, v)_X \to (u, v)_X \quad \text{as } n \to \infty, \quad \forall v \in X \implies u_n \rightharpoonup u \quad \text{in } X.$$

We conclude from this that a sequence $\{u_n\} \subset X$ converges weakly to $u \in X$ iff

$$(u_n, v)_X \to (u, v)_X \quad \text{as } n \to \infty, \quad \forall v \in X.$$

The Riesz representation theorem also allows to identify a Hilbert space with its dual and, therefore, with its bidual which, roughly speaking, shows that each Hilbert space is reflexive. Based on this result we have the following important property which represents a particular case of the well-known Eberlein–Smulyan theorem.

Theorem 1.19 *If X is a Hilbert space, then any bounded sequence in X has a weakly convergent subsequence.*

It follows that if X is a Hilbert space and the sequence $\{u_n\} \subset X$ is bounded, that is, $\sup_n \|u_n\|_X < \infty$, then there exists a subsequence $\{u_{n_k}\} \subset \{u_n\}$ and an element $u \in X$ such that $u_{n_k} \rightharpoonup u$ in X. Furthermore, if the limit u is independent of the subsequence, then the whole sequence $\{u_n\}$ converges weakly to u, as stated in the following result.

Theorem 1.20 *Let X be a Hilbert space and let $\{u_n\}$ be a bounded sequence of elements in X such that each weakly convergent subsequence of $\{u_n\}$ converges weakly to the same limit $u \in X$. Then $u_n \rightharpoonup u$ in X.*

Proof Arguing by contradiction, we assume in what follows that the sequence $\{u_n\}$ does not converge weakly to u in X. Then, there exists an element $v \in X$ such that $(u_n, v)_X \not\to (u, v)_X$ in $\mathbb{R}$, as $n \to \infty$. This, in turn, implies that there exists $\varepsilon_0 > 0$ such that for all $k \in \mathbb{N}$ there exists $u_{n_k} \in X$ which satisfies

$$|(u_{n_k}, v)_X - (u, v)_X| \geq \varepsilon_0 \qquad \forall\, k \in \mathbb{N}. \tag{1.43}$$

Since $\{u_{n_k}\}$ is a subsequence of the bounded sequence $\{u_n\}$ it follows that $\{u_{n_k}\}$ is bounded in X and, therefore, by Theorem 1.19 there exists a subsequence $\{u_{n_{k_p}}\} \subset \{u_{n_k}\}$ which is weakly convergent in X. Using the assumption of Theorem 1.20 it follows that $u_{n_{k_p}} \rightharpoonup u$ in X, which yields

$$(u_{n_{k_p}}, v)_X \to (u, v)_X \qquad \text{as} \quad p \to \infty. \tag{1.44}$$

On the other hand, $\{u_{n_{k_p}}\} \subset \{u_{n_k}\}$ and, therefore, by (1.43) we have

$$|(u_{n_{k_p}}, v)_X - (u, v)_X| \geq \varepsilon_0 \qquad \forall\, p \in \mathbb{N}. \tag{1.45}$$

Inequality (1.45) is in contradiction to the convergence (1.44), which concludes the proof. ■

1.3 Elements of nonlinear analysis

In the study of variational inequalities presented in Chapters 2 and 3 we need several results on nonlinear operators and convex functions that we introduce in this section.

1.3.1 Monotone operators

The projection operator on a convex subset K of a Hilbert space is, in general, a nonlinear operator on X. Its properties (1.23) and (1.24) can be extended as follows.

Definition 1.21 Let X be a space with inner product $(\cdot, \cdot)_X$ and norm $\|\cdot\|_X$ and let $A : X \to X$ be an operator. The operator A is said to be *monotone* if

$$(Au - Av, u - v)_X \geq 0 \quad \forall\, u, v \in X.$$

The operator A is *strictly monotone* if

$$(Au - Av, u - v)_X > 0 \quad \forall\, u, v \in X,\ u \neq v,$$

and *strongly monotone* if there exists a constant $m > 0$ such that

$$(Au - Av, u - v)_X \geq m\,\|u - v\|_X^2 \quad \forall\, u,\, v \in X. \tag{1.46}$$

The operator A is *nonexpansive* if

$$\|Au - Av\|_X \leq \|u - v\|_X \quad \forall\, u,\, v \in X$$

and *Lipschitz continuous* if there exists $M > 0$ such that

$$\|Au - Av\|_X \leq M\|u - v\|_X \quad \forall\, u,\, v \in X. \tag{1.47}$$

Finally, the operator A is *hemicontinuous* if the real-valued function

$$\theta \mapsto (A(u + \theta v), w)_X \quad \text{is continuous on } \mathbb{R}, \quad \forall\, u,\, v,\, w \in X$$

and, recalling Definition 1.8, A is *continuous* if

$$u_n \to u \ \text{ in } X \implies Au_n \to Au \ \text{ in } X.$$

It follows from the definition above that each strongly monotone operator is strictly monotone and a nonexpansive operator is Lipschitz continuous, with Lipschitz constant $M = 1$. Also, it is easy to check that a Lipschitz continuous operator is continuous and a continuous operator is hemicontinuous. Moreover, it follows from Proposition 1.13 that the projection operators are monotone and nonexpansive.

In many applications it is not necessary to define nonlinear operators on the entire space X. Indeed, in the study of the variational inequalities presented in Chapters 2 and 3 we shall consider strongly monotone Lipschitz continuous operators defined on a subset $K \subset X$. For this reason we complete Definition 1.21 with the following one.

Definition 1.22 Let X be a space with inner product $(\cdot,\cdot)_X$ and norm $\|\cdot\|_X$ and let $K \subset X$. An operator $A : K \to X$ is said to be *strongly monotone* if there exists a constant $m > 0$ such that

$$(Au - Av, u - v)_X \geq m\,\|u - v\|_X^2 \quad \forall\, u,\, v \in K. \tag{1.48}$$

The operator A is *Lipschitz continuous* if there exists $M > 0$ such that

$$\|Au - Av\|_X \leq M\,\|u - v\|_X \quad \forall\, u,\, v \in K. \tag{1.49}$$

The following result involving monotone operators will be used in Chapter 2, in the analysis of elliptic variational inequalities.

Proposition 1.23 *Let $(X, (\cdot,\cdot)_X)$ be an inner product space and let $A : X \to X$ be a monotone hemicontinuous operator. Assume that $\{u_n\}$ is a* *sequence of elements in X which converges weakly to the element $u \in X$, i.e.*

$$u_n \rightharpoonup u \quad \text{in } X \quad \text{as} \quad n \to \infty. \tag{1.50}$$

Moreover, assume that

$$\limsup_{n\to\infty}\,(Au_n, u_n - u)_X \le 0. \tag{1.51}$$

Then, for all $v \in X$, the following inequality holds:

$$\liminf_{n\to\infty}\,(Au_n, u_n - v)_X \ge (Au, u - v)_X. \tag{1.52}$$

Proof By the monotonicity of the operator A it follows that

$$(Au_n, u_n - u)_X \ge (Au, u_n - u)_X \qquad \forall\, n \in \mathbb{N}$$

and, passing to the lower limit as $n \to \infty$, by using (1.50) and (1.9) we find that

$$\liminf_{n\to\infty}\,(Au_n, u_n - u)_X \ge 0. \tag{1.53}$$

We combine the inequalities (1.51) and (1.53) to see that the sequence $(Au_n, u_n - u)_X$ converges as $n \to \infty$ and, moreover,

$$\lim_{n\to\infty}\,(Au_n, u_n - u)_X = 0. \tag{1.54}$$

Let $v \in X$, $\theta > 0$ and denote $w = (1-\theta)u + \theta v$. We use the monotonicity of the operator A to obtain

$$(Au_n - Aw, u_n - w)_X \ge 0 \qquad \forall\, n \in \mathbb{N}$$

which implies that

$$(Au_n - Aw, u_n - u + \theta(u - v))_X \ge 0 \qquad \forall\, n \in \mathbb{N}$$

or, equivalently,

$$\begin{aligned}(Au_n, u_n - u)_X &+ \theta(Au_n, u - v)_X \\ &\ge (Aw, u_n - u)_X + \theta(Aw, u - v)_X \qquad \forall\, n \in \mathbb{N}.\end{aligned} \tag{1.55}$$

We pass to the lower limit as $n \to \infty$ in (1.55), use (1.9) and (1.54) and then divide the resulting inequality by θ. We obtain

$$\liminf_{n\to\infty}\,(Au_n, u - v)_X \ge (Aw, u - v)_X$$

and, therefore,

$$\liminf_{n\to\infty}\,(Au_n, u - v)_X \ge (A(u + \theta(v - u)), u - v)_X.$$

Next, we pass to the limit in this last inequality as $\theta \to 0$ and use the hemicontinuity of the operator A to see that

$$\liminf_{n\to\infty} (Au_n, u - v)_X \geq (Au, u - v)_X. \tag{1.56}$$

We write now

$$(Au_n, u_n - v)_X = (Au_n, u_n - u)_X + (Au_n, u - v)_X \quad \forall\, n \in \mathbb{N},$$

then we pass to the lower limit as $n \to \infty$ in this equality and use (1.54) and (1.56). As a result we obtain (1.52), which concludes the proof. ■

An operator $A : X \to X$ for which (1.50) and (1.51) imply (1.52) for all $v \in X$ is called a *pseudomonotone* operator. We conclude from Proposition 1.23 that every monotone hemicontinuous operator on a Hilbert space is a pseudomonotone operator.

We proceed with the following existence and uniqueness result in the study of nonlinear equations involving monotone operators.

Theorem 1.24 *Let X be a Hilbert space and let $A : X \to X$ be a strongly monotone Lipschitz continuous operator. Then, for each $f \in X$ there exists a unique element $u \in X$ such that $Au = f$.*

Proof Since A is strongly monotone and Lipschitz continuous it follows from Definition 1.21 that there exist two constants $m > 0$ and $M > 0$ such that (1.46) and (1.47) hold. Moreover, we have

$$M \geq m. \tag{1.57}$$

Let $f \in X$ and let $\rho > 0$ be given. We consider the operator $S_\rho : X \to X$ defined by

$$S_\rho u = u - \rho(Au - f) \qquad \forall\, u \in X.$$

It follows from this definition that

$$\|S_\rho u - S_\rho v\|_X = \|(u - v) - \rho(Au - Av)\|_X \qquad \forall\, u,\, v \in X$$

and, using (1.46), (1.47) yields

$$\begin{aligned}
\|S_\rho u - S_\rho v\|_X^2 &= \|(u - v) - \rho(Au - Av)\|_X^2 \\
&= \|u - v\|_X^2 - 2\rho(Au - Av, u - v)_X + \rho^2\|Au - Av\|_X^2 \\
&\leq (1 - 2\rho m + \rho^2 M^2)\, \|u - v\|_X^2 \qquad \forall\, u,\, v \in X.
\end{aligned}$$

Next, using (1.57) it is easy to see that if $0 < \rho < \frac{2m}{M^2}$ then

$$0 \leq 1 - 2\rho m + \rho^2 M^2 < 1.$$

Therefore, with this choice of ρ, it follows that

$$\|S_\rho u - S_\rho v\|_X \leq k(\rho)\,\|u - v\|_X \qquad \forall\, u,\, v \in X, \tag{1.58}$$

where $k(\rho) = (1 - 2\rho m + \rho^2 M^2)^{\frac{1}{2}} \in [0, 1)$. Inequality (1.58) shows that S_ρ is a contraction on the space X and, using Theorem 1.5, we obtain that there exists $u \in X$ such that

$$S_\rho u = u - \rho(Au - f) = u. \tag{1.59}$$

Equality (1.59) yields $Au = f$, which proves the existence part of the theorem.

Next, consider two elements $u \in X$ and $v \in X$ such that $Au = f$ and $Av = f$. It follows that

$$(Au, u - v)_X = (f, u - v)_X, \qquad (Av, u - v)_X = (f, u - v)_X.$$

We subtract these equalities to obtain

$$(Au - Av, u - v)_X = 0,$$

then we use assumption (1.46) to find that $u = v$, which proves the uniqueness part. ■

Theorem 1.24 shows that if $A : X \to X$ is a strongly monotone Lipschitz continuous operator defined on a Hilbert space X, then A is invertible. The properties of its inverse, denoted A^{-1}, are given by the following result.

Proposition 1.25 *Let X be a Hilbert space and let $A : X \to X$ be a strongly monotone Lipschitz continuous operator. Then, $A^{-1} : X \to X$ is a strongly monotone Lipschitz continuous operator.*

Proof Let $u_1,\, u_2 \in X$ and denote

$$Au_1 = v_1, \qquad Au_2 = v_2. \tag{1.60}$$

It follows that

$$A^{-1}v_1 = u_1, \qquad A^{-1}v_2 = u_2. \tag{1.61}$$

We use now (1.46) and (1.60) to see that

$$(Au_1 - Au_2, u_1 - u_2)_X = (v_1 - v_2, u_1 - u_2)_X \geq m\,\|u_1 - u_2\|_X^2,$$

which implies that

$$\|u_1 - u_2\|_X \leq \frac{1}{m}\,\|v_1 - v_2\|_X.$$

This inequality and (1.61) lead to

$$\|A^{-1}v_1 - A^{-1}v_2\|_X \leq \frac{1}{m}\,\|v_1 - v_2\|_X,$$

which shows that A^{-1} is a Lipschitz continuous operator.

On the other hand, using (1.46), (1.47), (1.60) and (1.61) we have

$$(v_1 - v_2, A^{-1}v_1 - A^{-1}v_2)_X \geq m\,\|A^{-1}v_1 - A^{-1}v_2\|_X^2,$$

$$\frac{1}{M}\,\|v_1 - v_2\|_X \leq \|A^{-1}v_1 - A^{-1}v_2\|_X.$$

These inequalities imply that

$$(A^{-1}v_1 - A^{-1}v_2, v_1 - v_2)_X \geq \frac{m}{M^2}\,\|v_1 - v_2\|_X^2,$$

which shows that the operator A^{-1} is strongly monotone. ■

1.3.2 Convex lower semicontinuous functions

Convex lower semicontinuous functions represent a crucial ingredient in the study of variational inequalities. To introduce them, we start with the following definitions.

Definition 1.26 Let X be a linear space and let K be a nonempty convex subset of X. A function $\varphi : K \to \mathbb{R}$ is said to be *convex* if

$$\varphi((1-t)u + tv) \leq (1-t)\varphi(u) + t\varphi(v) \tag{1.62}$$

for all $u, v \in K$ and $t \in [0,1]$. The function φ is *strictly convex* if the inequality in (1.62) is strict for $u \neq v$ and $t \in (0,1)$.

We note that if $\varphi, \psi : K \to \mathbb{R}$ are convex and $\lambda \geq 0$, then the functions $\varphi + \psi$ and $\lambda\varphi$ are also convex.

Definition 1.27 Let $(X, \|\cdot\|_X)$ be a normed space and let K be a nonempty closed convex subset of X. A function $\varphi : K \to \mathbb{R}$ is said to be *lower semicontinuous (l.s.c.)* at $u \in K$ if

$$\liminf_{n\to\infty} \varphi(u_n) \geq \varphi(u) \tag{1.63}$$

for each sequence $\{u_n\} \subset K$ converging to u in X. The function φ is *l.s.c.* if it is l.s.c. at every point $u \in K$. When inequality (1.63) holds for each sequence $\{u_n\} \subset K$ that converges weakly to u, the function φ is said to be *weakly lower semicontinuous* at u. The function φ *is weakly l.s.c.* if it is weakly l.s.c. at every point $u \in K$.

We note that if $\varphi, \psi : K \to \mathbb{R}$ are l.s.c. functions and $\lambda \geq 0$, then the functions $\varphi + \psi$ and $\lambda\varphi$ are also lower semicontinuous. Moreover, if $\varphi : K \to \mathbb{R}$ is a continuous function then it is also lower semicontinuous. The converse is not true and a lower semicontinuous function can be discontinuous. Since strong convergence in X implies weak convergence, it follows that a weakly lower semicontinuous function is lower semicontinuous. Moreover, the following results hold.

Proposition 1.28 *Let $(X, \|\cdot\|_X)$ be a Banach space, K a nonempty closed convex subset of X and $\varphi : K \to \mathbb{R}$ a convex function. Then φ is lower semicontinuous if and only if it is weakly lower semicontinuous.*

Proposition 1.29 *Let $(X, \|\cdot\|_X)$ be a normed space, K a nonempty closed convex subset of X and $\varphi : K \to \mathbb{R}$ a convex lower semicontinuous function. Then φ is bounded from below by an affine function, i.e. there exist $\ell \in X'$ and $\alpha \in \mathbb{R}$ such that $\varphi(v) \geq \ell(v) + \alpha$ for all $v \in K$.*

The proof of Proposition 1.28 is a consequence of Mazur's theorem. It follows from this proposition that a convex continuous function $\varphi : X \to \mathbb{R}$ defined on the Banach space X is weakly lower semicontinuous. In particular, the norm function $v \mapsto \|v\|_X$ is weakly lower semicontinuous. A second example of a lower semicontinuous function is provided by the following result.

Proposition 1.30 *Let $(X, \|\cdot\|_X)$ be a normed space and let $a : X \times X \to \mathbb{R}$ be a bilinear symmetric continuous and positive form. Then the function $v \mapsto a(v, v)$ is strictly convex and lower semicontinuous.*

Proof The strict convexity is straightforward to show. To prove the lower semicontinuity consider a sequence $\{u_n\} \subset X$ such that $u_n \rightharpoonup u \in X$. Since a is positive it follows that

$$a(u_n - u, u_n - u) \geq 0 \qquad \forall n \in \mathbb{N}$$

and, therefore,

$$a(u_n, u_n) \geq a(u, u_n) + a(u_n, u) - a(u, u) \qquad \forall n \in \mathbb{N}. \tag{1.64}$$

For a fixed $v \in X$, the mappings $u \mapsto a(v, u)$ and $u \mapsto a(u, v)$ define linear continuous functionals on X and, since $\{u_n\}$ converges weakly to u, we have

$$\lim_{n\to\infty} a(u, u_n) = \lim_{n\to\infty} a(u_n, u) = a(u, u). \tag{1.65}$$

We pass to the lower limit in (1.64) and use (1.65) to see that

$$\liminf_{n\to\infty} a(u_n, u_n) \geq a(u, u),$$

which concludes the proof. ■

In particular, it follows from Proposition 1.30 that, if $(X, (\cdot, \cdot)_X)$ is an inner product space then the function $v \mapsto \|v\|_X^2 = (v, v)_X$ is strictly convex and lower semicontinuous.

We now recall the definition of Gâteaux differentiable functions.

Definition 1.31 Let $(X, (\cdot,\cdot)_X)$ be an inner product space, $\varphi : X \to \mathbb{R}$ and $u \in X$. Then φ is *Gâteaux differentiable* at u if there exists an element $\nabla\varphi(u) \in X$ such that

$$\lim_{t\to 0} \frac{\varphi(u+t\,v)-\varphi(u)}{t} = (\nabla\varphi(u), v)_X \quad \forall\, v \in X. \tag{1.66}$$

The element $\nabla\varphi(u)$ which satisfies (1.66) is unique and is called the *gradient of φ at u*. The function $\varphi : X \to \mathbb{R}$ is said to be *Gâteaux differentiable* if it is Gâteaux differentiable at every point of X. In this case the operator $\nabla\varphi : X \to X$ which maps every element $u \in X$ into the element $\nabla\varphi(u)$ is called the *gradient operator* of φ.

The convexity of Gâteaux differentiable functions can be characterized as follows.

Proposition 1.32 *Let $(X, (\cdot,\cdot)_X)$ be an inner product space and let $\varphi : X \to \mathbb{R}$ be a Gâteaux differentiable function. Then the following statements are equivalent:*

(i) φ is a convex function;

(ii) φ satisfies the inequality

$$\varphi(v) - \varphi(u) \ge (\nabla\varphi(u), v-u)_X \quad \forall\, u,\, v \in X; \tag{1.67}$$

(iii) the gradient of φ is a monotone operator, that is

$$(\nabla\varphi(u) - \nabla\varphi(v), u-v)_X \ge 0 \quad \forall\, u,\, v \in X. \tag{1.68}$$

Proof Proposition 1.32 is a direct consequence of the four implications we state and prove below.

(i) $\Longrightarrow$ *(ii)* If φ is a convex function, then for all $u,\, v \in X$ and $t \in (0,1)$, from (1.62) we deduce that

$$t(\varphi(v) - \varphi(u)) \ge \varphi(u + t(v-u)) - \varphi(u).$$

We divide both sides of the inequality by $t > 0$, pass to the limit as $t \to 0$ and use (1.66) to obtain (1.67).

(ii) $\Longrightarrow$ *(i)* Conversely, assume that (1.67) holds and let $u,\, v \in X$, $t \in [0,1]$. Let $w = (1-t)u + tv$; it follows from (1.67) that

$$\varphi(v) - \varphi(w) \ge (\nabla\varphi(w), v-w)_X,$$
$$\varphi(u) - \varphi(w) \ge (\nabla\varphi(w), u-w)_X.$$

We multiply the inequalities above by t and $(1-t)$, respectively, then add the results to obtain (1.62), which shows that φ is a convex function.

(ii) $\Longrightarrow$ *(iii)* Assume that (1.67) holds and let u, $v \in X$. We have

$$\varphi(u) - \varphi(v) \geq (\nabla\varphi(v), u - v)_X$$

and, adding this inequality to (1.67), we obtain (1.68).

(iii) $\Longrightarrow$ *(ii)* Assume that (1.68) holds and let u, $v \in X$. For all $t \in [0,1]$ we consider the element $w(t) \in X$ defined by $w(t) = (1-t)u + tv$ and let g be the real-valued function defined by $g(t) = \varphi(w(t))$. First, we see that $w(t) - u = t(v - u)$ and, therefore, we use (1.68) to write

$$(\nabla\varphi(w(t)), v - u)_X \geq (\nabla\varphi(u), v - u)_X \quad \forall\, t \in [0,1].$$

On the other hand $w(t+h) = w(t) + h(v-u)$ and, using the definition (1.66) of the gradient operator, it is easy to see that

$$\lim_{h\to 0} \frac{\varphi(w(t+h)) - \varphi(w(t))}{h} = (\nabla\varphi(w(t)), v - u)_X \quad \forall\, t \in [0,1],$$

where the limit is taken with respect to h with $t + h \in [0,1]$. We conclude from here that g is derivable on [0,1] and, moreover,

$$g'(t) \geq (\nabla\varphi(u), v - u)_X \quad \forall\, t \in [0,1].$$

We integrate this inequality on $[0,1]$ to obtain

$$g(1) - g(0) \geq (\nabla\varphi(u), v - u)_X. \tag{1.69}$$

Next, since $w(0) = u$ and $w(1) = v$ it follows that $g(0) = \varphi(u)$ and $g(1) = \varphi(v)$. We use these equalities in (1.69) to obtain (1.67). ■

From the previous proposition we easily deduce the following result.

Corollary 1.33 *Let $(X, (\cdot,\cdot)_X)$ be an inner product space and let $\varphi : X \to \mathbb{R}$ be a convex Gâteaux differentiable function. Then φ is lower semicontinuous.*

Proof Let $\{u_n\}$ be a sequence of elements of X converging to u in X. It follows from Proposition 1.32 that

$$\varphi(u_n) - \varphi(u) \geq (\nabla\varphi(u), u_n - u)_X \quad \forall\, n \in \mathbb{N}.$$

We pass to the lower limit as $n \to \infty$ in the previous inequality to obtain (1.63), which concludes the proof. ■

A large number of boundary value problems in Contact Mechanics lead to variational formulations in which the frictional term is associated with a continuous seminorm. For this reason we collect at the end of this subsection some results on continuous seminorms defined on a normed space, which we shall need in the rest of the book.

Given a linear space X, we recall that a *seminorm* j is a function from X to $\mathbb{R}$ satisfying the following properties:

(1) $j(u) \geq 0 \quad \forall\, u \in X$;

(2) $j(\alpha\, u) = |\alpha|\, j(u) \quad \forall\, u \in X, \forall\, \alpha \in \mathbb{R}$;

(3) $j(u + v) \leq j(u) + j(v) \quad \forall\, u, v \in X$.

It follows from the above that a seminorm satisfies the properties of a norm except that $j(u) = 0$ does not necessarily imply $u = 0_X$. In particular, note that $j(0_X) = 0$. The continuity of a seminorm defined on a normed space is characterized by the following result.

Proposition 1.34 *Let j be a seminorm on the normed space $(X, \|\cdot\|_X)$. Then j is continuous if and only if there exists $m > 0$ such that*

$$j(v) \leq m\, \|v\|_X \qquad \forall\, v \in X. \tag{1.70}$$

Proof Assume that j is continuous and, arguing by contradiction, assume that (1.70) does not hold. Then, for each $n \in \mathbb{N}$ there exists $v_n \in X$ such that $j(v_n) > n\, \|v_n\|_X$. It follows that $v_n \neq 0_X$ and, moreover,

$$j\Big(\frac{v_n}{n\, \|v_n\|_X}\Big) > 1. \tag{1.71}$$

For each $n \in \mathbb{N}$ denote

$$u_n = \frac{v_n}{n\, \|v_n\|_X}\,. \tag{1.72}$$

We use (1.72) and (1.71) to see that $u_n \to 0_X$ in X as $n \to \infty$ but $j(u_n) > 1$ for all $n \in \mathbb{N}$, which contradicts the continuity of j in 0_X.

Conversely, assume that (1.70) holds. Then, using the properties of the seminorm j we have

$$|j(u) - j(v)| \leq j(u - v) \leq m\, \|u - v\|_X \quad \forall\, u, v \in X,$$

which proves that j is continuous. ∎

It is easy to see that a seminorm defined on a linear space is a convex function. Therefore, a direct consequence of Proposition 1.34 is given by the following result.

Corollary 1.35 *Let $(X, \|\cdot\|_X)$ be a normed space and let j be a seminorm on X which satisfies (1.70) with some $m > 0$. Then j is a convex lower semicontinuous function.*

Note that in convex analysis it is usual to consider functions φ defined on a normed space X with values on $(-\infty, \infty]$. For such functions, concepts such as convexity and lower semicontinuity are introduced in Definitions 1.26 and 1.27, respectively, by using the convention that $\infty + \infty = \infty$ while an expression of the form $\infty - \infty$ is undefined. Nevertheless, in many applications in mechanics the domain of definition of various functions (like the energy function) is subjected to constraints. For this reason, above, we restricted ourselves to considering only functions $\varphi : K \to \mathbb{R}$ defined on a subset K of a normed or inner product space X. This choice is not restrictive. Indeed, assume that K is a nonempty closed convex subset of X, $\varphi : K \to \mathbb{R}$ is a given function and let $\widetilde{\varphi} : X \to (-\infty, \infty]$ be the function given by

$$\widetilde{\varphi}(v) = \begin{cases} \varphi(v) & \text{if} \quad v \in K, \\ \infty & \text{if} \quad v \notin K. \end{cases}$$

Then, it is easy to see that $\widetilde{\varphi}$ is convex and l.s.c. if and only if φ is convex and l.s.c.

1.3.3 Minimization problems

An important property of convex lower semicontinuous functions is given by the following well-known theorem.

Theorem 1.36 (The Weierstrass theorem) *Let X be a Hilbert space and K a nonempty closed convex subset of X. Let $J : K \to \mathbb{R}$ be a convex lower semicontinuous function. Then J is bounded from below and attains its infimum on K whenever one of the following two conditions hold:*

(i) K is bounded;

(ii) J is coercive, *i.e. $J(u) \to +\infty$ as $\|u\|_X \to +\infty$.*

Moreover, if J is a strictly convex function, then J attains its infimum on K at only one point.

Proof Let θ denote the infimum of J on K, that is

$$\theta = \inf_{v \in K} J(v) < \infty. \tag{1.73}$$

Consider a sequence $\{u_n\}$ of elements in K such that

$$J(u_n) \to \theta \qquad \text{as} \quad n \to \infty. \tag{1.74}$$

We claim that the sequence $\{u_n\}$ is bounded. Indeed, this claim is obviously satisfied if condition *(i)* holds. Assume now that *(ii)* holds, i.e. J is coercive. Arguing by contradiction, if the sequence $\{u_n\}$ is not bounded then, passing to a subsequence again denoted $\{u_n\}$, we can assume that

$$\|u_n\|_X \to \infty \qquad \text{as} \quad n \to \infty. \tag{1.75}$$

Using now the coercivity of J it follows that

$$J(u_n) \to \infty \qquad \text{as} \quad n \to \infty. \tag{1.76}$$

The convergences (1.76) and (1.74) are in contradiction to the inequality $\theta < \infty$ in (1.73). It follows from the above that the sequence $\{u_n\}$ is bounded and, therefore, our claim holds in this case, too.

We now use Theorem 1.19 to see that there exists an element $u \in K$ such that, passing to a subsequence again denoted $\{u_n\}$, we have

$$u_n \rightharpoonup u \quad \text{as} \quad n \to \infty. \tag{1.77}$$

Moreover, since K is a closed convex subset it follows from Theorem 1.4 that K is weakly closed and, therefore,

$$u \in K. \tag{1.78}$$

We use Proposition 1.28 to see that $J : K \to \mathbb{R}$ is a weakly lower semicontinuous function. Hence, (1.77) and (1.74) imply that $\theta \geq J(u)$. On the other hand, the converse inequality $\theta \leq J(u)$ is a consequence of (1.73) and (1.78) and, therefore, we obtain

$$\theta = J(u). \tag{1.79}$$

We now combine (1.78), (1.73) and (1.79) to see that

$$u \in K, \quad J(v) \geq J(u) \quad \forall\, v \in K, \tag{1.80}$$

which concludes the existence part of the theorem.

Assume now that J is strictly convex. Arguing by contradiction, assume that u_1 and u_2 satisfy (1.80) and, moreover, $u_1 \neq u_2$. Then, $J(u_1) = J(u_2) = \theta$ and by Definition 1.26 it follows that

$$J\Big(\frac{u_1 + u_2}{2}\Big) < \frac{1}{2}\, J(u_1) + \frac{1}{2}\, J(u_2) = \theta.$$

This inequality is in contradiction to the definition (1.73) and proves the uniqueness of the minimizer of the function J. ■

We use Theorem 1.36 to prove the following result.

Theorem 1.37 *Let X be a Hilbert space, K a nonempty closed convex subset of X, $a : X \times X \to \mathbb{R}$ a bilinear continuous symmetric and X-elliptic form, $j : K \to \mathbb{R}$ a convex lower semicontinuous function and $f \in X$. Denote by $J : K \to \mathbb{R}$ the function given by*

$$J(v) = \frac{1}{2}\, a(v, v) + j(v) - (f, v)_X \quad \forall\, v \in K \tag{1.81}$$

and consider the minimization problem

$$u \in K, \quad J(v) \geq J(u) \quad \forall\, v \in K. \tag{1.82}$$

Then:

(1) an element u is a solution to problem (1.82) iff

$$u \in K, \quad a(u, v-u) + j(v) - j(u) \geq (f, v-u)_X \quad \forall\, v \in K; \tag{1.83}$$

(2) there exists a unique solution to problem (1.82).

Proof (1) Assume that u satisfies (1.82) and let $v \in K$ and $t \in (0,1)$. It follows that

$$J(u + t(v-u)) \geq J(u)$$

and, using (1.81), the properties of a and the convexity of j, we deduce that

$$t\, a\,(u, v-u) + \frac{t^2}{2}\, a(v-u, v-u) + t\,(j(v) - j(u)) \geq t\,(f, v-u)_X.$$

Dividing both sides of the inequality by $t > 0$ and passing to the limit as $t \to 0$, we conclude that u satisfies inequality (1.83).

Conversely, suppose u is a solution of the inequality (1.83). Using this inequality and the properties of the form a, for each $v \in K$ we have

$$\begin{aligned} J(v) - J(u) &= a(u, v-u) + j(v) - j(u) - (f, v-u)_X \\ &\quad + \frac{1}{2}\, a(v-u, v-u) \geq \frac{1}{2}\, a(v-u, v-u) \geq 0. \end{aligned}$$

Thus, u is a solution of the minimization problem (1.82).

(2) We use Proposition 1.30 to see that the functional J is strictly convex and l.s.c. Therefore, in the case when K is bounded, Theorem 1.37 follows from Theorem 1.36.

Assume now that K is not bounded. Since the bilinear form a is X-elliptic, it follows that there exists $m > 0$ such that

$$J(v) \geq \frac{m}{2}\, \|v\|_X^2 + j(v) - (f, v)_X \quad \forall\, v \in K. \tag{1.84}$$

Next, Proposition 1.29 and Riesz's representation theorem guarantee that there exist $g \in X$ and $\alpha \in \mathbb{R}$ such that

$$j(v) \geq (g, v)_X + \alpha \quad \forall\, v \in K. \tag{1.85}$$

We combine inequalities (1.84) and (1.85) and then use the Cauchy–Schwarz inequality to see that the functional J is coercive. And, again, Theorem 1.37 is a consequence of Theorem 1.36. ■

We conclude from the first part of Theorem 1.37 that the minimization problem (1.82) and inequality (1.83) are equivalent. Since the existence of a unique minimizer for J is guaranteed by the second part of this theorem, we deduce the following existence and uniqueness result.

Corollary 1.38 *Let X be a Hilbert space, K a nonempty closed convex subset of X, $a : X \times X \to \mathbb{R}$ a bilinear continuous symmetric and X-elliptic form and $j : K \to \mathbb{R}$ a convex lower semicontinuous function. Then, for each $f \in X$ there exists a unique element u such that (1.83) holds.*

2
Elliptic variational inequalities

In this chapter we present theorems on the unique solvability of elliptic variational inequalities with strongly monotone Lipschitz continuous operators. We start with an existence and uniqueness result for elliptic variational inequalities of the first kind then we extend it to elliptic variational inequalities of the second kind, as well as to the elliptic quasivariational inequalities. We also present various convergence results. Besides their own interest, the results presented in this chapter represent crucial tools in the analysis of static frictionless and frictional contact problems with elastic materials. Even if most of the results we present in this chapter still remain valid for more general cases, we restrict ourselves to the framework of strongly monotone Lipschitz continuous operators in Hilbert spaces as this is sufficient for the applications we consider in Chapter 5. Everywhere in this chapter X denotes a real Hilbert space with inner product $(\cdot,\cdot)_X$ and norm $\|\cdot\|_X$.

2.1 Variational inequalities of the first kind

In this section we provide an extension of the existence and uniqueness result of Theorem 1.24. Thus, given an operator $A : X \to X$, a subset $K \subset X$ and an element $f \in X$, we consider the problem of finding an element u such that

$$u \in K, \quad (Au, v-u)_X \geq (f, v-u)_X \quad \forall\, v \in K. \tag{2.1}$$

An inequality of the form (2.1) is called an *elliptic variational inequality of the first kind.* An example is provided by inequality (1.22) which characterizes the projection on the nonempty closed convex subset $K \subset X$. Moreover, Theorem 1.10, Definition 1.11 and Proposition 1.12 provide, implicitly, a first existence result in the study of such inequalities.

2.1.1 Existence and uniqueness

The first result we present in this section is the following.

Theorem 2.1 *Let X be a Hilbert space and let $K \subset X$ be a nonempty closed convex subset. Assume that $A : K \to X$ is a strongly monotone Lipschitz continuous operator, i.e. it satisfies conditions (1.48) and (1.49). Then, for each $f \in X$ the variational inequality (2.1) has a unique solution.*

Proof Let $f \in X$ and let $\rho > 0$ be given. We consider the operator $S_\rho : K \to K$ defined by

$$S_\rho u = \mathcal{P}_K\big(u - \rho(Au - f)\big) \qquad \forall\, u \in K,$$

where $\mathcal{P}_K$ denotes the projection operator on K. Using (1.24) it follows that

$$\|S_\rho u - S_\rho v\|_X \le \|(u - v) - \rho(Au - Av)\|_X \qquad \forall\, u,\, v \in K.$$

Then, using arguments similar to those used in the proof of Theorem 1.24 we obtain that

$$\|S_\rho u - S_\rho v\|_X \le k(\rho)\, \|u - v\|_X \qquad \forall\, u,\, v \in K,$$

where $k(\rho) = (1 - 2\rho m + \rho^2 M^2)^{\frac{1}{2}}$ and m, M are the constants in (1.48) and (1.49), respectively. Also, with a convenient choice of ρ we may assume that $k(\rho) \in [0, 1)$. It follows now from Theorem 1.5 that there exists an element u such that

$$S_\rho u = \mathcal{P}_K\big(u - \rho(Au - f)\big) = u. \tag{2.2}$$

We now combine (2.2) and Proposition 1.12 to see that

$$u \in K, \quad (u, v - u)_X \ge (u - \rho(Au - f), v - u)_X \qquad v \in K.$$

Since $\rho > 0$ we conclude that u satisfies (2.1), which proves the existence part of the theorem.

Next, we consider two solutions u and v of (2.1). It follows that $u \in K$, $v \in K$ and, moreover,

$$(Au, v - u)_X \ge (f, v - u)_X, \qquad (Av, u - v)_X \ge (f, u - v)_X.$$

We add these inequalities to see that

$$(Au - Av, v - u)_X \le 0,$$

then we use assumption (1.48) to obtain $u = v$, which proves the uniqueness part. ■

Assume now that $K = X$. Then, taking $v = u \pm w$ it is easy to see that the variational inequality (2.1) is equivalent to the variational equation

$$(Au, w)_X = (f, w)_X \qquad \forall\, w \in X$$

which, in turn, is equivalent to the nonlinear equation $Au = f$. We conclude from above that Theorem 2.1 represents an extension of Theorem 1.24, as claimed at the beginning of this section.

2.1.2 *Penalization*

We now investigate the unique solvability of the variational inequality (2.1) by using a penalization method. To this end, we assume in what follows that $A : X \to X$ and we consider an operator G which satisfies the following conditions:

$$\left.\begin{array}{l} \text{(a) } G : X \to X \text{ is a monotone Lipschitz continuous operator;} \\ \text{(b) } (Gu, v - u)_X \le 0 \quad \forall\, u \in X,\ v \in K; \\ \text{(c) } Gu = 0_X \quad \text{iff} \quad u \in K. \end{array}\right\} \tag{2.3}$$

Note that such an operator G always exists, as shown in Proposition 1.14. For each $\mu > 0$ we consider the problem of finding an element u_μ such that

$$u_\mu \in X, \quad Au_\mu + \frac{1}{\mu} Gu_\mu = f. \tag{2.4}$$

We have the following existence, uniqueness and convergence result.

Theorem 2.2 *Let X be a Hilbert space and let $K \subset X$ be a nonempty closed convex subset. Assume that $A : X \to X$ is a strongly monotone Lipschitz continuous operator, G is an operator which satisfies (2.3) and $f \in X$. Then:*

(1) *for each $\mu > 0$ there exists a unique element u_μ which solves the nonlinear equation (2.4);*

(2) *there exists a unique element u which solves the variational inequality (2.1);*

(3) *the solution u_μ of (2.4) converges strongly to the solution u of (2.1), i.e.*

$$u_\mu \to u \quad \textit{in } X \quad \textit{as} \quad \mu \to 0. \tag{2.5}$$

Note that the convergence (2.5) above is understood in the following sense: for every sequence $\{\mu_n\} \subset \mathbb{R}_+$ converging to 0 as $n \to \infty$ one has $u_{\mu_n} \to u$ in X as $n \to \infty$. We shall use such notation to indicate various convergences in the rest of the book, see for instance (2.35) and (2.75).

The proof of Theorem 2.2 will be carried out in several steps, based on arguments of compactness and monotonicity. We suppose in what follows that the assumptions of Theorem 2.2 hold. We use c to denote a positive constant which does not depend on μ, and whose value may change from line to line. We start with the unique solvability of the nonlinear equation (2.4).

Lemma 2.3 *For each $\mu > 0$ there exists a unique element which solves the nonlinear equation (2.4).*

Proof Let $\mu > 0$. Since A is strongly monotone and Lipschitz continuous, by assumption (2.3)(a) it follows that the operator $v \mapsto Av + \frac{1}{\mu} Gv$ is also a strongly monotone Lipschitz continuous operator on X. Lemma 2.3 is now a consequence of Theorem 1.24. ■

Next, we perform *a priori* estimates on the solution of equation (2.4), which imply the following convergence result.

Lemma 2.4 *There exists an element $u \in X$ and a subsequence of the sequence $\{u_\mu\}$, again denoted $\{u_\mu\}$, which converges weakly to u, i.e.*

$$u_\mu \rightharpoonup u \quad in\ X \quad as \quad \mu \to 0. \tag{2.6}$$

Proof Let $\mu > 0$ and let $v_0 \in K$. We use (2.4) to obtain

$$(Au_\mu, v_0 - u_\mu)_X + \frac{1}{\mu}(Gu_\mu, v_0 - u_\mu)_X = (f, v_0 - u_\mu)_X$$

and, therefore,

$$(Au_\mu - Av_0, u_\mu - v_0)_X = (Av_0, v_0 - u_\mu)_X + \frac{1}{\mu}(Gu_\mu, v_0 - u_\mu)_X + (f, u_\mu - v_0)_X.$$

Then, by using (1.46), (2.3)(b) and the Cauchy–Schwarz inequality it follows that

$$m\,\|u_\mu - v_0\|_X^2 \le \|Av_0\|_X \|u_\mu - v_0\|_X + \|f\|_X \|u_\mu - v_0\|_X$$

and, therefore,

$$\|u_\mu - v_0\|_X \le \frac{1}{m}\Big(\|Av_0\|_X + \|f\|_X\Big). \tag{2.7}$$

Since

$$\|u_\mu\|_X \le \|u_\mu - v_0\|_X + \|v_0\|_X$$

it follows from (2.7) that there exists $c > 0$ such that

$$\|u_\mu\|_X \le c \tag{2.8}$$

and, therefore, the sequence $\{u_\mu\}$ is bounded in X. Lemma 2.4 is now a consequence of Theorem 1.19. ■

Next, we investigate the properties of the element u defined in Lemma 2.4.

Lemma 2.5 *The element u satisfies the variational inequality (2.1) and, moreover, it is the unique solution of this inequality.*

Proof Following Lemma 2.4 we consider a subsequence of the sequence $\{u_\mu\}$, again denoted $\{u_\mu\}$, such that (2.6) holds. Let $\mu > 0$ and let v be an arbitrary element in X. We use (2.4) to see that

$$(Au_\mu, v - u_\mu)_X + \frac{1}{\mu}(Gu_\mu, v - u_\mu)_X = (f, v - u_\mu)_X,$$

which implies that

$$\frac{1}{\mu}(Gu_\mu, u_\mu - v)_X = (Au_\mu, v - u_\mu)_X + (f, u_\mu - v)_X.$$

We write

$$Au_\mu = Au_\mu - A0_X + A0_X,$$

then we use the Lipschitz continuity of the operator A, (1.47), to obtain

$$\begin{aligned}\frac{1}{\mu}(Gu_\mu, u_\mu - v)_X &\le M\,\|u_\mu\|_X\|v - u_\mu\|_X \\ &\quad + \|A0_X\|_X\|v - u_\mu\|_X + \|f\|_X\|v - u_\mu\|_X.\end{aligned}$$

Next, we use (2.8) and the previous inequality to see that there exists a positive constant c which depends on v but is independent of μ, such that

$$(Gu_\mu, u_\mu - v)_X \le c\,\mu. \tag{2.9}$$

We now take $v = u$ in (2.9), then we pass to the upper limit as $\mu \to 0$ in the resulting inequality to obtain

$$\limsup_{\mu\to 0}\,(Gu_\mu, u_\mu - u)_X \le 0.$$

Therefore, using assumption (2.3)(a), the convergence (2.6) and Proposition 1.23 we deduce that

$$\liminf_{\mu\to 0}\,(Gu_\mu, u_\mu - v)_X \ge (Gu, u - v)_X \quad \forall\, v \in X. \tag{2.10}$$

On the other hand, inequality (2.9) implies that

$$\liminf_{\mu\to 0}\,(Gu_\mu, u_\mu - v)_X \le 0 \quad \forall\, v \in X. \tag{2.11}$$

We combine the inequalities (2.10) and (2.11) to see that

$$(Gu, u - v)_X \le 0 \quad \forall\, v \in X$$

and taking $v = u - Gu$ in this inequality yields $\|Gu\|_X^2 \le 0$. We conclude that $Gu = 0_X$ and, using assumption (2.3)(c), it follows that

$$u \in K. \tag{2.12}$$

Next, we use equation (2.4) to see that

$$(Au_\mu, v - u_\mu)_X + \frac{1}{\mu}\,(Gu_\mu, v - u_\mu)_X = (f, v - u_\mu)_X \quad \forall\, v \in K$$

and, using assumption (2.3)(b), we find that

$$\frac{1}{\mu}\,(Gu_\mu, v - u_\mu)_X = (f, v - u_\mu)_X - (Au_\mu, v - u_\mu)_X \le 0 \quad \forall\, v \in K.$$

We conclude from the above that

$$(Au_\mu, u_\mu - v)_X \le (f, u_\mu - v)_X \quad \forall\, v \in K. \tag{2.13}$$

We use now (2.12) and take $v = u$ in (2.13), then we pass to the upper limit as $\mu \to 0$ in the resulting inequality and use the weak convergence (2.6). As a result we obtain

$$\limsup_{\mu\to 0}\,(Au_\mu, u_\mu - u)_X \le 0 \tag{2.14}$$

and, using Proposition 1.23, it follows that

$$\liminf_{\mu\to 0}\,(Au_\mu, u_\mu - v)_X \ge (Au, u - v)_X \qquad \forall\, v \in X. \tag{2.15}$$

On the other hand, passing to the lower limit as $\mu \to 0$ in (2.13) yields

$$\liminf_{\mu\to 0}\,(Au_\mu, u_\mu - v)_X \le (f, u - v)_X \qquad \forall\, v \in K. \tag{2.16}$$

We combine now the inequalities (2.15) and (2.16) to see that

$$(Au, u - v)_X \le (f, u - v)_X \qquad \forall\, v \in K,$$

and, therefore,

$$(Au, v - u)_X \ge (f, v - u)_X \qquad \forall\, v \in K. \tag{2.17}$$

It follows now from (2.12) and (2.17) that u satisfies the variational inequality (2.1), which concludes the first part of the lemma. The uniqueness part follows by using the strong monotonicity of the operator A and was already presented in the proof of Theorem 2.1. ■

We proceed our analysis with the following weak convergence result.

Lemma 2.6 *The whole sequence* $\{u_\mu\}$ *converges weakly in* X *to the element* u.

Proof A careful examination of the proof of Lemma 2.5 shows that if u is the weak limit of a weakly convergent subsequence of the sequence $\{u_\mu\} \subset X$, then u satisfies the variational inequality (2.1). Recall also that this variational inequality has a unique solution and, moreover, it results from the proof of Lemma 2.4 that the sequence $\{u_\mu\}$ is bounded in X. Lemma 2.6 is now a consequence of Theorem 1.20. ■

The last step is provided by the following strong convergence result.

Lemma 2.7 *The sequence* $\{u_\mu\}$ *converges strongly in* X *to the element* u, *that is*

$$u_\mu \to u \quad in\ X \quad as \quad \mu \to 0. \tag{2.18}$$

Proof Let $\mu > 0$. We take $v = u$ in (2.15) to see that

$$\liminf_{\mu \to 0} (Au_\mu, u_\mu - u)_X \geq 0,$$

then we combine this inequality with (2.14) to obtain that

$$\lim_{\mu \to 0} (Au_\mu, u_\mu - u)_X = 0. \tag{2.19}$$

On the other hand, from the weak convergence of the sequence $\{u_\mu\}$ to u, guaranteed by Lemma 2.6, it follows that

$$\lim_{\mu \to 0} (Au, u_\mu - u)_X = 0. \tag{2.20}$$

Next, from the strong monotonicity of the operator A, (1.46), it follows that

$$\begin{aligned} m\,\|u_\mu - u\|_X^2 &\leq (Au_\mu - Au, u_\mu - u)_X \\ &= (Au_\mu, u_\mu - u)_X - (Au, u_\mu - u)_X. \end{aligned} \tag{2.21}$$

The convergence (2.18) is now a consequence of (2.19)–(2.21). ■

We can now easily provide the proof of Theorem 2.2.

Proof The points (1), (2) and (3) of Theorem 2.2 are direct consequences of Lemmas 2.3, 2.5 and 2.7, respectively. ■

The interest in Theorem 2.2 is twofold; first, it provides the existence and uniqueness of the solution to the variational inequality (2.1); second, it shows that the solution of (2.1) represents the strong limit of the sequence of solutions u_μ of the problem (2.4), as $\mu \to 0$.

2.2 Variational inequalities of the second kind

Given a set $K \subset X$, an operator $A : K \to X$, a function $j : K \to \mathbb{R}$ and an element $f \in X$, in this section we consider the problem of finding an element u such that

$$u \in K, \quad (Au, v-u)_X + j(v) - j(u) \geq (f, v-u)_X \quad \forall v \in K. \qquad (2.22)$$

A variational inequality of the form (2.22) is called an *elliptic variational inequality of the second kind.* An example is provided by inequality (1.83) which characterizes the minimizers of the functional (1.81) on K. Moreover, Corollary 1.38 provides a first existence and uniqueness result in the study of this kind of inequality. Finally, note that in the particular case when $j \equiv 0$, the variational inequality (2.22) represents an elliptic variational inequality of the form (2.1), i.e. an elliptic variational inequality of the first kind.

2.2.1 Existence and uniqueness

In the study of (2.22) we consider the following assumptions:

$$K \text{ is a nonempty closed convex subset of } X; \qquad (2.23)$$

$$\begin{aligned} &A : K \to X \text{ is a strongly monotone} \\ &\text{Lipschitz continuous operator, i.e. it} \\ &\text{satisfies conditions (1.48) and (1.49);} \end{aligned} \qquad (2.24)$$

$$j : K \to \mathbb{R} \text{ is a convex l.s.c. function.} \qquad (2.25)$$

The main result of this section is the following.

Theorem 2.8 *Let X be a Hilbert space and assume that (2.23)–(2.25) hold. Then, for each $f \in X$ the variational inequality (2.22) has a unique solution.*

The proof of Theorem 2.8 will be carried out in two steps. The first step consists of solving (2.22) in the case when A is the identity operator and obtaining an estimated result.

Lemma 2.9 *Assume (2.23) and (2.25). Then, for each $f \in X$ there exists a unique element u such that*

$$u \in K, \quad (u, v-u)_X + j(v) - j(u) \geq (f, v-u)_X \quad \forall v \in K. \qquad (2.26)$$

Moreover, if u_1 and u_2 denote the solutions of the inequality (2.26) for $f = f_1 \in X$ and $f = f_2 \in X$, respectively, then

$$\|u_1 - u_2\|_X \leq \|f_1 - f_2\|_X. \qquad (2.27)$$

Proof The first part of the lemma is a direct consequence of Corollary 1.38. Let now $f_1, f_2 \in X$ and let $u_1,\, u_2 \in K$ be such that

$$(u_1, v - u_1)_X + j(v) - j(u_1) \geq (f_1, v - u_1)_X \quad \forall\, v \in K, \tag{2.28}$$
$$(u_2, v - u_2)_X + j(v) - j(u_2) \geq (f_2, v - u_2)_X \quad \forall\, v \in K. \tag{2.29}$$

Taking $v = u_2$ in (2.28), $v = u_1$ in (2.29) and adding the resulting inequalities we find that

$$\|u_1 - u_2\|_X^2 \leq (f_1 - f_2, u_1 - u_2)_X, \tag{2.30}$$

which implies (2.27). ■

Lemma 2.9 allows us to introduce the following definition.

Definition 2.10 Let X be a Hilbert space, $K \subset X$ a nonempty closed convex subset and $j : K \to \mathbb{R}$ a convex lower semicontinuous function. Then, for each $f \in X$, the solution u of the variational inequality (2.26) is called the *proximal element* of f with respect to the function j and it is usually denoted $\text{prox}_j f$. The operator $\text{prox}_j : X \to K$ defined by $f \mapsto \text{prox}_j f$ is called the *proximity operator* of the function j.

The proximity operators were first introduced in [112]. Note that Lemma 2.9 states that prox_j is a nonexpansive operator, i.e.

$$\|\text{prox}_j f_1 - \text{prox}_j f_2\|_X \leq \|f_1 - f_2\|_X \quad \forall\, f_1, f_2 \in X. \tag{2.31}$$

Moreover, it follows from (2.30) that prox_j is a monotone operator, since

$$(\text{prox}_j f_1 - \text{prox}_j f_2, f_1 - f_2)_X \geq 0 \quad \forall\, f_1, f_2 \in X.$$

Finally, we remark that if K is a nonempty closed convex subset of X, we may consider the proximity operator of the zero function on K, denoted prox_K. It follows from Definition 2.10 that $u = \text{prox}_K f$ if and only if

$$u \in K, \quad (u, v - u)_X \geq (f, v - u)_X \quad \forall\, v \in K.$$

Therefore, using Proposition 1.12 we obtain that $\text{prox}_K = \mathcal{P}_K$, where $\mathcal{P}_K$ denotes the projection operator on K. We conclude from the above that the projection operators represent a particular type of proximity operator.

We have now all the ingredients to provide the proof of Theorem 2.8.

Proof Let $f \in X$ and let $\rho > 0$ be a parameter to be chosen later. Since $\rho j : K \to \mathbb{R}$ is again a convex lower semicontinuous function, we can define an operator $S_\rho : K \to K$ by

$$S_\rho(v) = \text{prox}_{\rho j}(\rho f - \rho A v + v) \quad \forall\, v \in K. \tag{2.32}$$

Moreover, using (2.32) and (2.31) we find that

$$\|S_\rho u - S_\rho v\|_X \leq \|(u-v) - \rho(Au - Av)\|_X \qquad \forall\, u,\, v \in K.$$

Then, using arguments similar to those used in the proof of Theorem 1.24 we obtain that, with a convenient choice of ρ, we have

$$\|S_\rho u - S_\rho v\|_X \leq k(\rho)\, \|u - v\|_X \qquad \forall\, u,\, v \in K,$$

where $k(\rho) = (1 - 2\rho m + \rho^2 M^2)^{\frac{1}{2}} \in [0, 1)$ and m, M are the constants in (1.48) and (1.49), respectively. Next, we use the Banach fixed point argument to see that there exists $u \in K$ such that

$$S_\rho u = \mathrm{prox}_{\rho j}(\rho f - \rho A u + u) = u.$$

Then, by Definition 2.10 we obtain

$$(u, v-u)_X + \rho j(v) - \rho j(u) \geq (\rho f - \rho A u + u, v - u)_X \quad \forall\, v \in K,$$

i.e.,

$$\rho\,[(Au, v-u)_X + j(v) - j(u)] \geq \rho\,(f, v-u)_X \quad \forall\, v \in K.$$

Since $\rho > 0$, we deduce from the above inequality that u is a solution of the variational inequality (2.22), which proves the existence part of the theorem.

To show the uniqueness, assume there are two solutions u_1, $u_2 \in K$ of the variational inequality (2.22). Then,

$$(Au_1, v - u_1)_X + j(v) - j(u_1) \geq (f, v - u_1)_X \quad \forall\, v \in K,$$
$$(Au_2, v - u_2)_X + j(v) - j(u_2) \geq (f, v - u_2)_X \quad \forall\, v \in K.$$

Taking $v = u_2$ in the first inequality, $v = u_1$ in the second one and adding the resulting inequalities, we find that

$$(Au_1 - Au_2, u_1 - u_2)_X \leq 0.$$

We combine this inequality with (1.48) to obtain $u_1 = u_2$, which proves the uniqueness part of Theorem 2.8. ∎

Recall that for $j \equiv 0$ the variational inequality (2.22) reduces to the variational inequality (2.1) and, moreover, for $K = X$ and $j \equiv 0$ it reduces to the nonlinear equation $Au = f$. We conclude that Theorems 1.24 and 2.1 can be recovered by Theorem 2.8.

2.2.2 A convergence result

We proceed with a result concerning the dependence of the solution of the variational inequality (2.22) with respect to the function j. To this end, we assume in what follows that (2.23)–(2.25) hold, $f \in X$ and, for each $\rho > 0$, let j_ρ be a perturbation of j which satisfies (2.25). We consider the problem of finding an element u_ρ such that

$$u_\rho \in K, \quad (Au_\rho, v - u_\rho)_X + j_\rho(v) - j_\rho(u_\rho) \geq (f, v - u_\rho)_X \quad \forall v \in K. \quad (2.33)$$

We deduce from Theorem 2.8 that inequality (2.22) has a unique solution $u \in K$ and, for each $\rho > 0$, inequality (2.33) has a unique solution $u_\rho \in K$.

Consider now the following assumption.

$$\left.\begin{array}{l} \text{There exists } G : \mathbb{R}_+ \to \mathbb{R}_+ \text{ such that:} \\ \text{(a) } j_\rho(u_1) - j_\rho(u_2) + j(u_2) - j(u_1) \leq G(\rho)\,\|u_1 - u_2\|_X \\ \quad \forall\, u_1, u_2 \in K, \text{ for each } \rho > 0; \\ \text{(b) } \lim_{\rho \to 0} G(\rho) = 0. \end{array}\right\} \quad (2.34)$$

Then the behavior of the solution u_ρ as ρ converges to zero is given in the following theorem.

Theorem 2.11 *Under the assumption (2.34), the solution u_ρ of problem (2.33) converges to the solution u of problem (2.22), i.e.*

$$u_\rho \to u \quad \text{in } X \quad \text{as} \quad \rho \to 0. \quad (2.35)$$

Note that, as in the case of (2.5), the convergence (2.35) is understood in the following sense: for every sequence $\{\rho_n\} \subset \mathbb{R}_+$ converging to 0 as $n \to \infty$ one has $u_{\rho_n} \to u$ in X as $n \to \infty$.

Proof Let $\rho > 0$. We take $v = u_\rho$ in (2.22) and $v = u$ in (2.33) and add the resulting inequalities to obtain

$$(Au_\rho - Au, u_\rho - u)_X \leq j_\rho(u) - j_\rho(u_\rho) + j(u_\rho) - j(u).$$

We now use (1.48) and assumption (2.34)(a) to find that

$$m\,\|u_\rho - u\|_X \leq G(\rho). \quad (2.36)$$

The convergence result (2.35) is now a consequence of the inequality (2.36) combined with assumption (2.34)(b). ■

2.2.3 Regularization

We consider now the variational inequality (2.22) in the case when $K = X$ and we investigate its unique solvability by using a regularization method. Thus, we assume in what follows that $A : X \to X$, $j : X \to \mathbb{R}$ and

$f \in X$ are given and we consider the problem of finding an element u such that

$$u \in X, \quad (Au, v-u)_X + j(v) - j(u) \geq (f, v-u)_X \quad \forall v \in X. \tag{2.37}$$

Let $\rho > 0$ be a parameter. We also assume that there exists a family of functionals (j_ρ) which satisfies:

$$\left.\begin{array}{l} \text{(a) } j_\rho : X \to \mathbb{R} \text{ is a convex Gâteaux differentiable function,} \\ \quad\ \text{for each } \rho > 0; \\ \text{(b) } \nabla j_\rho : X \to X \text{ is a Lipschitz continuous operator,} \\ \quad\ \text{for each } \rho > 0. \end{array}\right\} \tag{2.38}$$

$$\left.\begin{array}{l} \text{There exists } F : \mathbb{R}_+ \to \mathbb{R}_+ \text{ such that} \\ \text{(a) } |j_\rho(v) - j(v)| \leq F(\rho) \quad \forall v \in X, \text{ for each } \rho > 0; \\ \text{(b) } \lim\limits_{\rho \to 0} F(\rho) = 0. \end{array}\right\} \tag{2.39}$$

For each $\rho > 0$ we consider the problem of finding an element u_ρ such that

$$u_\rho \in X, \quad Au_\rho + \nabla j_\rho(u_\rho) = f. \tag{2.40}$$

We have the following existence, uniqueness and convergence result.

Theorem 2.12 *Let X be a Hilbert space. Assume that $A : X \to X$ is a strongly monotone Lipschitz continuous operator, $j : X \to \mathbb{R}$, (j_ρ) is a family of functionals which satisfies (2.38), (2.39) and $f \in X$. Then:*

(1) *for each $\rho > 0$ there exists a unique element u_ρ which solves the nonlinear equation (2.40);*

(2) *there exists a unique element u which solves the variational inequality (2.37);*

(3) *The solution u_ρ of (2.40) converges strongly to the solution u of (2.37), i.e.*

$$u_\rho \to u \quad \text{in } X \quad \text{as} \quad \rho \to 0.$$

The proof of Theorem 2.12 will be carried out in several steps, based on arguments of compactness and monotonicity. We suppose in what follows that the assumptions of Theorem 2.12 hold; moreover, we use c to denote a positive constant which may depend on A and f, but is independent of ρ, and whose value may change from line to line. We start with the unique solvability of the nonlinear equation (2.40).

Lemma 2.13 *For each $\rho > 0$ there exists a unique element which solves the nonlinear equation (2.40).*

Proof Let $\rho > 0$. Using Proposition 1.32 and assumption (2.38) it follows that the gradient operator ∇j_ρ is a monotone Lipschitz continuous operator on X. Therefore, since the operator A is strongly monotone and Lipschitz continuous, it follows that $A + \nabla j_\rho$ is also a strongly monotone Lipschitz continuous operator on X. Lemma 2.13 is now a consequence of Theorem 1.24. ■

Next, we use the uniform convergence assumption (2.39) to derive the following result on the functional j.

Lemma 2.14 *The functional j is convex and lower semicontinuous.*

Proof Let $u, v \in X$ and let $t \in [0, 1]$. We use the convexity of the function j_ρ to see that

$$j_\rho(u + t(v - u)) \leq (1 - t) j_\rho(u) + t\, j_\rho(v),$$

for each $\rho > 0$. We pass to the limit as $\rho \to 0$ in this inequality and use (2.39) to obtain

$$j(u + t(v - u)) \leq (1 - t) j(u) + t\, j(v),$$

which proves the convexity of j.

Next, assume that $u \in X$ and let $\{u_n\}$ be a sequence of elements of X such that $u_n \to u$ in X. Let $\rho > 0$ and $n \in \mathbb{N}$ be fixed. We write

$$j(u) = (j(u) - j_\rho(u)) + (j_\rho(u) - j_\rho(u_n)) + (j_\rho(u_n) - j(u_n)) + j(u_n).$$

Then, we use assumption (2.39)(a) to obtain

$$j(u) \leq j_\rho(u) - j_\rho(u_n) + j(u_n) + 2\, F(\rho)$$

or, equivalently,

$$j(u) + j_\rho(u_n) \leq j_\rho(u) + j(u_n) + 2\, F(\rho).$$

We pass to the lower limit in this inequality as $n \to \infty$. As a result we obtain

$$j(u) + \liminf_n j_\rho(u_n) \leq j_\rho(u) + \liminf_n j(u_n) + 2\, F(\rho). \tag{2.41}$$

On the other hand, recall the inequality

$$j_\rho(u) \leq \liminf_n j_\rho(u_n), \tag{2.42}$$

which results from the lower semicontinuity of the function j_ρ, guaranteed by assumption (2.38) and Corollary 1.33. We now combine (2.41) and (2.42) to see that

$$j(u) \leq \liminf_n j(u_n) + 2\, F(\rho) \tag{2.43}$$

and recall that this inequality holds for each $\rho > 0$. Finally, we pass to the limit in (2.43) as $\rho \to 0$ and use assumption (2.39)(b) to deduce that j is l.s.c. ■

Next, we perform *a priori* estimates on the solution of the equation (2.40), which imply the following convergence result.

Lemma 2.15 *There exists an element $u \in X$ and a subsequence of the sequence $\{u_\rho\}$, again denoted $\{u_\rho\}$, which converges weakly to u, i.e.*

$$u_\rho \rightharpoonup u \quad \text{in } X \quad \text{as} \quad \rho \to 0. \tag{2.44}$$

Proof Let $\rho > 0$. We use (2.40) to obtain

$$(Au_\rho, v - u_\rho)_X + (\nabla j_\rho(u_\rho), v - u_\rho)_X = (f, v - u_\rho)_X \quad \forall\, v \in X$$

and, keeping in mind (1.67) we deduce that

$$(Au_\rho, v - u_\rho)_X + j_\rho(v) - j_\rho(u_\rho) \geq (f, v - u_\rho)_X \quad \forall\, v \in X. \tag{2.45}$$

We test in (2.45) with $v = 0_X$ to see that

$$(Au_\rho, u_\rho)_X + j_\rho(u_\rho) \leq j_\rho(0_X) + (f, u_\rho)_X.$$

This inequality, combined with inequalities

$$j(u_\rho) - F(\rho) \leq j_\rho(u_\rho), \quad j_\rho(0_X) \leq j(0_X) + F(\rho),$$

guaranteed by (2.39)(a), implies that

$$\begin{aligned}(Au_\rho &- A0_X, u_\rho)_X + j(u_\rho)\\ &\leq (f, u_\rho)_X - (A0_X, u_\rho)_X + j(0_X) + 2\,F(\rho).\end{aligned} \tag{2.46}$$

On the other hand, we use Lemma 2.14, Proposition 1.29 and Riesz's representation theorem to see that there exist $\omega \in X$ and $\alpha \in \mathbb{R}$ such that

$$j(u_\rho) \geq (\omega, u_\rho)_X + \alpha. \tag{2.47}$$

Therefore, combining (2.46), (2.47), by using (1.46) and the Cauchy–Schwarz inequality it follows that

$$m\,\|u_\rho\|_X^2 \leq \big(\|f\|_X + \|A0_X\|_X + \|\omega\|_X\big)\|u_\rho\|_X + |j(0_X)| + 2\,F(\rho) + |\alpha|.$$

We now use the elementary inequality

$$x,\ a,\ b \geq 0 \ \text{ and } \ x^2 \leq ax + b \implies x^2 \leq a^2 + 2b$$

and assumption (2.39)(b) to deduce that the sequence $\{u_\rho\}$ is bounded in X as $\rho \to 0$. Lemma 2.15 is now a consequence of Theorem 1.19. ■

Next, we investigate the properties of the element u defined in Lemma 2.15.

Lemma 2.16 *The element u satisfies the variational inequality (2.37) and, moreover, it is the unique solution of this inequality.*

Proof Following Lemma 2.15 we consider a subsequence of the sequence $\{u_\rho\}$, again denoted $\{u_\rho\}$, such that (2.44) holds. Let $\rho > 0$. We use (2.45) with $v = u$ to see that

$$(Au_\rho, u_\rho - u)_X \le j_\rho(u) - j_\rho(u_\rho) + (f, u_\rho - u)_X. \qquad (2.48)$$

We write

$$j_\rho(u) - j_\rho(u_\rho) = \big(j_\rho(u) - j(u)\big) + \big(j(u) - j(u_\rho)\big) + \big(j(u_\rho) - j_\rho(u_\rho)\big)$$

and use assumption (2.39)(a) to see that

$$j_\rho(u) - j_\rho(u_\rho) \le 2F(\rho) + j(u) - j(u_\rho).$$

Next, we pass to the upper limit as $\rho \to 0$ in the previous inequality and use assumption (2.39)(b) to obtain

$$\limsup_{\rho\to 0} \big(j_\rho(u) - j_\rho(u_\rho)\big) \le j(u) - \liminf_{\rho\to 0} j(u_\rho).$$

Therefore, using the lower semicontinuity of j we deduce that

$$\limsup_{\rho\to 0} \big(j_\rho(u) - j_\rho(u_\rho)\big) \le 0. \qquad (2.49)$$

And, finally, passing to the upper limit as $\rho \to 0$ in (2.48) and using (2.49) yields

$$\limsup_{\rho\to 0} (Au_\rho, u_\rho - u)_X \le 0. \qquad (2.50)$$

Using now (2.44), (2.50) and Proposition 1.23 we deduce that

$$\liminf_{\rho\to 0} (Au_\rho, u_\rho - v)_X \ge (Au, u - v)_X \quad \forall\, v \in X. \qquad (2.51)$$

Next, we use inequality (2.45) to see that

$$j_\rho(v) - j_\rho(u_\rho) \ge (f, v - u_\rho)_X + (Au_\rho, u_\rho - v)_X \quad \forall\, v \in X$$

and, therefore,

$$\begin{aligned} &j(v) + \big(j_\rho(v) - j(v)\big) + \big(j(u_\rho) - j_\rho(u_\rho)\big) \\ &\qquad \ge j(u_\rho) + (f, v - u_\rho)_X + (Au_\rho, u_\rho - v)_X \quad \forall\, v \in X. \end{aligned} \qquad (2.52)$$

We pass to the lower limit as $\rho \to 0$ in (2.52) and use the assumption (2.39), the lower semicontinuity of j and inequality (2.51). As a result we obtain

$$j(v) \ge j(u) + (f, v - u)_X + (Au, u - v)_X \quad \forall\, v \in X. \qquad (2.53)$$

It follows now from (2.53) that u satisfies the variational inequality (2.37), which concludes the first part of the lemma. The uniqueness part follows by using the strong monotonicity of the operator A and was already presented in the proof of Theorem 2.8. ■

We proceed through our analysis with the following weak convergence result.

Lemma 2.17 *The whole sequence $\{u_\rho\}$ converges weakly in X to the element u.*

Proof A careful examination of the proof of Lemma 2.16 shows that if u is the weak limit of a weakly convergent subsequence of the sequence $\{u_\rho\} \subset X$, then u satisfies the variational inequality (2.37). Recall also that this variational inequality has a unique solution and, moreover, it was shown in the proof of Lemma 2.15 that the sequence $\{u_\rho\}$ is bounded in X. Lemma 2.17 is now a consequence of Theorem 1.20. ■

The last step is provided by the following strong convergence result.

Lemma 2.18 *The sequence $\{u_\rho\}$ converges strongly in X to the element u, that is*

$$u_\rho \to u \quad \text{in } X \quad \text{as} \quad \rho \to 0. \tag{2.54}$$

Proof Let $\rho > 0$. We take $v = u$ in (2.51) to see that

$$\liminf_{\rho \to 0} (Au_\rho, u_\rho - u)_X \geq 0,$$

then we combine this inequality with (2.50) to obtain that

$$\lim_{\rho \to 0} (Au_\rho, u_\rho - u)_X = 0. \tag{2.55}$$

On the other hand, from the weak convergence of the sequence $\{u_\rho\}$ to u, guaranteed by Lemma 2.17, it follows that

$$\lim_{\rho \to 0} (Au, u_\rho - u)_X = 0. \tag{2.56}$$

Next, from the strong monotonicity of the operator A, (1.46), it follows that

$$\begin{aligned} m\,\|u_\rho - u\|_X^2 &\leq (Au_\rho - Au, u_\rho - u)_X \\ &= (Au_\rho, u_\rho - u)_X - (Au, u_\rho - u)_X. \end{aligned} \tag{2.57}$$

The convergence (2.54) is now a consequence of (2.55)–(2.57). ■

We can now easily provide the proof of Theorem 2.12.

Proof The points (1), (2) and (3) of Theorem 2.12 are direct consequences of Lemmas 2.13, 2.16 and 2.18, respectively. ■

The interest in Theorem 2.12 is twofold; first, it provides the existence and uniqueness of the solution to the variational inequality (2.37); second, it shows that the solution of (2.37) represents the strong limit of the sequence of solutions u_ρ of the problems (2.40), as $\rho \to 0$.

2.3 Quasivariational inequalities

For the variational inequalities studied in this section we allow the function j to depend on the solution. Therefore, given a subset $K \subset X$, an operator $A : K \to X$, a function $j : K \times K \to \mathbb{R}$ and an element $f \in X$, in this section we consider the problem of finding an element u such that

$$u \in K, \quad (Au, v-u)_X + j(u,v) - j(u,u) \geq (f, v-u)_X \quad \forall\, v \in K. \tag{2.58}$$

A problem of the form (2.58) is called an *elliptic quasivariational inequality.* Its solvability is obtained, usually, by using fixed point arguments.

2.3.1 The Banach fixed point argument

In the study of (2.58), besides (2.23) and (2.24) we consider the following assumption on the function j.

$$\left.\begin{array}{l} \text{(a) For all } \eta \in K,\ j(\eta,\cdot) : K \to \mathbb{R} \text{ is convex and l.s.c.} \\ \text{(b) There exists } \alpha \geq 0 \text{ such that} \\ \quad j(\eta_1, v_2) - j(\eta_1, v_1) + j(\eta_2, v_1) - j(\eta_2, v_2) \\ \qquad \leq \alpha\, \|\eta_1 - \eta_2\|_X \|v_1 - v_2\|_X \quad \forall\, \eta_1, \eta_2, v_1, v_2 \in K. \end{array}\right\} \tag{2.59}$$

Our first result in this section is the following.

Theorem 2.19 *Let X be a Hilbert space and assume that (2.23), (2.24), and (2.59) hold. Moreover, assume that $m > \alpha$ where $m > 0$ is the constant defined in (1.48). Then, for each $f \in X$ the quasivariational inequality (2.58) has a unique solution.*

The proof of Theorem 2.19 will be carried out in three steps. We assume in what follows that (2.23), (2.24) and (2.59) hold and let $f \in X$. In the first step, for each $\eta \in K$ we consider the auxiliary problem of finding u_η which solves the elliptic variational inequality

$$\begin{aligned} u_\eta \in K, \quad & (Au_\eta, v - u_\eta)_X + j(\eta, v) - j(\eta, u_\eta) \\ & \geq (f, v - u_\eta)_X \quad \forall\, v \in K. \end{aligned} \tag{2.60}$$

We use Theorem 2.8 to obtain the following existence and uniqueness result.

Lemma 2.20 *For each $\eta \in K$ there exists a unique solution u_η of inequality (2.60).*

Next, we consider the operator $\Lambda : K \to K$ defined by

$$\Lambda\eta = u_\eta \quad \forall\, \eta \in K, \tag{2.61}$$

where $u_\eta \in K$ is the unique solution of (2.60), guaranteed by Lemma 2.20. We have the following fixed point result.

Lemma 2.21 *If $m > \alpha$ then Λ has a unique fixed point $\eta^* \in K$.*

Proof Let $\eta_1,, \eta_2 \in K$ and let u_i denote the solution of (2.60) for $\eta = \eta_i$, i.e. $u_i = u_{\eta_i}$, $i = 1, 2$. We have

$$\begin{aligned}(Au_1, v - u_1)_X + j(\eta_1, v) - j(\eta_1, u_1) &\geq (f, v - u_1)_X \quad \forall\, v \in K,\\ (Au_2, v - u_2)_X + j(\eta_2, v) - j(\eta_2, u_2) &\geq (f, v - u_2)_X \quad \forall\, v \in K.\end{aligned}$$

We take $v = u_2$ in the first inequality, $v = u_1$ in the second one and add the resulting inequalities to obtain

$$(Au_1 - Au_2, u_1 - u_2)_X \leq j(\eta_1, u_2) - j(\eta_1, u_1) + j(\eta_2, u_1) - j(\eta_2, u_2).$$

We now use the properties (1.48) and (2.59)(b) of A and j, respectively, to see that

$$\|u_1 - u_2\|_X \leq \frac{\alpha}{m}\,\|\eta_1 - \eta_2\|_X. \tag{2.62}$$

Since $m > \alpha$ the inequality (2.62) shows that the operator Λ given by (2.61) is a contraction on K and, therefore, Lemma 2.21 follows from Theorem 1.5. ■

We can now proceed with the proof of Theorem 2.19.

Proof Let η^* be the fixed point of the operator Λ obtained in Lemma 2.21. Since $\eta^* = \Lambda\eta^* = u_{\eta^*}$ it follows from (2.60) that η^* is a solution of the quasivariational inequality (2.58), which concludes the proof of the existence part.

Next, let u be a solution of (2.58). It follows that u is a solution of the variational inequality (2.60) with $\eta = u$ and, since by Lemma 2.20 this inequality has a unique solution denoted u_η, we have $u = u_\eta$. This equality shows that $\eta = u_\eta$ and, keeping in mind the definition (2.61) of the operator Λ, we deduce that $\eta = \Lambda\eta$. Since Lemma 2.21 guarantees that the operator Λ has a unique fixed point, denoted η^*, we find that $\eta = \eta^*$. Therefore, $u = \eta^*$, which concludes the proof of the uniqueness part. ■

2.3.2 *The Schauder fixed point argument*

Next, we investigate the quasivariational inequality (2.58) by using the Schauder fixed point argument. To this end we assume in what follows that (2.23) and (2.24) hold and, in addition,

$$0_X \in K. \tag{2.63}$$

Moreover, we assume that $j : K \times K \to \mathbb{R}$ satisfies the following conditions.

$$\left.\begin{array}{l}\text{For all } \eta \in K, \text{ the function } v \mapsto j(\eta, v) : K \to \mathbb{R} \\ \text{is convex, } j(\eta, v) \geq 0 \text{ for all } v \in K \text{ and } j(\eta, 0_X) = 0.\end{array}\right\} \tag{2.64}$$

$$\left.\begin{array}{l}\text{For all sequences } \{\eta_n\} \subset K \text{ and } \{u_n\} \subset K \text{ such that} \\ \eta_n \rightharpoonup \eta \text{ in } X, \ u_n \rightharpoonup u \text{ in } X \text{ and for all } v \in K, \\ \text{the inequality below holds:} \\ \limsup_{n\to\infty} \left[j(\eta_n, v) - j(\eta_n, u_n)\right] \leq j(\eta, v) - j(\eta, u).\end{array}\right\} \tag{2.65}$$

Our second result in this section is the following.

Theorem 2.22 *Let X be a Hilbert space and assume that (2.23), (2.24), (2.63), (2.64) and (2.65) hold. Then, for each $f \in X$ there exists at least one solution to the quasivariational inequality (2.58).*

The proof of Theorem 2.22 will be carried out in five steps. We assume in what follows that (2.23), (2.24), (2.63), (2.64) and (2.65) hold and let $f \in X$. As in the proof of Theorem 2.19, we start with the study of the elliptic variational inequality (2.60), defined for each $\eta \in K$.

Lemma 2.23 *For each $\eta \in K$, the inequality (2.60) has a unique solution $u_\eta \in K$. Moreover, the solution satisfies*

$$\|u_\eta\|_X \leq \frac{1}{m}\left(\|f\|_X + \|A0_X\|_X\right). \tag{2.66}$$

Proof Let $\eta \in K$. It follows from (2.64) that $j(\eta, \cdot) : K \to \mathbb{R}$ is a convex function; moreover, choosing $\eta_n = \eta$ in (2.65) yields

$$\liminf_{n\to\infty} j(\eta, u_n) \geq j(\eta, u)$$

whenever $u_n \rightharpoonup u$ in X, i.e. $j(\eta, \cdot)$ is an l.s.c. function. The existence and uniqueness part in Lemma 2.23 is now a consequence of Theorem 2.8.

Next, we choose $v = 0_X$ in (2.60) to obtain

$$(Au_\eta, u_\eta)_X + j(\eta, u_\eta) \leq (f, u_\eta)_X + j(\eta, 0_X)$$

and, using (2.64) we deduce that

$$(Au_\eta, u_\eta)_X \le (f, u_\eta)_X.$$

Then we write $Au_\eta = Au_\eta - A0_X + A0_X$ and use the strong monotonicity of the operator A to find that

$$m\,\|u_\eta\|_X^2 \le (f, u_\eta)_X - (A0_X, u_\eta)_X$$

which implies (2.66) and concludes the proof. ■

Lemma 2.23 allows us, again, to consider the operator Λ defined by (2.61). The properties of this operator are investigated in the next three steps.

Lemma 2.24 *The operator $\Lambda : K \to K$ is weakly continuous, i.e. $\{\eta_n\} \subset K$, $\eta_n \rightharpoonup \eta$ in X imply $\Lambda\eta_n \rightharpoonup \Lambda\eta$ in X.*

Proof Let $\{\eta_n\} \subset K$ be a sequence such that

$$\eta_n \rightharpoonup \eta \quad \text{in } X \tag{2.67}$$

and note that (2.23) implies that $\eta \in K$. For simplicity, we denote $\Lambda\eta_n = u_n$ and, using (2.66), we deduce that $\{u_n\}$ is a bounded sequence in K. Therefore there exists a subsequence, again denoted $\{u_n\}$, and an element $w \in K$ such that

$$u_n \rightharpoonup w \quad \text{in } X. \tag{2.68}$$

We write (2.60) for η_n and test in the resulting inequality with $v = w$ to deduce

$$(Au_n, u_n - w)_X \le (f, u_n - w)_X + j(\eta_n, w) - j(\eta_n, u_n) \quad \forall\, n \in \mathbb{N}.$$

Using (2.68) it follows that

$$\limsup_{n\to\infty}\,(Au_n, u_n - w)_X \le \limsup_{n\to\infty}\,[j(\eta_n, w) - j(\eta_n, u_n)]$$

and, keeping in mind (2.67), (2.68) and assumption (2.65), we obtain

$$\limsup_{n\to\infty}\,(Au_n, u_n - w)_X \le 0. \tag{2.69}$$

We now use (2.68), (2.69) and arguments similar to those used in the proof of Proposition 1.23 to see that

$$\liminf_{n\to\infty}\,(Au_n, u_n - v)_X \ge (Aw, w - v)_X \quad \forall\, v \in K. \tag{2.70}$$

On the other hand, using (2.60) again, we obtain

$$(Au_n, u_n - v)_X \le (f, u_n - v)_X + j(\eta_n, v) - j(\eta_n, u_n) \quad \forall\, v \in K,\ n \in \mathbb{N}$$

and, combining this inequality with (2.67), (2.68) and assumption (2.65), we find

$$\limsup_{n\to\infty} (Au_n, u_n - v)_X \le (f, w - v)_X + j(\eta, v) - j(\eta, w) \quad \forall v \in K. \tag{2.71}$$

It follows from (2.70) and (2.71) that

$$(Aw, w - v)_X \le (f, w - v)_X + j(\eta, v) - j(\eta, w) \quad \forall v \in K,$$

which shows that w solves (2.60) and, therefore, using the definition of the operator Λ it follows that $w = \Lambda\eta$. A careful examination based on the arguments above shows that $\Lambda\eta$ is the unique weak limit of any subsequence of the bounded sequence $\{\Lambda\eta_n\} \subset K$. Lemma 2.24 is now a consequence of Theorem 1.20. ■

Lemma 2.25 *The operator $\Lambda : K \to K$ maps weak convergent sequences into strong convergent sequences, i.e. $\{\eta_n\} \subset K$, $\eta_n \rightharpoonup \eta$ in X imply $\Lambda\eta_n \to \Lambda\eta$ in X.*

Proof Let $\{\eta_n\} \subset K$ be a sequence such that $\eta_n \rightharpoonup \eta$ in X and recall that (2.23) implies that $\eta \in K$. Denote $\Lambda\eta_n = u_n$ and $\Lambda\eta = u$. Using Lemma 2.24 we deduce that

$$u_n \rightharpoonup u \quad \text{in } X. \tag{2.72}$$

Moreover, writing (2.60) for $\eta = \eta_n$ and taking $v = u$ in the resulting inequality it follows that

$$(Au_n, u_n - u)_X \le (f, u_n - u)_X + j(\eta_n, u) - j(\eta_n, u_n) \quad \forall n \in \mathbb{N}.$$

Next, using assumption (1.48) we find that

$$\begin{aligned} m\,\|u_n - u\|_X^2 &\le (Au_n - Au, u_n - u)_X \\ &= (Au_n, u_n - u)_X - (Au, u_n - u)_X \\ &\le (f, u_n - u)_X + j(\eta_n, u) - j(\eta_n, u_n) - (Au, u_n - u)_X. \end{aligned}$$

This inequality combined with the convergence (2.72) and assumption (2.65) implies that

$$\limsup_{n\to\infty} m\,\|u_n - u\|_X^2 \le \limsup_{n\to\infty} \,[j(\eta_n, u) - j(\eta_n, u_n)] = 0.$$

We obtain from the previous inequality that $u_n \to u$ in X which shows that $\Lambda\eta_n \to \Lambda\eta$ in X and concludes the proof. ■

Lemma 2.26 *The operator Λ is compact, i.e. it is continuous and maps bounded sets into relatively compact sets.*

Proof Let $\{\eta_n\}$ be a sequence in K such that $\eta_n \to \eta$ in X. It follows that $\eta_n \rightharpoonup \eta$ in X and using Lemma 2.25 we deduce that $\Lambda\eta_n \to \Lambda\eta$ in X. We conclude from here that the operator Λ is continuous.

Next, let $\{\eta_n\}$ be a bounded sequence in K. Then Theorem 1.19 implies that there exists a subsequence $\{\eta_{n_k}\}$ of $\{\eta_n\}$ and an element $\eta \in K$ such that $\eta_{n_k} \rightharpoonup \eta$ in X. This convergence combined with Lemma 2.25 shows that $\Lambda\eta_{n_k} \to \Lambda\eta$ in X. We deduce from this and the comments on page 10 that Λ is a compact operator, which concludes the proof. ■

We have now all the ingredients needed to prove Theorem 2.22.

Proof Let $f \in X$ and let $\widetilde{K}$ denote the set given by

$$\widetilde{K} = \{\, v \in K \,:\, \|v\|_X \le \frac{1}{m}\,(\|f\|_X + \|A0_X\|_X)\,\}.$$

Clearly, $\widetilde{K}$ is a nonempty closed convex bounded subset of X. Consider Λ, the operator given by (2.61). It follows from Lemma 2.23 that $\Lambda\eta \in \widetilde{K}$ for all $\eta \in K$ and, therefore, considering the restriction of Λ to $\widetilde{K}$, we have $\Lambda(\widetilde{K}) \subset \widetilde{K}$. Recall also that, as proved in Lemma 2.26, the operator Λ is compact. Therefore we can use Theorem 1.9 to deduce that there exists an element $\eta^* \in \widetilde{K}$ such that $\Lambda\eta^* = \eta^*$. By (2.61) it follows now that $u_{\eta^*} = \eta^*$ and, writing (2.60) for $\eta = \eta^*$, it is easy to see that η^* is a solution of the quasivariational inequality (2.58). ■

Theorems 2.19 and 2.22 provide existence results in the study of the elliptic quasivariational inequality (2.58). However, besides the fact that the arguments used in their proofs are different, we note that the statements of these theorems are different too, since the assumptions on the functional j in the two theorems are different. Moreover, in contrast with Theorem 2.19, Theorem 2.22 does not provide the uniqueness of the solution of the quasivariational inequality (2.58).

2.3.3 A convergence result

We proceed with a result concerning the dependence of the solution of the quasivariational inequality (2.58) with respect to the function j. To this end, we assume in what follows that (2.23), (2.24) and (2.59) hold and, moreover, $m > \alpha$. Let $f \in X$ and, for each $\rho > 0$, let j_ρ be a perturbation of j which satisfies (2.59) with $0 < \alpha_\rho < m$. We consider the problem of finding an element u_ρ such that

$$\begin{aligned} u_\rho \in K, \quad & (Au_\rho, v - u_\rho)_X + j_\rho(u_\rho, v) - j_\rho(u_\rho, u_\rho) \\ & \ge (f, v - u_\rho)_X \qquad \forall\, v \in K. \end{aligned} \tag{2.73}$$

We deduce from Theorem 2.19 that inequality (2.58) has a unique solution $u \in K$ and, for each $\rho > 0$, inequality (2.73) has a unique solution $u_\rho \in K$.

Consider now the following assumption:

$$\left.\begin{array}{l}\text{There exists } G : \mathbb{R}_+ \to \mathbb{R}_+ \text{ such that:}\\[4pt] \text{(a) } j_\rho(u_1, u_2) - j_\rho(u_1, u_1) + j(u_2, u_1) - j(u_2, u_2)\\ \qquad \le G(\rho)\|u_1 - u_2\|_X + \alpha\,\|u_1 - u_2\|_X^2\\ \qquad \forall\, u_1, u_2 \in K, \ \text{for each } \rho > 0;\\[4pt] \text{(b) } \lim\limits_{\rho \to 0} G(\rho) = 0.\end{array}\right\} \tag{2.74}$$

Then the behavior of the solution u_ρ as $\rho \to 0$ is given in the following theorem.

Theorem 2.27 *Under the assumption (2.74), the solution u_ρ of problem (2.73) converges to the solution u of problem (2.58), i.e.,*

$$u_\rho \to u \quad \textit{in } X \quad \textit{as} \quad \rho \to 0. \tag{2.75}$$

Proof Let $\rho > 0$. We take $v = u_\rho$ in (2.58), $v = u$ in (2.73) and add the resulting inequalities to obtain

$$(Au_\rho - Au, u_\rho - u)_X \le j_\rho(u_\rho, u) - j_\rho(u_\rho, u_\rho) + j(u, u_\rho) - j(u, u).$$

We now use (1.48) and assumption (2.74)(a) to find that

$$m\,\|u_\rho - u\|_X^2 \le G(\rho)\|u_\rho - u\|_X + \alpha\,\|u_\rho - u\|_X^2$$

and, since $m > \alpha$, it follows that

$$\|u_\rho - u\|_X \le \frac{G(\rho)}{m - \alpha}. \tag{2.76}$$

The convergence result (2.75) is now a consequence of the inequality (2.76) combined with assumption (2.74)(b). ■

We end this section with the remark that the results presented in this section represent extensions of the results presented in Section 2.2, in the study of elliptic variational inequalities. Indeed, assume the particular case when j does not depend on the first variable. Then it is easy to see that (2.59)(a) implies (2.25) and (2.59)(b) holds with $\alpha = 0$. It follows from this that the smallness assumption $\alpha < m$ is satisfied and, therefore, Theorem 2.19 can be used in this case and allows us to recover the existence and uniqueness result in Theorem 2.8. Moreover, assumption (2.74) implies assumption (2.34) and, therefore, Theorem 2.27 implies Theorem 2.11.

3 History-dependent variational inequalities

In this chapter we deal with the study of variational inequalities which involve the so-called history-dependent operators. We start with some preliminary material on spaces of vector-valued functions. Then, in order to illustrate the main ideas which are developed in this chapter, we analyze two nonlinear equations which lead to history-dependent operators. We proceed by introducing the concept of the history-dependent quasivariational inequalities together with a theorem on their unique solvability. Also, we state and prove a general convergence result in the study of these inequalities. Next, we particularize our results in the study of evolutionary variational and quasivariational inequalities. Besides their own interest, the results presented in this chapter represent crucial tools in deriving the existence of a unique weak solution to frictionless and frictional contact problems with viscoelastic and viscoplastic materials. Even if most of the results we present in this chapter still remain valid for more general cases, we restrict ourselves to the framework of strongly monotone Lipschitz continuous operators in Hilbert spaces as this is sufficient for the applications we consider in Chapter 6. Everywhere in this chapter X denotes a real Hilbert space with inner product $(\cdot,\cdot)_X$ and norm $\|\cdot\|_X$.

3.1 Nonlinear equations with history-dependent operators

In this section we consider two nonlinear equations in Hilbert spaces which lead to history-dependent operators. This will allow us to illustrate the main idea of this chapter which consists of unifying the study of various types of evolutionary equations or inequalities, by using a convenient choice of the variables and operators. To this end we need to introduce spaces of vector-valued continuous and continuously differentiable functions.

3.1.1 Spaces of vector-valued functions

Let $T > 0$. We denote by $C([0,T];X)$ the space of continuous functions defined on $[0,T]$ with values in X. It is well known that $C([0,T];X)$ is a Banach space with the norm

$$\|v\|_{C([0,T];X)} = \max_{t\in[0,T]} \|v(t)\|_X.$$

We also recall that a function $v : [0,T] \to X$ is said to be *differentiable* at $t_0 \in [0,T]$ if there exists an element in X, denoted $\dot{v}(t_0)$ and called the *derivative* of v at t_0, such that

$$\lim_{h\to 0} \left\| \frac{1}{h}(v(t_0+h) - v(t_0)) - \dot{v}(t_0) \right\|_X = 0,$$

where the limit is taken with respect to h with $t_0+h \in [0,T]$. The derivative at $t_0 = 0$ is defined as a right-sided limit, and that at $t_0 = T$ as a left-sided limit. The function v is said to be *differentiable* on $[0,T]$ if it is differentiable at every $t_0 \in [0,T]$. In this case the function $\dot{v} : [0,T] \to X$ is called the *derivative* of v. The function v is said to be *continuously differentiable* on $[0,T]$ if it is differentiable and its derivative is continuous.

We denote by $C^1([0,T];X)$ the space of continuously differentiable functions on $[0,T]$ with values in X and we recall that this is a Banach space with the norm

$$\|v\|_{C^1([0,T];X)} = \max_{t\in[0,T]} \|v(t)\|_X + \max_{t\in[0,T]} \|\dot{v}(t)\|_X.$$

Using the properties of the integral it is easy to see that if $f \in C([0,T];X)$ then the function $g : [0,T] \to X$ given by

$$g(t) = \int_0^t f(s)\,\mathrm{d}s \quad \forall\, t \in [0,T]$$

belongs to $C^1([0,T];X)$ and, moreover, $\dot{g} = f$. In addition, we recall that for a function $v \in C^1([0,T];X)$ the following equality holds:

$$v(t) = \int_0^t \dot{v}(s)\,\mathrm{d}s + v(0) \quad \forall\, t \in [0,T]. \tag{3.1}$$

Finally, for a subset $K \subset X$ we still use the notation $C([0,T];K)$ and $C^1([0,T];K)$ for the set of continuous and continuously differentiable functions defined on $[0,T]$ with values in K, respectively.

We now present a fixed point result which is useful for proving the solvability of nonlinear equations and variational inequalities with history-dependent operators.

Proposition 3.1 *Let $\Lambda : C([0,T];X) \to C([0,T];X)$ be an operator which satisfies the following property: there exist $k \in [0,1)$ and $c \geq 0$ such that*

$$\|\Lambda\eta_1(t) - \Lambda\eta_2(t)\|_X \leq k\,\|\eta_1(t) - \eta_2(t)\|_X$$
$$+\,c\int_0^t \|\eta_1(s) - \eta_2(s)\|_X\,\mathrm{d}s \quad \forall\,\eta_1,\,\eta_2 \in C([0,T];X), \quad t \in [0,T]. \tag{3.2}$$

Then there exists a unique element $\eta^* \in C([0,T];X)$ such that $\Lambda\eta^* = \eta^*$.

Proof Denote

$$\|\eta\|_\beta = \max_{t\in[0,T]} e^{-\beta t}\,\|\eta(t)\|_X \quad \forall\,\eta \in C([0,T];X), \tag{3.3}$$

with $\beta > 0$ to be chosen later. Clearly $\|\cdot\|_\beta$ defines a norm on the space $C([0,T];X)$ that is equivalent to the usual norm $\|\cdot\|_{C([0,T];X)}$. As a consequence it results that $C([0,T];X)$ is a Banach space with the norm $\|\cdot\|_\beta$, too. Let $t \in [0,T]$. Using (3.2) and (3.3) it follows that

$$\begin{aligned}
&\mathrm{e}^{-\beta t}\,\|\Lambda\eta_1(t) - \Lambda\eta_2(t)\|_X \\
&\leq k\mathrm{e}^{-\beta t}\,\|\eta_1(t) - \eta_2(t)\|_X + c\,\mathrm{e}^{-\beta t}\int_0^t \|\eta_1(s) - \eta_2(s)\|_X\,\mathrm{d}s \\
&= k\mathrm{e}^{-\beta t}\,\|\eta_1(t) - \eta_2(t)\|_X + c\,\mathrm{e}^{-\beta t}\int_0^t (\mathrm{e}^{-\beta s}\|\eta_1(s) - \eta_2(s)\|_X)\mathrm{e}^{\beta s}\,\mathrm{d}s \\
&\leq k\,\|\eta_1 - \eta_2\|_\beta + c\,\mathrm{e}^{-\beta t}\|\eta_1 - \eta_2\|_\beta\int_0^t \mathrm{e}^{\beta s}\,\mathrm{d}s \\
&= k\,\|\eta_1 - \eta_2\|_\beta + \frac{c}{\beta}\,\|\eta_1 - \eta_2\|_\beta(1 - \mathrm{e}^{-\beta t})
\end{aligned}$$

for all $\eta_1,\,\eta_2 \in C([0,T];X)$ and, therefore,

$$\|\Lambda\eta_1 - \Lambda\eta_2\|_\beta \leq \Big(k + \frac{c}{\beta}\Big)\|\eta_1 - \eta_2\|_\beta \quad \forall\,\eta_1,\,\eta_2 \in C([0,T];X).$$

Next, we choose β such that $\beta > \dfrac{c}{1-k}$ and note that this choice is possible since $k \in [0,1)$. Then

$$k + \frac{c}{\beta} < 1$$

and, therefore, the operator Λ is a contraction on the space $C([0,T];X)$ endowed with the norm $\|\cdot\|_\beta$. By Theorem 1.5 it follows now that Λ has a unique fixed point $\eta^* \in C([0,T];X)$, which concludes the proof. ∎

We use below the notation $C([0,T])$ for the space of real-valued continuous functions defined on the compact interval $[0,T] \subset \mathbb{R}$, that is $C([0,T]) = C([0,T],\mathbb{R})$. The following inequality is useful for obtaining uniqueness results in the study of nonlinear equations and variational inequalities with history-dependent operators.

Lemma 3.2 (The Gronwall inequality) *Let $f, g \in C([0,T])$ and assume that there exists $c > 0$ such that*

$$f(t) \le g(t) + c \int_0^t f(s)\,\mathrm{d}s \quad \forall\, t \in [0,T]. \tag{3.4}$$

Then

$$f(t) \le g(t) + c \int_0^t g(s)\,\mathrm{e}^{c\,(t-s)}\,\mathrm{d}s \quad \forall\, t \in [0,T]. \tag{3.5}$$

Moreover, if g is nondecreasing, then

$$f(t) \le g(t)\,\mathrm{e}^{c\,t} \quad \forall\, t \in [0,T].$$

Proof Define

$$F(s) = \int_0^s f(r)\,\mathrm{d}r \quad \forall\, s \in [0,T]. \tag{3.6}$$

Then $F'(s) = f(s)$. From assumption (3.4) we get

$$F'(s) \le g(s) + c\,F(s) \quad \forall\, s \in [0,T].$$

Thus,

$$\big(\mathrm{e}^{-c\,s}F(s)\big)' \le g(s)\,\mathrm{e}^{-c\,s} \quad \forall\, s \in [0,T].$$

Let $t \in [0,T]$. We integrate this inequality from 0 to t and, since $F(0) = 0$, we find that

$$\mathrm{e}^{-c\,t}F(t) \le \int_0^t g(s)\,\mathrm{e}^{-c\,s}\mathrm{d}s.$$

So,

$$F(t) \le \int_0^t g(s)\,\mathrm{e}^{c\,(t-s)}\mathrm{d}s. \tag{3.7}$$

We combine now (3.4) with (3.6) to find that $f(t) \le g(t) + c\,F(t)$ and then, using (3.7), we obtain (3.5).

If g is nondecreasing, then (3.5) implies that

$$f(t) \le g(t) + g(t)\,c \int_0^t \mathrm{e}^{c\,(t-s)}\,\mathrm{d}s = g(t)\,\mathrm{e}^{c\,t},$$

which concludes the proof. ∎

3.1.2 Two examples

Let $A : X \to X$ and $B : X \to X$ be given nonlinear operators, let $f : [0,T] \to X$ be a continuous function and let u_0 be the initial data. The first nonlinear equation we are interested in is the following: find a function $u : [0,T] \to X$ such that

$$A\dot{u}(t) + Bu(t) = f(t) \quad \forall\, t \in [0,T], \tag{3.8}$$

$$u(0) = u_0. \tag{3.9}$$

Alternatively, assume that $B \in C([0,T]; \mathcal{L}(X))$ where, recall, $\mathcal{L}(X)$ denotes the space of linear continuous operators from X to X. Then, the second nonlinear equation we are interested in is the following: find a function $u : [0,T] \to X$ such that

$$Au(t) + \int_0^t B(t-s)u(s)\,\mathrm{d}s = f(t) \quad \forall\, t \in [0,T]. \tag{3.10}$$

The second term in the equation above is called the *Volterra integral term* and for this reason we say that (3.10) represents a *nonlinear equation with Volterra integral term* or a *Volterra-type nonlinear equation.*

A brief comparison of the two problems above leads to the conclusion that there are several major differences between them. First, in contrast with equation (3.8), the nonlinear equation (3.10) does not involve the derivative of the unknown u and, in turn, involves an integral term. Next, in the first problem we have an initial condition for the unknown u which is not included in the second problem. Despite these differences, we shall see in this section that the two problems above have a common feature. More precisely, they can be treated in the same way, provided that we operate a change of variable in the problem (3.8)–(3.9).

Before operating this change of variable, we present the following existence and uniqueness result in the study of the Cauchy problem (3.8)–(3.9), which has some interest of its own.

Theorem 3.3 *Let X be a Hilbert space and assume that $A : X \to X$ is a strongly monotone Lipschitz continuous operator and $B : X \to X$ is a Lipschitz continuous operator. Then, for each $f \in C([0,T]; X)$ and $u_0 \in X$ there exists a unique solution $u \in C^1([0,T]; X)$ which satisfies (3.8)–(3.9).*

Proof It follows from Definition 1.21 that there exist three positive constants m_A, L_A and L_B, such that

$$(Au - Av, u - v)_X \geq m_A \|u - v\|_X^2, \tag{3.11}$$

$$\|Au - Av\|_X \leq L_A \|u - v\|_X, \tag{3.12}$$

$$\|Bu - Bv\|_X \leq L_B \|u - v\|_X, \tag{3.13}$$

for all $u,\, v \in X$. Also, it follows from Proposition 1.25 that A is invertible and, moreover, $A^{-1} : X \to X$ satisfies

$$\|A^{-1}u - A^{-1}v\|_X \le \frac{1}{m_A}\,\|u - v\|_X \quad \forall\, u,\, v \in X. \tag{3.14}$$

Let $\eta \in C([0,T];X)$. We use the properties of A^{-1}, B and f to see that $s \mapsto A^{-1}(f(s) - B\eta(s))$ is a continuous function from $[0,T]$ to X. Therefore, we may consider the function $\Lambda\eta : [0,T] \to X$ given by

$$\Lambda\eta(t) = \int_0^t A^{-1}(f(s) - B\eta(s))\,\mathrm{d}s + u_0 \quad \forall\, t \in [0,T] \tag{3.15}$$

and it follows that $\Lambda\eta \in C^1([0,T];X)$. Moreover, $\eta \mapsto \Lambda\eta$ defines an operator on the space $C([0,T];X)$ with values in the space $C^1([0,T];X) \subset C([0,T];X)$.

Using (3.1) it is easy to see that a function $u \in C^1([0,T];X)$ is a solution to problem (3.8)–(3.9) if and only if $u \in C([0,T];X)$ and u is a fixed point of the operator Λ, i.e.

$$u(t) = \Lambda u(t) \quad \forall\, t \in [0,T].$$

For this reason we investigate in what follows the properties of the operator $\Lambda : C([0,T];X) \to C([0,T];X)$.

Let $\eta_1,\, \eta_2 \in C([0,T];X)$. We use the definition (3.15) and the properties (3.14) and (3.13) of the operators A^{-1} and B, respectively, to see that

$$\begin{aligned}
&\|\Lambda\eta_1(t) - \Lambda\eta_2(t)\|_X \\
&\quad \le \int_0^t \|A^{-1}(f(s) - B\eta_1(s)) - A^{-1}(f(s) - B\eta_2(s))\|_X\,\mathrm{d}s \\
&\quad \le \frac{1}{m_A}\int_0^t \|B\eta_1(s) - B\eta_2(s)\|_X\,\mathrm{d}s \\
&\quad \le \frac{L_B}{m_A}\int_0^t \|\eta_1(s) - \eta_2(s)\|_X\,\mathrm{d}s \quad \forall\, t \in [0,T].
\end{aligned}$$

We conclude from the above that

$$\|\Lambda\eta_1(t) - \Lambda\eta_2(t)\|_X \le c\int_0^t \|\eta_1(s) - \eta_2(s)\|_X\,\mathrm{d}s \quad \forall\, t \in [0,T],$$

where $c = L_B/m_A > 0$. Therefore, by using Proposition 3.1 it follows that Λ has a unique fixed point $u^* \in C([0,T];X)$. We have

$$u^*(t) = \int_0^t A^{-1}(f(s) - Bu^*(s))\,\mathrm{d}s + u_0 \quad \forall\, t \in [0,T],$$

which shows that u^* is a solution of the problem (3.8)–(3.9) and, moreover, $u^* \in C^1([0,T];X)$. This concludes the proof of the existence part of Theorem 3.3.

The uniqueness part is a consequence of the uniqueness of the fixed point of the operator Λ. Nevertheless, a direct proof can be obtained as follows. Assume that $u_1, u_2 \in C^1([0,T];X)$ are two solutions of problem (3.8)–(3.9). Then,

$$A\dot{u}_1(t) - A\dot{u}_2(t) + Bu_1(t) - Bu_2(t) = 0 \quad \forall\, t \in [0,T],$$

which implies that

$$(A\dot{u}_1(t) - A\dot{u}_2(t), \dot{u}_1(t) - \dot{u}_2(t))_X = -(Bu_1(t) - Bu_2(t), \dot{u}_1(t) - \dot{u}_2(t))_X$$

for all $t \in [0,T]$. We use assumptions (3.11) and (3.13) and the previous equality to obtain

$$m_A \|\dot{u}_1(t) - \dot{u}_2(t)\|_X \le L_B \|u_1(t) - u_2(t)\|_X \quad \forall\, t \in [0,T]. \tag{3.16}$$

On the other hand, since $u_1(0) = u_2(0) = u_0$, we have

$$u_1(t) = \int_0^t \dot{u}_1(s)\,\mathrm{d}s + u_0, \quad u_2(t) = \int_0^t \dot{u}_2(s)\,\mathrm{d}s + u_0 \quad \forall\, t \in [0,T],$$

which implies that

$$\|u_1(t) - u_2(t)\|_X \le \int_0^t \|\dot{u}_1(s) - \dot{u}_2(s)\|_X\,\mathrm{d}s \quad \forall\, t \in [0,T]. \tag{3.17}$$

We now combine inequalities (3.16) and (3.17) to see that

$$\|\dot{u}_1(t) - \dot{u}_2(t)\|_X \le \frac{L_B}{m_A} \int_0^t \|\dot{u}_1(s) - \dot{u}_2(s)\|_X\,\mathrm{d}s.$$

An application of the Gronwall inequality shows that $\dot{u}_1(t) = \dot{u}_2(t)$ for all $t \in [0,T]$. Therefore, inequality (3.17) implies that $u_1 = u_2$, which concludes the proof of the uniqueness part of Theorem 3.3. ■

Assume in what follows that the hypotheses of Theorem 3.3 hold. We denote $\dot{u} = w$ and we remark that, using the initial condition (3.9), it follows that

$$u(t) = \int_0^t w(s)\,\mathrm{d}s + u_0 \quad \forall\, t \in [0,T]. \tag{3.18}$$

We also denote by $\mathcal{S} : C([0,T];X) \to C([0,T];X)$ the operator given by

$$\mathcal{S}v(t) = B\Big(\int_0^t v(s)\,\mathrm{d}s + v_0\Big) \quad \forall\, v \in C([0,T];X),\ \forall\, t \in [0,T]. \tag{3.19}$$

We note that, using the Lipschitz continuity of B, it follows that $\mathcal{S}$ satisfies the inequality

$$\|\mathcal{S}u_1(t) - \mathcal{S}u_2(t)\|_X \le L_B \int_0^t \|u_1(s) - u_2(s)\|_X \, ds \tag{3.20}$$

for all $u_1,\ u_2 \in C([0,T];X)$ and for all $t \in [0,T]$. Moreover, with the notation above it is easy to see that the function $u \in C^1([0,T];X)$ is a solution to the Cauchy problem (3.8)–(3.9) if and only if the function w belongs to the space $C([0,T];X)$ and satisfies the equation

$$Aw(t) + \mathcal{S}w(t) = f(t) \quad \forall\, t \in [0,T]. \tag{3.21}$$

To avoid any confusion, we note that here and below, the notations $Au(t)$ and $\mathcal{S}u(t)$ are short-hand notation for $A(u(t))$ and $(\mathcal{S}u)(t)$, i.e. $Au(t) = A(u(t))$ and $\mathcal{S}u(t) = (\mathcal{S}u)(t)$, for all $t \in [0,T]$.

Assume now that $B \in C([0,T];\mathcal{L}(X))$ and consider the *Volterra operator* $\mathcal{S} : C([0,T];X) \to C([0,T];X)$ given by

$$\mathcal{S}v(t) = \int_0^t B(t-s)\, v(s)\, ds \quad \forall\, v \in C([0,T];X),\ \forall t \in [0,T]. \tag{3.22}$$

Then it is easy to see that the operator $\mathcal{S}$ satisfies the inequality

$$\|\mathcal{S}u_1(t) - \mathcal{S}u_2(t)\|_X \le \|B\| \int_0^t \|u_1(s) - u_2(s)\|_X \, ds \tag{3.23}$$

for all $u_1,\ u_2 \in C([0,T];X)$ and for all $t \in [0,T]$, where

$$\|B\| = \|B\|_{C([0,T];\mathcal{L}(X))} = \max_{s\in[0,T]} \|B(s)\|_{\mathcal{L}(X)}$$

represents the norm of the operator B in the space $C([0,T];\mathcal{L}(X))$. Moreover, the nonlinear equation (3.10) can be written, equivalently, as follows:

$$Au(t) + \mathcal{S}u(t) = f(t) \quad \forall\, t \in [0,T]. \tag{3.24}$$

We conclude from the above that the nonlinear equations (3.8)–(3.9) and (3.10) can be cast in the same form given by (3.21) and (3.24), respectively, in which the operators $\mathcal{S}$ are defined by (3.19) and (3.22), respectively. Moreover, these operators satisfy similar inequalities, given by (3.20) and (3.23), respectively. The only difference arises from the fact that in the case of the problem (3.8)–(3.9) the unknown is the velocity field that is $w = \dot{u}$.

The previous remark motivates us to study equation (3.24) in a general framework. In this way we will obtain an alternative for the proof of Theorem 3.3 and an existence and uniqueness result for the Volterra-type nonlinear equation (3.10).

3.1.3 The general case

Let X be a real Hilbert space, $A : X \to X$, $\mathcal{S} : C([0,T];X) \to C([0,T];X)$ and $f : [0,T] \to X$. We assume that operator $\mathcal{S}$ satisfies:

$$\left.\begin{array}{l}\text{there exists } L_S > 0 \text{ such that} \\[2mm] \|\mathcal{S}u_1(t) - \mathcal{S}u_2(t)\|_X \le L_S \displaystyle\int_0^t \|u_1(s) - u_2(s)\|_X \, \mathrm{d}s \\[2mm] \forall\, u_1,\, u_2 \in C([0,T];X),\ \forall\, t \in [0,T]. \end{array}\right\} \tag{3.25}$$

Note that this condition is satisfied for the operators given by (3.19) and (3.22), as shown above. Clearly, in the case of these operators the current value $\mathcal{S}v(t)$ at the moment t depends on the history of the values of v at the moments $0 \le s \le t$ and, therefore, we refer to the operators of the form (3.19) or (3.22) as *history-dependent operators.* We extend this definition to all the operators $\mathcal{S} : C([0,T];X) \to C([0,T];X)$ which satisfy condition (3.25) and, for this reason, we say that the nonlinear equation (3.24) is a *history-dependent nonlinear equation.* Our main result in its study is the following.

Theorem 3.4 *Let X be a Hilbert space, assume that $A : X \to X$ is a strongly monotone Lipschitz continuous operator and, moreover, $\mathcal{S} : C([0,T];X) \to C([0,T];X)$ is an operator which satisfies condition (3.25). Then, for each $f \in C([0,T];X)$ there exists a unique solution of (3.24), which satisfies $u \in C([0,T];X)$.*

Proof The proof of Theorem 3.4 is based on a fixed point argument and will be established in several steps.

(i) In the first step let $\eta \in C([0,T];X)$ be given and denote by $y_\eta \in C([0,T];X)$ the function

$$y_\eta(t) = \mathcal{S}\eta(t) \qquad \forall\, t \in [0,T]. \tag{3.26}$$

Consider the problem of finding a function $u_\eta : [0,T] \to X$ such that

$$Au_\eta(t) + y_\eta(t) = f(t) \qquad \forall\, t \in [0,T]. \tag{3.27}$$

Using Theorem 1.24 we obtain that there exists a unique solution $u_\eta(t)$ to (3.27), at each $t \in [0,T]$. Moreover, using the strong monotonicity of A, (1.46), it follows that

$$\|u_\eta(t_1) - u_\eta(t_2)\|_X \le \frac{1}{m}\left(\|y_\eta(t_1) - y_\eta(t_2)\|_X + \|f(t_1) - f(t_2)\|_X\right)$$

for all $t_1,\, t_2 \in [0,T]$. Therefore, since y_η and f belong to $C([0,T];X)$, we deduce from the above that $u_\eta : [0,T] \to X$ is a continuous function, i.e. $u_\eta \in C([0,T];X)$. To clarify, we have proved in this step that equation (3.27) has a unique solution $u_\eta \in C([0,T];X)$.

(ii) In the next step we define the operator $\Lambda : C([0,T];X) \to C([0,T];X)$ by the equality

$$\Lambda\eta = u_\eta \qquad \forall\, \eta \in C([0,T];X). \tag{3.28}$$

We prove that the operator Λ has a unique fixed point $\eta^* \in C([0,T];X)$. Indeed, let $\eta_1, \eta_2 \in C([0,T];X)$, and let y_i be the function defined by (3.26) for $\eta = \eta_i$, i.e. $y_i = y_{\eta_i}$, for $i = 1,2$. We also denote by u_i the solution of the equation (3.27) for $\eta = \eta_i$, i.e. $u_i = u_{\eta_i}$, $i = 1,2$. Let $t \in [0,T]$. From definition (3.28) we have

$$\|\Lambda\eta_1(t) - \Lambda\eta_2(t)\|_X = \|u_1(t) - u_2(t)\|_X. \tag{3.29}$$

Moreover, using the properties of the operator A, from (3.27) it follows that

$$m\,\|u_1(t) - u_2(t)\|_X \le \|y_1(t) - y_2(t)\|_X. \tag{3.30}$$

Next, we use (3.26) and the property (3.25) of the operator $\mathcal{S}$ to see that

$$\|y_1(t) - y_2(t)\|_X = \|\mathcal{S}\eta_1(t) - \mathcal{S}\eta_2(t)\|_X \le L_{\mathcal{S}} \int_0^t \|\eta_1(s) - \eta_2(s)\|_X\, ds$$

and using this inequality in (3.30) yields

$$\|u_1(t) - u_2(t)\|_X \le \frac{L_{\mathcal{S}}}{m} \int_0^t \|\eta_1(s) - \eta_2(s)\|_X\, ds. \tag{3.31}$$

We now combine (3.29) and (3.31) to see that

$$\|\Lambda\eta_1(t) - \Lambda\eta_2(t)\|_X \le \frac{L_{\mathcal{S}}}{m} \int_0^t \|\eta_1(s) - \eta_2(s)\|_X\, ds. \tag{3.32}$$

Finally, we use (3.32) and Proposition 3.1 to find that the operator Λ has a unique fixed point $\eta^* \in C([0,T];X)$.

(iii) *Existence* Let $\eta^* \in C([0,T];X)$ be the fixed point of the operator Λ, that is $\Lambda\eta^* = \eta^*$. It follows from (3.26) and (3.28) that, for all $t \in [0,T]$, the following equalities hold:

$$y_{\eta^*}(t) = \mathcal{S}\eta^*(t), \qquad u_{\eta^*}(t) = \eta^*(t). \tag{3.33}$$

We now write equality (3.27) for $\eta = \eta^*$ and then use (3.33) to conclude that the function $\eta^* \in C([0,T];X)$ is a solution to the nonlinear equation (3.24).

(iv) *Uniqueness* The uniqueness part is a consequence of the uniqueness of the fixed point of the operator Λ and can be proved as follows. Denote by $\eta^* \in C([0,T];X)$ the solution of the time-dependent equation (3.24) obtained above, and let $\eta \in C([0,T];X)$ be a different solution of this equation. Also, consider the function $y_\eta \in C([0,T];X)$ defined by (3.26). Then, it follows that η is solution to the equation (3.27) and, since step (i) implies that this inequality has a unique solution, denoted by u_η, we conclude that

$$\eta = u_\eta. \tag{3.34}$$

Equality (3.34) shows that $\Lambda\eta = \eta$, where Λ is the operator defined by (3.28). Therefore, by step (ii) it follows that $\eta = \eta^*$, which concludes the proof of the uniqueness part. ■

Theorem 3.4 allows us to obtain an alternative proof of Theorem 3.3. Indeed, under the assumption of Theorem 3.3 it follows from Theorem 3.4 that there exists a unique function $w \in C([0,T];X)$ which satisfies the nonlinear equation (3.21) where, we recall, the operator $\mathcal{S}$ is given by (3.19). Consider now the function $u : [0,T] \to X$ given by (3.18). Then it follows from the discussion on page 64 that $u \in C^1([0,T;X)$ is the unique solution of the Cauchy problem (3.8)–(3.9). In addition, Theorem 3.4 allows us to obtain the following existence and uniqueness results in the study of equation (3.10).

Corollary 3.5 *Let X be a Hilbert space, assume that $A : X \to X$ is a strongly monotone Lipschitz continuous operator and, moreover, let $B \in C([0,T];\mathcal{L}(X))$. Then for each $f \in C([0,T];X)$ there exists a unique solution $u \in C([0,T];X)$ which satisfies (3.10).*

The proof of this corollary follows from Theorem 3.4 and is based on the fact that the Volterra operator (3.22) satisfies condition (3.25).

3.2 History-dependent quasivariational inequalities

In this section we introduce the concept of history-dependent quasivariational inequalities for which we provide an existence and uniqueness result. We complete it with a general convergence result.

3.2.1 A basic existence and uniqueness result

Let X be a real Hilbert space with inner product $(\cdot,\cdot)_X$ and associated norm $\|\cdot\|_X$. Let also Y be a normed space with the norm $\|\cdot\|_Y$ and recall that $\mathcal{L}(X,Y)$ denotes the space of linear continuous operators from X to Y with the usual norm $\|\cdot\|_{\mathcal{L}(X,Y)}$. We also denote by $C([0,T];\mathcal{L}(X,Y))$ the space of continuous functions defined on the bounded interval $[0,T]$ with values in $\mathcal{L}(X,Y)$.

Let K be a subset of X and consider the operators $A : K \to X$, $\mathcal{S} : C([0,T];X) \to C([0,T];Y)$, the functions $\varphi : Y \times K \to \mathbb{R}$, $j : K \times K \to \mathbb{R}$ and $f : [0,T] \to X$. We are interested in the problem of finding a function $u \in C([0,T],X)$ such that, for all $t \in [0,T]$, the inequality below holds:

$$\begin{aligned} u(t) \in K, \quad & (Au(t), v - u(t))_X + \varphi(\mathcal{S}u(t), v) - \varphi(\mathcal{S}u(t), u(t)) \\ & + j(u(t), v) - j(u(t), u(t)) \geq (f(t), v - u(t))_X \qquad \forall\, v \in K. \end{aligned} \tag{3.35}$$

Note that (3.35) represents a *time-dependent variational inequality* governed by two functions φ and j which depend on the solution and, therefore, using a terminology similar to that introduced on page 49, it represents a *quasivariational inequality.*

In the study of (3.35) we assume that

$$K \text{ is a nonempty closed convex subset of } X \tag{3.36}$$

and $A : K \to X$ is a strongly monotone Lipschitz continuous operator, i.e.

$$\left.\begin{array}{l} \text{(a) there exists } m > 0 \text{ such that} \\ \qquad (Au_1 - Au_2, u_1 - u_2)_X \geq m\, \|u_1 - u_2\|_X^2 \quad \forall\, u_1, u_2 \in K; \\ \text{(b) there exists } M > 0 \text{ such that} \\ \qquad \|Au_1 - Au_2\|_X \leq M\, \|u_1 - u_2\|_X \quad \forall\, u_1, u_2 \in K. \end{array}\right\} \tag{3.37}$$

The functions $\varphi : Y \times K \to \mathbb{R}$ and $j : K \times K \to \mathbb{R}$ satisfy

$$\left.\begin{array}{l} \text{(a) for all } y \in Y,\ \varphi(y, \cdot) : K \to \mathbb{R} \text{ is convex and l.s.c.;} \\ \text{(b) there exists } \beta \geq 0 \text{ such that} \\ \qquad \varphi(y_1, u_2) - \varphi(y_1, u_1) + \varphi(y_2, u_1) - \varphi(y_2, u_2) \\ \qquad \leq \beta\, \|y_1 - y_2\|_Y\, \|u_1 - u_2\|_X \quad \forall\, y_1,\, y_2 \in Y,\ \forall\, u_1,\, u_2 \in K; \end{array}\right\} \tag{3.38}$$

$$\left.\begin{array}{l} \text{(a) for all } u \in K,\ j(u, \cdot) : K \to \mathbb{R} \text{ is convex and l.s.c.;} \\ \text{(b) there exists } \alpha \geq 0 \text{ such that} \\ \qquad j(u_1, v_2) - j(u_1, v_1) + j(u_2, v_1) - j(u_2, v_2) \\ \qquad \leq \alpha\, \|u_1 - u_2\|_X\, \|v_1 - v_2\|_X \quad \forall\, u_1,\, u_2 \in K,\ \forall\, v_1,\, v_2 \in K. \end{array}\right\} \tag{3.39}$$

Moreover, we assume that

$$m > \alpha, \tag{3.40}$$

where m and α are the constants in (3.37) and (3.39), respectively. The operator $\mathcal{S} : C([0,T];X) \to C([0,T];Y)$ satisfies:

$$\left.\begin{array}{l} \text{there exists } L_{\mathcal{S}} > 0 \text{ such that} \\ \qquad \|\mathcal{S}u_1(t) - \mathcal{S}u_2(t)\|_Y \leq L_{\mathcal{S}} \displaystyle\int_0^t \|u_1(s) - u_2(s)\|_X \,\mathrm{d}s \\ \qquad \forall\, u_1,\, u_2 \in C([0,T];X),\ \forall\, t \in [0,T] \end{array}\right\} \tag{3.41}$$

and, finally, we assume that

$$f \in C([0,T];X). \tag{3.42}$$

Note that examples of operators S which satisfy condition (3.41) in the case $Y = X$ were already considered in Section 3.1. These examples can be easily extended in the general case. For instance, condition (3.41) is satisfied for the operator $\mathcal{S} : C([0,T];X) \to C([0,T];Y)$ given by

$$\mathcal{S}v(t) = R\left(\int_0^t v(s)\,\mathrm{d}s + v_0\right) \quad \forall\, v \in C([0,T];X),\ \forall\, t \in [0,T], \tag{3.43}$$

where $R : X \to Y$ is a Lipschitz continuous operator and $v_0 \in X$. It is also satisfied for *Volterra's operator* $\mathcal{S} : C([0,T];X) \to C([0,T];Y)$ given by

$$\mathcal{S}v(t) = \int_0^t R(t-s)\,v(s)\,\mathrm{d}s \quad \forall v \in C([0,T];X),\ \forall t \in [0,T], \tag{3.44}$$

where now $R \in C([0,T];\mathcal{L}(X,Y))$. Indeed, in the case of the operator (3.43), inequality (3.41) holds with L_S being the Lipschitz constant of the operator R, and in the case of the operator (3.44) it holds with

$$L_S = \|R\|_{C([0,T];\mathcal{L}(X,Y))} = \max_{s\in[0,T]} \|R(s)\|_{\mathcal{L}(X,Y)}.$$

Clearly, in the case of the operators (3.43) and (3.44) the current value $\mathcal{S}v(t)$ at the moment t depends on the history of the values of v at the moments $0 \le s \le t$ and, therefore, we refer to the operators of the form (3.43) or (3.44) as *history-dependent operators.* We extend this definition to all the operators $\mathcal{S} : C([0,T];X) \to C([0,T];Y)$ which satisfy condition (3.41) and, for this reason, we say that the quasivariational inequalities of the form (3.35) are *history-dependent quasivariational inequalities.* Their main feature consists of the fact that at each moment $t \in [0,T]$ the function φ depends on $\mathcal{S}u(t)$ and, therefore, it depends on the history of the solution up to the moment t. This feature makes a difference with respect to the quasivariational inequalities studied in the literature in which, usually, φ is assumed to depend on the current value of the solution, $u(t)$.

The main result of this section is the following.

Theorem 3.6 *Assume that (3.36)–(3.42) hold. Then the variational inequality (3.35) has a unique solution $u \in C([0,T];K)$.*

The proof of Theorem 3.6 is based on a fixed point argument similar to that used in the proof of Theorem 3.4 and will be established in several steps. We assume in what follows that (3.36)–(3.42) hold. In the first step let $\eta \in C([0,T];X)$ be given and denote by $y_\eta \in C([0,T];Y)$ the function

$$y_\eta(t) = \mathcal{S}\eta(t) \qquad \forall t \in [0,T]. \tag{3.45}$$

Consider now the problem of finding a function $u_\eta : [0,T] \to X$ such that for all $t \in [0,T]$ the inequality below holds:

$$u_\eta(t) \in K, \quad (Au_\eta(t), v - u_\eta(t))_X + \varphi(y_\eta(t), v) - \varphi(y_\eta(t), u_\eta(t)) \\ +j(\eta(t), v) - j(\eta(t), u_\eta(t)) \geq (f(t), v - u_\eta(t))_X \qquad \forall v \in K. \tag{3.46}$$

We have the following existence and uniqueness result.

Lemma 3.7 *There exists a unique solution $u_\eta \in C([0,T]; K)$ to the problem (3.46).*

Proof Using assumptions (3.36), (3.37), (3.38)(a), (3.39)(a) and (3.42) it follows from Theorem 2.8 that there exists a unique element $u_\eta(t)$ that solves (3.46), for each $t \in [0,T]$. Let us show that $u_\eta : [0,T] \to K$ is continuous and, to this end, consider $t_1, t_2 \in [0,T]$. For the sake of simplicity in writing we denote $u_\eta(t_i) = u_i$, $\eta(t_i) = \eta_i$, $y(t_i) = y_i$, $f(t_i) = f_i$ for $i = 1, 2$. Using (3.46) we obtain

$$u_1 \in K, \quad (Au_1, v - u_1)_X + \varphi(y_1, v) - \varphi(y_1, u_1) \\ +j(\eta_1, v) - j(\eta_1, u_1) \geq (f_1, v - u_1)_X \quad \forall v \in K, \tag{3.47}$$

$$u_2 \in K, \quad (Au_2, v - u_2)_X + \varphi(y_2, v) - \varphi(y_2, u_2) \\ +j(\eta_2, v) - j(\eta_2, u_2) \geq (f_2, v - u_2)_X \quad \forall v \in K. \tag{3.48}$$

We take $v = u_2$ in (3.47) and $v = u_1$ in (3.48), then we add the resulting inequalities to find that

$$(Au_1 - Au_2, u_1 - u_2)_X \leq \varphi(y_1, u_2) - \varphi(y_1, u_1) + \varphi(y_2, u_1) - \varphi(y_2, u_2) \\ +j(\eta_1, u_2) - j(\eta_1, u_1) + j(\eta_2, u_1) - j(\eta_2, u_2) + (f_1 - f_2, u_1 - u_2)_X.$$

Next, we use assumptions (3.37)(a), (3.38)(b) and (3.39)(b) to obtain

$$m\, \|u_1 - u_2\|_X \leq \beta\, \|y_1 - y_2\|_Y + \alpha\, \|\eta_1 - \eta_2\|_X + \|f_1 - f_2\|_X. \tag{3.49}$$

We deduce from (3.49) that $t \mapsto u_\eta(t) : [0,T] \to K$ is a continuous function, which concludes the proof. ■

In the second step we use Lemma 3.7 to consider the operator

$$\Lambda : C([0,T]; X) \to C([0,T]; K) \subset C([0,T]; X)$$

defined by the equality

$$\Lambda\eta = u_\eta \quad \forall \eta \in C([0,T]; X). \tag{3.50}$$

We have the following fixed point result.

Lemma 3.8 *The operator Λ has a unique fixed point $\eta^* \in C([0,T];K)$.*

Proof Let $\eta_1, \eta_2 \in C([0,T];X)$, and let y_i be the function defined by (3.45) for $\eta = \eta_i$, i.e. $y_i = y_{\eta_i}$, for $i = 1, 2$. We also denote by u_i the solution of the variational inequality (3.46) for $\eta = \eta_i$, i.e. $u_i = u_{\eta_i}$, $i = 1, 2$. Let $t \in [0,T]$. From definition (3.50) we have

$$\|\Lambda\eta_1(t) - \Lambda\eta_2(t)\|_X = \|u_1(t) - u_2(t)\|_X. \tag{3.51}$$

Moreover, an argument similar to that in the proof of (3.49) shows that

$$m\,\|u_1(t) - u_2(t)\|_X \le \beta\,\|y_1(t) - y_2(t)\|_Y + \alpha\,\|\eta_1(t) - \eta_2(t)\|_X. \tag{3.52}$$

Next, we use (3.45) and the property (3.41) of the operator $\mathcal{S}$ to see that

$$\|y_1(t) - y_2(t)\|_Y = \|\mathcal{S}\eta_1(t) - \mathcal{S}\eta_2(t)\|_Y \le L_{\mathcal{S}} \int_0^t \|\eta_1(s) - \eta_2(s)\|_X\, ds$$

and using this inequality in (3.52) yields

$$\|u_1(t) - u_2(t)\|_X \le \frac{\beta L_{\mathcal{S}}}{m} \int_0^t \|\eta_1(s) - \eta_2(s)\|_X\, ds + \frac{\alpha}{m}\,\|\eta_1(t) - \eta_2(t)\|_X. \tag{3.53}$$

We now combine (3.51) and (3.53) to see that

$$\|\Lambda\eta_1(t) - \Lambda\eta_2(t)\|_X \le \frac{\beta L_{\mathcal{S}}}{m} \int_0^t \|\eta_1(s) - \eta_2(s)\|_X\, ds + \frac{\alpha}{m}\,\|\eta_1(t) - \eta_2(t)\|_X. \tag{3.54}$$

Finally, we use (3.54), the smallness assumption (3.40) and Proposition 3.1 to see that the operator Λ has a unique fixed point $\eta^* \in C([0,T];X)$. And, since Λ has values on $C([0,T];K)$, we deduce that $\eta^* \in C([0,T];K)$, which concludes the proof. ■

We have now all the ingredients to prove Theorem 3.6.

Proof *Existence* Let $\eta^* \in C([0,T];K)$ be the fixed point of the operator Λ, i.e. $\Lambda\eta^* = \eta^*$. It follows from (3.45) and (3.50) that, for all $t \in [0,T]$, the following equalities hold:

$$y_{\eta^*}(t) = \mathcal{S}\eta^*(t), \qquad u_{\eta^*}(t) = \eta^*(t). \tag{3.55}$$

We now write the inequality (3.46) for $\eta = \eta^*$ and then use the equalities (3.55) to conclude that the function $\eta^* \in C([0,T];K)$ is a solution to the quasivariational inequality (3.35).

Uniqueness The uniqueness part is a consequence of the uniqueness of the fixed point of the operator Λ and can be proved as follows. Denote by $\eta^* \in C([0,T];K)$ the solution of the quasivariational inequality (3.35) obtained above, and let $\eta \in C([0,T];K)$ be a different solution of this inequality. Also, consider the function $y_\eta \in C([0,T];Y)$ defined by (3.45). Then it follows from (3.35) that η is solution to the variational inequality (3.46) and, since by Lemma 3.7 this inequality has a unique solution, denoted u_η, we conclude that

$$\eta = u_\eta. \tag{3.56}$$

Equality (3.56) shows that $\Lambda\eta = \eta$, where Λ is the operator defined by (3.50). Therefore, by Lemma 3.8 it follows that $\eta = \eta^*$, which concludes the first proof of the uniqueness part.

A direct proof of the uniqueness part can be obtained by using the Gronwall argument and is as follows. Assume that u_1, u_2 are two solutions of the variational inequality (3.35) with regularity $C([0,T];K)$ and let $t \in [0,T]$. We use (3.35) to see that

$$\begin{aligned}(Au_1(t) - Au_2(t), u_1(t) - u_2(t))_X &\le \varphi(\mathcal{S}u_1(t), u_2(t)) - \varphi(\mathcal{S}u_1(t), u_1(t))\\ &+\varphi(\mathcal{S}u_2(t), u_1(t)) - \varphi(\mathcal{S}u_2(t), u_2(t)) + j(u_1(t), u_2(t)) - j(u_1(t), u_1(t))\\ &+j(u_2(t), u_1(t)) - j(u_2(t), u_2(t))\end{aligned}$$

and then, using assumptions (3.37)(a), (3.38)(b) and (3.39)(b) yields

$$\begin{aligned}m\,\|u_1(t) - u_2(t)\|_X^2 &\le \beta\,\|\mathcal{S}u_1(t) - \mathcal{S}u_2(t)\|_Y \|u_1(t) - u_2(t)\|_X\\ &+\alpha\,\|u_1(t) - u_2(t)\|_X^2.\end{aligned}$$

We use this inequality and assumption (3.41) on the operator $\mathcal{S}$ to find that

$$\|u_1(t) - u_2(t)\|_X \le \frac{\beta L_{\mathcal{S}}}{m}\int_0^t \|u_1(s) - u_2(s)\|_X\,ds + \frac{\alpha}{m}\,\|u_1(t) - u_2(t)\|_X.$$

We now combine the smallness assumption (3.40) with the previous inequality to see that there exists a positive constant c such that

$$\|u_1(t) - u_2(t)\|_X \le c\int_0^t \|u_1(s) - u_2(s)\|_X\,ds. \tag{3.57}$$

We now use Lemma 3.2 on page 60 and inequality (3.57) to see that $u_1(t) = u_2(t)$ for all $t \in [0,T]$, which concludes the second proof of the uniqueness part of the theorem. ■

We end this subsection with the remark that in the particular case when $K = X = Y$, $\varphi(u,v) = (u,v)_X$, $j \equiv 0$, inequality (3.35) is equivalent to equation (3.24). We conclude from here that Theorem 3.6 represents an extension of Theorem 3.4 on page 65.

3.2.2 A convergence result

We proceed with a result concerning the dependence of the solution of the history-dependent quasivariational inequality (3.35) with respect to the operator $\mathcal{S}$ and the functions φ and j. To this end, we assume in what follows that (3.36)–(3.42) hold. For each $\rho > 0$, let $\varphi_\rho : Y \times K \to \mathbb{R}$, $j_\rho : K \times K \to \mathbb{R}$ and $\mathcal{S}_\rho : C([0,T];X) \to C([0,T];Y)$ be perturbations of φ, j and $\mathcal{S}$, respectively, which satisfy conditions (3.38)–(3.41) with constants β_ρ, α_ρ and $L_{\mathcal{S}_\rho}$ respectively. We consider the problem of finding a function $u_\rho \in C([0,T];X)$ such that for all $t \in [0,T]$, the inequality below holds:

$$\begin{aligned} &u_\rho(t) \in K, \quad (Au_\rho(t), v - u_\rho(t))_X \\ &\quad +\varphi_\rho(\mathcal{S}_\rho u_\rho(t), v) - \varphi_\rho(\mathcal{S}_\rho u_\rho(t), u_\rho(t)) \\ &\qquad +j_\rho(u_\rho(t), v) - j_\rho(u_\rho(t), u_\rho(t)) \geq (f(t), v - u_\rho(t))_X \quad \forall\, v \in K. \end{aligned} \tag{3.58}$$

We deduce from Theorem 3.6 that inequality (3.35) has a unique solution $u \in C([0,T];K)$ and, for each $\rho > 0$, inequality (3.58) has a unique solution $u_\rho \in C([0,T];K)$.

Consider now the following assumptions.

$$\left.\begin{array}{l} \text{There exist } F : \mathbb{R}_+ \to \mathbb{R}_+ \text{ and } c > 0 \text{ such that} \\[2mm] \text{(a) } \varphi_\rho(y_1, u_2) - \varphi_\rho(y_1, u_1) + \varphi(y_2, u_1) - \varphi(y_2, u_2) \\ \qquad \leq F(\rho)\|u_1 - u_2\|_X + c\,\|y_1 - y_2\|_Y \|u_1 - u_2\|_X \\ \qquad \text{for all } y_1, y_2 \in Y,\ u_1, u_2 \in K, \text{ for each } \rho > 0; \\[2mm] \text{(b) } \lim\limits_{\rho \to 0} F(\rho) = 0. \end{array}\right\} \tag{3.59}$$

$$\left.\begin{array}{l} \text{There exists } G : \mathbb{R}_+ \to \mathbb{R}_+ \text{ such that} \\[2mm] \text{(a) } j_\rho(u_1, u_2) - j_\rho(u_1, u_1) + j(u_2, u_1) - j(u_2, u_2) \\ \qquad \leq G(\rho)\|u_1 - u_2\|_X + \alpha\,\|u_1 - u_2\|_X^2 \\ \qquad \text{for all } u_1, u_2 \in K, \text{ for each } \rho > 0; \\[2mm] \text{(b) } \lim\limits_{\rho \to 0} G(\rho) = 0. \end{array}\right\} \tag{3.60}$$

$$\left.\begin{array}{l} \text{There exist } H : \mathbb{R}_+ \to \mathbb{R}_+,\ J : C([0,T];K) \to \mathbb{R}_+ \\ \text{and } \widetilde{L} > 0 \text{ such that} \\[2mm] \text{(a) } \|\mathcal{S}_\rho u(t) - \mathcal{S}u(t)\|_Y \leq H(\rho)J(u) \quad \forall\, u \in C([0,T];K), \\ \qquad \forall\, t \in [0,T], \text{ for each } \rho > 0; \\[2mm] \text{(b) } \lim\limits_{\rho \to 0} H(\rho) = 0; \\[2mm] \text{(c) } L_{\mathcal{S}_\rho} \leq \widetilde{L}, \text{ for each } \rho > 0. \end{array}\right\} \tag{3.61}$$

Then the behavior of the solution u_ρ as $\rho \to 0$ is given by the following theorem.

Theorem 3.9 *Under the assumptions (3.59)–(3.61), the solution u_ρ of the history-dependent quasivariational inequality (3.58) converges to the solution u of the history-dependent quasivariational inequality (3.35), i.e.,*

$$u_\rho \to u \quad in \ C([0,T];X) \quad as \quad \rho \to 0. \tag{3.62}$$

Proof Let $\rho > 0$ and let $t \in [0,T]$. We take $v = u(t)$ in (3.58) and $v = u_\rho(t)$ in (3.35). Then we add the resulting inequalities to obtain

$$\begin{aligned}(Au_\rho(t) - Au(t), u_\rho(t) - u(t))_X &\le \varphi_\rho(\mathcal{S}_\rho u_\rho(t), u(t)) \\ -\varphi_\rho(\mathcal{S}_\rho u_\rho(t), u_\rho(t)) &+ \varphi(\mathcal{S}u(t), u_\rho(t)) - \varphi(\mathcal{S}u(t), u(t)) \\ +j_\rho(u_\rho(t), u(t)) - j_\rho(u_\rho(t), u_\rho(t)) &+ j(u(t), u_\rho(t)) - j(u(t), u(t)).\end{aligned}$$

Next we use assumptions (3.37)(a), (3.59)(a) and (3.60)(a) to see that

$$\begin{aligned}m\,\|u_\rho(t) - u(t)\|_X^2 &\le F(\rho)\|u_\rho(t) - u(t)\|_X \\ &+c\,\|\mathcal{S}_\rho u_\rho(t) - \mathcal{S}u(t)\|_Y \|u_\rho(t) - u(t)\|_X \\ &+G(\rho)\|u_\rho(t) - u(t)\|_X + \alpha\,\|u_\rho(t) - u(t)\|_X^2\end{aligned}$$

and, therefore,

$$(m - \alpha)\|u_\rho(t) - u(t)\|_X \le F(\rho) + G(\rho) + c\,\|\mathcal{S}_\rho u_\rho(t) - \mathcal{S}u(t)\|_Y. \tag{3.63}$$

On the other hand, using the triangle inequality it follows that

$$\|\mathcal{S}_\rho u_\rho(t) - \mathcal{S}u(t)\|_Y \le \|\mathcal{S}_\rho u_\rho(t) - \mathcal{S}_\rho u(t)\|_Y + \|\mathcal{S}_\rho u(t) - \mathcal{S}u(t)\|_Y.$$

We combine the previous inequality with (3.41) and (3.61)(a) to see that

$$\|\mathcal{S}_\rho u_\rho(t) - \mathcal{S}u(t)\|_Y \le L_{\mathcal{S}_\rho} \int_0^t \|u_\rho(s) - u(s)\|_X \, ds + H(\rho)J(u). \tag{3.64}$$

It follows from (3.63) and (3.64) that

$$\begin{aligned}&(m - \alpha)\|u_\rho(t) - u(t)\|_X \\ &\quad \le F(\rho) + G(\rho) + c\,H(\rho)J(u) + c\,L_{\mathcal{S}_\rho} \int_0^t \|u_\rho(s) - u(s)\|_X \, ds\end{aligned}$$

and, using Gronwall's inequality together with (3.61)(c), we find that

$$\|u_\rho(t) - u(t)\|_X \le \widetilde{c}\big(F(\rho) + G(\rho) + H(\rho)J(u)\big), \tag{3.65}$$

where $\widetilde{c} > 0$ does not depend on ρ. The convergence (3.62) is now a consequence of (3.65), (3.59)(b), (3.60)(b) and (3.61)(b). ■

3.3 Evolutionary variational inequalities

In this section we specialize the results on history-dependent quasivariational inequalities presented in Section 3.2 in order to obtain existence, uniqueness and convergence results in the study of other types of variational inequality which arise in Contact Mechanics. For simplicity, we consider only the case $K = Y = X$, since this will be enough for the study of the examples we present in Chapters 6 and 7.

3.3.1 Existence and uniqueness

Given the operators $A : X \to X$ and $B : X \to X$, the functions $j : X \times X \to \mathbb{R}$ and $f : [0,T] \to X$, and the initial data $u_0 \in X$, we start by considering the problem of finding a function $u : [0,T] \to X$ such that

$$(A\dot{u}(t), v - \dot{u}(t))_X + (Bu(t), v - \dot{u}(t))_X + j(u(t), v) \\ -j(u(t), \dot{u}(t)) \geq (f(t), v - \dot{u}(t))_X \quad \forall\, v \in X,\ t \in [0,T], \tag{3.66}$$

$$u(0) = u_0. \tag{3.67}$$

Alternatively, we consider the problem of finding a function $u : [0,T] \to X$ such that

$$(A\dot{u}(t), v - \dot{u}(t))_X + (Bu(t), v - \dot{u}(t))_X + j(\dot{u}(t), v) \\ -j(\dot{u}(t), \dot{u}(t)) \geq (f(t), v - \dot{u}(t))_X \quad \forall\, v \in X,\ t \in [0,T], \tag{3.68}$$

$$u(0) = u_0. \tag{3.69}$$

And, finally, when $j : X \to \mathbb{R}$, we consider the problem of finding a function $u : [0,T] \to X$ such that

$$(A\dot{u}(t), v - \dot{u}(t))_X + (Bu(t), v - \dot{u}(t))_X + j(v) - j(\dot{u}(t)) \\ \geq (f(t), v - \dot{u}(t))_X \quad \forall\, v \in X,\ t \in [0,T], \tag{3.70}$$

$$u(0) = u_0. \tag{3.71}$$

We remark that one of the common features of the problems (3.66)–(3.67), (3.68)–(3.69) and (3.70)–(3.71) above is that they involve the time derivative of the unknown function u. As a consequence, an initial condition is needed. For this reason we refer to such problems as *evolutionary problems*.

The term $(A\dot{u}(t), v - \dot{u}(t))_X$ is called the *viscosity term*; the origin of this terminology arises from mechanics and will be explained in Chapter 6. Note that this term simplifies considerably the analysis of the problem, since it produces a regularization effect; its presence allows us to prove the unique solvability of the problems (3.66)–(3.67), (3.68)–(3.69) and (3.70)–(3.71) by using arguments of history-dependent variational inequalities, as we shall see below in this section.

Finally, we remark that the functional j in (3.66) and (3.68) depends on the solution. Therefore, we refer to a problem of the form (3.66)–(3.67) or (3.68)–(3.69) as an *evolutionary quasivariational inequality.* In contrast, we note that the functional j in (3.70) does not depend on the solution. For this reason, a problem of the form (3.70)–(3.71) is called an *evolutionary variational inequality.*

We assume in what follows that the functional j satisfies

$$\left.\begin{array}{l}\text{(a) for all } u \in X,\ j(u,\cdot) : X \to \mathbb{R} \text{ is convex and l.s.c.;}\\[6pt] \text{(b) there exists } \alpha \geq 0 \text{ such that}\\ \qquad j(u_1,v_2) - j(u_1,v_1) + j(u_2,v_1) - j(u_2,v_2)\\ \qquad \leq \alpha \,\|u_1 - u_2\|_X \,\|v_1 - v_2\|_X \ \forall\, u_1,\, u_2 \in X,\ \forall\, v_1,\, v_2 \in X.\end{array}\right\} \tag{3.72}$$

Moreover, the initial condition is such that

$$u_0 \in X. \tag{3.73}$$

The main result in the study of problem (3.66)–(3.67) is the following.

Theorem 3.10 *Let X be a Hilbert space and assume that (3.11)–(3.13), (3.42), (3.72) and (3.73) hold. Then there exists a unique solution u to problem (3.66)–(3.67). Moreover, the solution satisfies $u \in C^1([0,T];X)$.*

Proof Define the functional $\varphi : X \times X \to \mathbb{R}$ and the operator $\mathcal{S}$: $C([0,T];X) \to C([0,T];X)$ by the equalities

$$\varphi(u,v) = (Bu,v)_X + j(u,v) \qquad \forall\, u,v \in X, \tag{3.74}$$

$$\mathcal{S}w(t) = \int_0^t w(s)\mathrm{d}s + u_0 \qquad \forall\, w \in C([0,T];X),\ t \in [0,T]. \tag{3.75}$$

And, for a regular function $w : [0,T] \to X$, let $u : [0,T] \to X$ be the function defined by

$$u(t) = \mathcal{S}w(t) \quad \forall\, t \in [0,T]. \tag{3.76}$$

Then it is easy to see that the function u belongs to the space $C^1([0,T];X)$ and is a solution to the Cauchy problem (3.68)–(3.69) if and only if the function w belongs to the space $C([0,T];X)$ and is a solution to the history-dependent quasivariational inequality

$$\begin{aligned}(Aw(t), v - w(t))_X + \varphi(\mathcal{S}w(t), v) - \varphi(\mathcal{S}w(t), w(t))\\ \geq (f(t), v - w(t))_X \quad \forall\, v \in X.\end{aligned} \tag{3.77}$$

Next, we note that inequality (3.77) represents an inequality of the form (3.35) in which $K = Y = X$ and $j \equiv 0$. Using (3.13) and (3.72) it follows that φ satisfies condition (3.38) and using (3.75) it follows that (3.41) holds. The rest of the assumptions in Theorem 3.6 are clearly satisfied. Therefore, using this theorem it follows that the history-dependent quasivariational inequality (3.77) has a unique solution $w \in C([0,T];X)$, which concludes the proof. ■

The main result in the study of problem (3.68)–(3.69) is the following.

Theorem 3.11 *Let X be a Hilbert space, assume that (3.11)–(3.13), (3.42), (3.72) and (3.73) hold and $m_A > \alpha$. Then there exists a unique solution u to problem (3.68)–(3.69). Moreover, the solution satisfies $u \in C^1([0,T];X)$.*

Proof Let $\varphi : X \times X \to \mathbb{R}$ be the form given by

$$\varphi(u,v) = (Bu,v)_X \quad \forall\, u,v \in X$$

and consider the operator $\mathcal{S}$ defined by (3.75). Moreover, for a regular function $w : [0,T] \to X$, let $u : [0,T] \to X$ be the function defined by (3.76). Then, it is easy to see that the function u belongs to the space $C^1([0,T];X)$ and is a solution to the Cauchy problem (3.68)–(3.69) if and only if the function w belongs to the space $C([0,T];X)$ and is a solution to the history-dependent quasivariational inequality

$$\begin{aligned}(Aw(t), v - w(t))_X + \varphi(\mathcal{S}w(t), v) - \varphi(\mathcal{S}w(t), w(t))\\ +j(w(t), v) - j(w(t), w(t)) \geq (f(t), v - w(t))_X \quad \forall\, t \in [0,T].\end{aligned} \tag{3.78}$$

Note that inequality (3.78) represents an inequality of the form (3.35) in which $K = Y = X$. Theorem 3.11 is now a consequence of the unique solvability of the inequality (3.78), guaranteed by Theorem 3.6. ■

A brief comparison between Theorems 3.10 and 3.11 shows that, besides the common assumptions (3.11)–(3.13), (3.42), (3.72) and (3.73), in solving problem (3.68)–(3.69) we need the additional assumption $m_A > \alpha$. This represents a smallness assumption on the coefficient α similar to that used in Theorem 2.19.

Consider now the case when the functional j does not depend on the solution. In this case we replace assumption (3.72) by the following one:

$$j : X \to \mathbb{R} \quad \text{is convex and l.s.c.} \tag{3.79}$$

We have the following existence and uniqueness result.

Corollary 3.12 *Let X be a Hilbert space and assume that (3.11)–(3.13), (3.42), (3.73) and (3.79) hold. Then there exists a unique solution u to problem (3.70)–(3.71). Moreover, the solution satisfies $u \in C^1([0,T];X)$.*

Proof Let $J : X \times X \to \mathbb{R}$ be the functional given by $J(u,v) = j(v)$ for all u, $v \in X$. Since j satisfies condition (3.79) is easy to see that J satisfies condition (3.72) with $\alpha = 0$. Corollary 3.12 is now a direct consequence of Theorem 3.10 or 3.11. ■

3.3.2 Convergence results

We proceed with some convergence results concerning the evolutionary variational inequalities presented above. These results can be easily obtained by specializing the general convergence result in Theorem 3.9. To avoid repetition, we restrict ourselves to the case of the Cauchy problems (3.68)–(3.69) and (3.70)–(3.71).

We assume in what follows that the hypotheses of Theorem 3.11 hold and we denote by u the solution of problem (3.68)–(3.69). Also, for each $\rho > 0$ we consider a perturbation $B_\rho : X \to X$ of B as well as a perturbation $j_\rho : X \times X \to \mathbb{R}$ of j which satisfy conditions (3.13) and (3.72), respectively, with constants L_{B_ρ} and α_ρ such that $m_A > \alpha_\rho$. We consider the problem of finding a function $u_\rho : [0,T] \to X$ such that

$$\begin{aligned} &(A\dot{u}_\rho(t), v - \dot{u}_\rho(t))_X + (B_\rho u_\rho(t), v - \dot{u}_\rho(t))_X + j_\rho(\dot{u}_\rho(t), v) \\ &\quad -j_\rho(\dot{u}_\rho(t), \dot{u}_\rho(t)) \geq (f(t), v - \dot{u}_\rho(t))_X \quad \forall\, v \in X,\ t \in [0,T], \end{aligned} \tag{3.80}$$

$$u_\rho(0) = u_0. \tag{3.81}$$

We deduce from Theorem 3.11 that problem (3.80)–(3.81) has a unique solution $u_\rho \in C^1([0,T]; X)$.

Consider now the following assumptions.

$$\left.\begin{array}{l} \text{There exists } F : \mathbb{R}_+ \to \mathbb{R}_+ \text{ such that:} \\ \text{(a) } \|B_\rho(v) - B(v)\|_X \leq F(\rho) \quad \forall\, v \in X, \text{ for each } \rho > 0; \\ \text{(b) } \lim\limits_{\rho \to 0} F(\rho) = 0. \end{array}\right\} \tag{3.82}$$

$$\left.\begin{array}{l} \text{There exists } G : \mathbb{R}_+ \to \mathbb{R}_+ \text{ such that} \\ \text{(a) } j_\rho(u_1, u_2) - j_\rho(u_1, u_1) + j(u_2, u_1) - j(u_2, u_2) \\ \qquad \leq G(\rho)\|u_1 - u_2\|_X + \alpha\, \|u_1 - u_2\|_X^2 \\ \qquad \text{for all } u_1, u_2 \in X, \text{ for each } \rho > 0; \\ \text{(b) } \lim\limits_{\rho \to 0} G(\rho) = 0. \end{array}\right\} \tag{3.83}$$

The behavior of the solution u_ρ as $\rho \to 0$ is given by the following theorem.

Theorem 3.13 *Under the assumptions (3.82)–(3.83) the solution u_ρ of problem (3.80)–(3.81) converges to the solution u of problem (3.68)–(3.69), i.e.,*

$$u_\rho \to u \quad \textit{in } C^1([0,T]; X) \quad \textit{as} \quad \rho \to 0.$$

Proof Let $\rho > 0$ and denote by w_ρ and w the functions defined by

$$w_\rho(t) = \dot{u}_\rho(t), \quad w(t) = \dot{u}(t) \quad \forall\, t \in [0,T]. \tag{3.84}$$

Also, denote by $\mathcal{S} : C([0,T];X) \to C([0,T];X)$ the operator

$$\mathcal{S}v(t) = \int_0^t v(s)\,ds + u_0 \quad \forall\, v \in C([0,T];X),\ t \in [0,T]$$

and, finally, let $\varphi_\rho : X \times X \to \mathbb{R}$, $\varphi : X \times X \to \mathbb{R}$ be the functionals

$$\varphi_\rho(u,v) = (B_\rho u, v)_X, \quad \varphi(u,v) = (Bu,v)_X \quad \forall\, u,v \in X. \tag{3.85}$$

Then from the proof of Theorem 3.11 we deduce that the following inequalities hold, for all $t \in [0,T]$:

$$\begin{aligned}
&(Aw_\rho(t), v - w_\rho(t))_X + \varphi_\rho(\mathcal{S}w_\rho(t), v) - \varphi_\rho(\mathcal{S}w_\rho(t), w_\rho(t))\\
&\quad + j_\rho(w_\rho(t), v) - j_\rho(w_\rho(t), w_\rho(t)) \geq (f(t), v - w_\rho(t))_X,\\
&(Aw(t), v - w(t))_X + \varphi(\mathcal{S}w(t), v) - \varphi(\mathcal{S}w(t), w(t))\\
&\quad + j(w(t), v) - j(w(t), w(t)) \geq (f(t), v - w(t))_X.
\end{aligned}$$

These inequalities lead us to apply the convergence result in Theorem 3.9 with $K = Y = X$, $\mathcal{S}_\rho = \mathcal{S}$ and, to this end, we have to check the validity of conditions (3.59)–(3.61). First, we use (3.85) to see that

$$\begin{aligned}
&\varphi_\rho(y_1, u_2) - \varphi_\rho(y_1, u_1) + \varphi(y_2, u_1) - \varphi(y_2, u_2)\\
&\quad = (B_\rho y_1 - By_2, u_2 - u_1)_X \leq \|B_\rho y_1 - By_2\|_X \|u_2 - u_1\|_X\\
&\quad \leq (\|B_\rho y_1 - By_1\|_X + \|By_1 - By_2\|_X)\|u_2 - u_1\|_X
\end{aligned}$$

for all $y_1, y_2 \in X$, $u_1, u_2 \in X$, for each $\rho > 0$. Next, we use assumption (3.82)(a) and (3.13) to obtain

$$\begin{aligned}
&\varphi_\rho(y_1, u_2) - \varphi_\rho(y_1, u_1) + \varphi(y_2, u_1) - \varphi(y_2, u_2)\\
&\quad \leq F(\rho)\|u_1 - u_2\|_X + L_B\|y_1 - y_2\|_X\|u_1 - u_2\|_X
\end{aligned}$$

for all $y_1, y_2 \in X$, $u_1, u_2 \in X$, for each $\rho > 0$. This inequality combined with (3.82)(b) shows that condition (3.59) holds. Also, we remark that condition (3.83) implies (3.60) and, finally, (3.61) holds since $\mathcal{S}_\rho = \mathcal{S}$.

We are now in position to use Theorem 3.9 to deduce that

$$w_\rho \to w \quad \text{in} \quad C([0,T];X) \quad \text{as} \quad \rho \to 0. \tag{3.86}$$

We now use (3.84) and the initial conditions $u_\rho(0) = u(0) = u_0$ to see that

$$u_\rho(t) - u(t) = \int_0^t (w_\rho(s) - w(s))\,ds \quad \forall\, t \in [0,T]. \tag{3.87}$$

Equalities (3.84), (3.87) combined with the convergence (3.86) show that

$$\dot{u}_\rho \to \dot{u} \quad \text{in } C([0,T];X), \quad u_\rho \to u \quad \text{in } C([0,T];X) \quad \text{as} \quad \rho \to 0,$$

which concludes the proof. ■

We proceed with a result concerning the dependence of the solution of the variational inequality (3.70)–(3.71) with respect to the operator B and the functional j. To this end we assume in what follows that (3.11)–(3.13), (3.42), (3.73) and (3.79) hold, and we denote by $u \in C^1([0,T];X)$ the solution of (3.70)–(3.71) provided by Corollary 3.12. For each $\rho > 0$ we consider a perturbation $B_\rho : X \to X$ of B which satisfies condition (3.13) with $L_{B_\rho} > 0$ and a perturbation j_ρ of j which satisfies (3.79). We also consider the problem of finding a function $u_\rho : [0,T] \to X$ such that

$$\begin{aligned}(A\dot{u}_\rho(t), v - \dot{u}_\rho(t))_X + (B_\rho u_\rho(t), v - \dot{u}_\rho(t))_X + j_\rho(v) - j_\rho(\dot{u}_\rho(t))\\ \geq (f(t), v - \dot{u}_\rho(t))_X \quad \forall\, v \in X,\ t \in [0,T],\end{aligned} \tag{3.88}$$

$$u_\rho(0) = u_0. \tag{3.89}$$

We deduce from Corollary 3.12 that problem (3.88)–(3.89) has a unique solution $u_\rho \in C^1([0,T];X)$.

Consider now the following assumption.

$$\left.\begin{aligned}&\text{There exists } G : \mathbb{R}_+ \to \mathbb{R}_+ \text{ such that}\\ &\text{(a) } j_\rho(u_2) - j_\rho(u_1) + j(u_1) - j(u_2) \leq G(\rho)\|u_1 - u_2\|_X\\ &\qquad \text{for all } u_1, u_2 \in X, \text{ for each } \rho > 0;\\ &\text{(b) } \lim_{\rho \to 0} G(\rho) = 0.\end{aligned}\right\} \tag{3.90}$$

The behavior of the solution u_ρ as ρ converges to zero is given in the following result.

Corollary 3.14 *Under the assumptions (3.82) and (3.90), the solution u_ρ of problem (3.88)–(3.89) converges to the solution u of problem (3.70)–(3.71), i.e.*

$$u_\rho \to u \quad \textit{in } C^1([0,T];X) \quad \textit{as} \quad \rho \to 0.$$

Proof Let $J_\rho : X \times X \to \mathbb{R}$ and $J : X \times X \to \mathbb{R}$ be the functionals given by $J_\rho(u,v) = j_\rho(v)$ and $J(u,v) = j(v)$ for all u, $v \in X$. Since j_ρ and j satisfy condition (3.90) it is easy to see that J_ρ and J satisfy condition (3.83) with $\alpha = 0$. Therefore, Corollary 3.14 is a consequence of Theorem 3.13. ∎

Part II

Modelling and analysis of contact problems

4
Modelling of contact problems

In this chapter we present notation and preliminary material which is necessary in the study of the boundary value problems presented in Chapters 5–7. We start by introducing some function spaces that will be relevant in the study of contact problems. Then we provide a general description of the mathematical modelling of the processes involved in contact between an elastic, viscoelastic or viscoplastic body and an obstacle, say a foundation. We describe the physical setting, the variables which determine the state of the system, the material's behavior which is reflected in the constitutive law, the input data, the equation of equilibrium for the state of the system and the boundary conditions for the system variables. Finally, we extend this description to the contact of piezoelectric bodies for which we consider both the case when the foundation is conductive and the case when it is an insulator.

Everywhere in the rest of the book we assume that $\Omega \subset \mathbb{R}^d$ $(d = 1, 2, 3)$ is open, connected, bounded, and has a Lipschitz continuous boundary Γ. We denote by $\overline{\Omega} = \Omega \cup \Gamma$ the closure of Ω in $\mathbb{R}^d$. We use bold face letters for vectors and tensors, such us the outward unit normal on Γ, denoted by $\boldsymbol{\nu}$. A typical point in $\mathbb{R}^d$ is denoted by $\boldsymbol{x} = (x_i)$. The indices i, j, k, l run between 1 and d, and, unless stated otherwise, the summation convention over repeated indices is used. Also, the index that follows a comma indicates a partial derivative with respect to the corresponding component of the spatial variable $\boldsymbol{x}$. Finally, Γ is decomposed into three parts $\overline{\Gamma}_1$, $\overline{\Gamma}_2$ and $\overline{\Gamma}_3$, with Γ_1, Γ_2 and Γ_3 being relatively open and mutually disjoint and, moreover, meas $(\Gamma_1) > 0$.

4.1 Function spaces in contact mechanics

In order to introduce the mathematical models that describe various contact processes we need to describe the functional spaces to which the data and the unknown belong. The aim of this section is to introduce these spaces together with their basic properties.

4.1.1 Preliminaries

We denote by $\mathbb{R}^d$ the d-dimensional real linear space, and the symbol $\mathbb{S}^d$ stands for the space of second order symmetric tensors on $\mathbb{R}^d$ or, equivalently, the space of symmetric matrices of order d. The canonical inner products and the corresponding norms on $\mathbb{R}^d$ and $\mathbb{S}^d$ are given by

$$\boldsymbol{u}\cdot\boldsymbol{v} = u_i v_i, \quad \|\boldsymbol{v}\| = (\boldsymbol{v}\cdot\boldsymbol{v})^{1/2} \quad \forall\, \boldsymbol{u} = (u_i),\, \boldsymbol{v} = (v_i) \in \mathbb{R}^d,$$
$$\boldsymbol{\sigma}\cdot\boldsymbol{\tau} = \sigma_{ij}\tau_{ij}, \quad \|\boldsymbol{\tau}\| = (\boldsymbol{\tau}\cdot\boldsymbol{\tau})^{1/2} \quad \forall\, \boldsymbol{\sigma} = (\sigma_{ij}),\, \boldsymbol{\tau} = (\tau_{ij}) \in \mathbb{S}^d,$$

respectively. The notation $\boldsymbol{I}_d$ will represent the identity operator on $\mathbb{R}^d$ or, equivalently, the unit matrix of order d. And, as usual, the zero elements of $\mathbb{R}^d$ and $\mathbb{S}^d$ will be denoted by $\mathbf{0}_{\mathbb{R}^d}$ and $\mathbf{0}_{\mathbb{S}^d}$, respectively.

We use standard notation for various spaces of real-valued functions defined on Ω and Γ, and we assume that the reader has some basic knowledge of the properties of these spaces. For this reason we restrict ourselves to recalling the following list of the spaces we shall use in the rest of the book:

- $C^m(\overline{\Omega})$ – the space of functions whose derivatives up to and including order m are continuous up to the boundary Γ;
- $C^\infty(\overline{\Omega})$ – the space of infinitely differentiable functions up to the boundary Γ;
- $C_0^\infty(\Omega)$ – the space of infinitely differentiable functions with compact support in Ω;
- $L^p(\Omega)$ – the Lebesgue space of p-integrable functions on Ω, with the usual modification if $p = \infty$;
- $L^1_{\mathrm{loc}}(\Omega)$ – the space of locally integrable functions on Ω;
- $W^{m,p}(\Omega)$ – the Sobolev space of functions whose weak derivatives of orders m or less are p-integrable on Ω;
- $H^m(\Omega) \equiv W^{m,2}(\Omega)$, for positive integer m;
- $L^p(\Gamma)$ – the Lebesgue space of p-integrable functions on Γ, with the usual modification if $p = \infty$;
- $L^p(\Gamma_i)$ – the Lebesgue space of p-integrable functions on Γ_i $(i = 1, 2, 3)$, with the usual modification if $p = \infty$.

If X represents one of the above spaces we use the notation X^d and $X_s^{d\times d}$ for the spaces

$$\begin{aligned} X^d &= \{\, \boldsymbol{x} = (x_i) \;:\; x_i \in X,\ 1 \le i \le d \,\},\\ X_s^{d\times d} &= \{\, \boldsymbol{x} = (x_{ij}) \;:\; x_{ij} = x_{ji} \in X,\ 1 \le i,j \le d \,\} \end{aligned}$$

and, in particular, we will frequently use the spaces

$$L^2(\Omega)^d = \{\boldsymbol{v} = (v_i) \;:\; v_i \in L^2(\Omega),\ 1 \le i \le d\}, \tag{4.1}$$

$$Q = L^2(\Omega)_s^{d\times d} = \{\boldsymbol{\tau} = (\tau_{ij}) : \tau_{ij} = \tau_{ji} \in L^2(\Omega),\ 1 \le i,\, j \le d\}. \tag{4.2}$$

These are Hilbert spaces with the canonical inner products

$$(\boldsymbol{u},\boldsymbol{v})_{L^2(\Omega)^d} = \int_\Omega u_i\, v_i \,\mathrm{d}x = \int_\Omega \boldsymbol{u}\cdot\boldsymbol{v}\,\mathrm{d}x,$$

$$(\boldsymbol{\sigma},\boldsymbol{\tau})_Q = \int_\Omega \sigma_{ij}\,\tau_{ij}\,\mathrm{d}x = \int_\Omega \boldsymbol{\sigma}\cdot\boldsymbol{\tau}\,\mathrm{d}x,$$

and the associated norms denoted by $\|\cdot\|_{L^2(\Omega)^d}$ and $\|\cdot\|_Q$, respectively.

Besides the function spaces introduced above, we need specific function spaces for the displacement and the stress field.

4.1.2 Spaces for the displacement field

In the contact problems presented in Chapters 5–7, displacements are sought in the space

$$H^1(\Omega)^d = \{\boldsymbol{v} = (v_i) : v_i \in H^1(\Omega),\ 1 \le i \le d\}$$

or its subspaces or subsets, depending on prescribed boundary conditions. The space $H^1(\Omega)^d$ is a Hilbert space with the canonical inner product

$$(\boldsymbol{u},\boldsymbol{v})_{H^1(\Omega)^d} = \int_\Omega (u_i v_i + u_{i,j} v_{i,j})\,\mathrm{d}x$$

and the corresponding norm

$$\|\boldsymbol{v}\|_{H^1(\Omega)^d} = \Big(\int_\Omega (v_i\,v_i + v_{i,j}\,v_{i,j})\,\mathrm{d}x\Big)^{1/2}. \tag{4.3}$$

Let $\boldsymbol{\varepsilon} : H^1(\Omega)^d \to Q$ be the *deformation* operator defined by

$$\boldsymbol{\varepsilon}(\boldsymbol{u}) = (\varepsilon_{ij}(\boldsymbol{u})), \quad \varepsilon_{ij}(\boldsymbol{u}) = \frac{1}{2}\,(u_{i,j} + u_{j,i}).$$

The quantity $\boldsymbol{\varepsilon}(\boldsymbol{u})$ is the *linearized* (or *small*) *strain tensor* associated with the displacement $\boldsymbol{u}$. Define

$$((\boldsymbol{u},\boldsymbol{v}))_{H^1(\Omega)^d} = (\boldsymbol{u},\boldsymbol{v})_{L^2(\Omega)^d} + (\boldsymbol{\varepsilon}(\boldsymbol{u}),\boldsymbol{\varepsilon}(\boldsymbol{v}))_Q \quad \forall\, \boldsymbol{u},\boldsymbol{v} \in H^1(\Omega)^d, \tag{4.4}$$

and the corresponding norm

$$|||\boldsymbol{v}|||_{H^1(\Omega)^d} = \left(\|\boldsymbol{v}\|^2_{L^2(\Omega)^d} + \|\boldsymbol{\varepsilon}(\boldsymbol{v})\|^2_Q\right)^{1/2} \quad \forall\, \boldsymbol{v} \in H^1(\Omega)^d. \tag{4.5}$$

It can be proved (see, e.g., [75, p. 106]) that there exists a positive constant c such that

$$\int_\Omega (v_i v_i + v_{i,j} v_{i,j})\,\mathrm{d}x \le c \int_\Omega (v_i v_i + \varepsilon_{ij}(\boldsymbol{v})\,\varepsilon_{ij}(\boldsymbol{v}))\,\mathrm{d}x \quad \forall\, \boldsymbol{v} \in H^1(\Omega)^d.$$

It follows that $((\cdot,\cdot))_{H^1(\Omega)^d}$, defined in (4.4), is an inner product in $H^1(\Omega)^d$, and the corresponding norm $|||\cdot|||_{H^1(\Omega)^d}$ is equivalent to the canonical norm (4.3). We conclude from here that $(H^1(\Omega)^d, |||\cdot|||_{H^1(\Omega)^d})$ is a real Hilbert space.

Next, we note that by the Sobolev trace theorem we can define the *trace* $\gamma\boldsymbol{v}$ of a function $\boldsymbol{v} \in H^1(\Omega)^d$ on the boundary Γ, such that $\gamma\boldsymbol{v} = \boldsymbol{v}|_\Gamma$ if $\boldsymbol{v} \in H^1(\Omega)^d \cap C(\overline{\Omega})^d$. And, for simplicity, for an element $\boldsymbol{v} \in H^1(\Omega)^d$ we still denote by $\boldsymbol{v}$ its trace $\gamma\boldsymbol{v}$ on Γ. The trace operator $\gamma : H^1(\Omega)^d \to L^2(\Gamma)^d$ is a linear continuous operator, i.e. there exists a positive constant c, depending only on Ω, such that

$$\|\boldsymbol{v}\|_{L^2(\Gamma)^d} \le c\,|||\boldsymbol{v}|||_{H^1(\Omega)^d} \quad \forall\, \boldsymbol{v} \in H^1(\Omega)^d. \tag{4.6}$$

The continuity of the trace operator shows that the convergence in $H^1(\Omega)^d$ implies the convergence of the traces in $L^2(\Gamma)^d$, i.e.

$$\boldsymbol{u}_n \to \boldsymbol{u} \quad \text{in} \quad H^1(\Omega)^d \implies \boldsymbol{u}_n \to \boldsymbol{u} \quad \text{in} \quad L^2(\Gamma)^d.$$

Moreover, in Section 5.3 we need a stronger result which states that the weak convergence in $H^1(\Omega)^d$ is enough to imply the strong convergence of the traces in $L^2(\Gamma)^d$, i.e.

$$\boldsymbol{u}_n \rightharpoonup \boldsymbol{u} \quad \text{in} \quad H^1(\Omega)^d \implies \boldsymbol{u}_n \to \boldsymbol{u} \quad \text{in} \quad L^2(\Gamma)^d. \tag{4.7}$$

The convergence result (4.7) shows that the trace operator $\gamma : H^1(\Omega)^d \to L^2(\Gamma)^d$ is a compact operator.

Besides its trace, for an element $\boldsymbol{v} \in H^1(\Omega)^d$ we also consider its *normal* component and *tangential* part on the boundary, denoted by v_ν and $\boldsymbol{v}_\tau$, respectively, defined by

$$v_\nu = \boldsymbol{v} \cdot \boldsymbol{\nu}, \qquad \boldsymbol{v}_\tau = \boldsymbol{v} - v_\nu \boldsymbol{\nu}. \tag{4.8}$$

In the study of contact problems, we will frequently use the subspace V of $H^1(\Omega)^d$ given by

$$V = \{\, \boldsymbol{v} \in H^1(\Omega)^d : \boldsymbol{v} = \boldsymbol{0} \ \text{ a.e. on } \Gamma_1 \,\}. \tag{4.9}$$

Here, the condition "$\boldsymbol{v} = \mathbf{0}$ a.e. on Γ_1" is understood in the sense of trace, i.e., $\gamma\boldsymbol{v} = \mathbf{0}$ a.e. on Γ_1. Since the trace operator is a linear continuous operator, it follows that V is a closed subspace of $H^1(\Omega)^d$. Moreover, since meas $(\Gamma_1) > 0$, Korn's inequality holds, i.e. there exists a positive constant c_K which depends only on Ω and Γ_1 such that

$$\|\boldsymbol{\varepsilon}(\boldsymbol{v})\|_Q \geq c_K \,|||\boldsymbol{v}|||_{H^1(\Omega)^d} \quad \forall\, \boldsymbol{v} \in V. \tag{4.10}$$

A proof of this inequality can be found in [116, p. 79]. We define the inner product $(\cdot\,,\cdot)_V$ on V by

$$(\boldsymbol{u}, \boldsymbol{v})_V = (\boldsymbol{\varepsilon}(\boldsymbol{u}), \boldsymbol{\varepsilon}(\boldsymbol{v}))_Q, \tag{4.11}$$

and it induces the norm

$$\|\boldsymbol{v}\|_V = \|\boldsymbol{\varepsilon}(\boldsymbol{v})\|_Q. \tag{4.12}$$

It follows from (4.5), (4.10) and (4.12) that $|||\cdot|||_{H^1(\Omega)^d}$ and $\|\cdot\|_V$ are equivalent norms on V and, therefore, $(V, \|\cdot\|_V)$ is a real Hilbert space. Moreover, the deformation operator $\boldsymbol{\varepsilon} : V \to Q$ is a linear continuous operator. Finally, note that by (4.6), (4.10) and (4.12) there exists a positive constant c_0, depending on Ω, Γ_1 and Γ_3, such that

$$\|\boldsymbol{v}\|_{L^2(\Gamma_3)^d} \leq c_0 \,\|\boldsymbol{v}\|_V \quad \forall\, \boldsymbol{v} \in V. \tag{4.13}$$

We now proceed with the following consequence of Korn's inequality.

Theorem 4.1 *Assume that meas* $(\Gamma_1) > 0$ *and denote by* $\boldsymbol{\varepsilon}(V)$ *the range of the deformation operator* $\boldsymbol{\varepsilon} : V \to Q$, *i.e.*

$$\boldsymbol{\varepsilon}(V) = \{\, \boldsymbol{\varepsilon}(\boldsymbol{v}) \,:\, \boldsymbol{v} \in V \,\}. \tag{4.14}$$

Then $\boldsymbol{\varepsilon}(V)$ *is a closed subspace of* Q.

Proof Let $\{\boldsymbol{\tau}_n\}$ be a sequence of elements of $\boldsymbol{\varepsilon}(V)$ which converges in Q to an element $\boldsymbol{\tau} \in Q$, i.e.

$$\boldsymbol{\tau}_n \to \boldsymbol{\tau} \quad \text{in } Q \quad \text{as} \quad n \to \infty. \tag{4.15}$$

Then there exists a sequence $\{\boldsymbol{v}_n\} \subset V$ such that

$$\boldsymbol{\tau}_n = \boldsymbol{\varepsilon}(\boldsymbol{v}_n) \quad \text{for all} \quad n \in \mathbb{N}. \tag{4.16}$$

It follows from (4.15) that $\{\boldsymbol{\tau}_n\}$ is a Cauchy sequence in Q and, therefore, (4.12) implies that $\{\boldsymbol{v}_n\}$ is a Cauchy sequence in V. Next, since V is complete, there exists an element $\boldsymbol{v} \in V$ such that

$$\boldsymbol{v}_n \to \boldsymbol{v} \quad \text{in } V \quad \text{as} \quad n \to \infty. \tag{4.17}$$

We now use the convergence (4.17) to see that

$$\boldsymbol{\varepsilon}(\boldsymbol{v}_n) \to \boldsymbol{\varepsilon}(\boldsymbol{v}) \quad \text{in } Q \quad \text{as} \quad n \to \infty. \tag{4.18}$$

Then we combine (4.15), (4.16) and (4.18) to deduce that $\boldsymbol{\tau} = \boldsymbol{\varepsilon}(\boldsymbol{v})$. It follows from this that $\boldsymbol{\tau} \in \boldsymbol{\varepsilon}(V)$, which concludes the proof. ∎

In addition to the space V defined in (4.9), we will also need its subspace

$$V_1 = \{\, \boldsymbol{v} \in V : v_\nu = 0 \ \text{ a.e. on } \Gamma_3 \,\}. \tag{4.19}$$

Over V_1, we use the inner product $(\cdot,\cdot)_V$ and the associated norm $\|\cdot\|_V$ of the space V. Since V_1 is a closed subspace of V, we see that $(V_1, \|\cdot\|_V)$ is itself a real Hilbert space.

4.1.3 Spaces for the stress field

To introduce a function space for stress fields which is useful in the study of contact problems we need to extend the definition of the divergence of a regular tensor field. A general approach to such an extension is to use the derivatives in the distribution sense, see e.g., [66]. Here, we choose to introduce the concept of the weak divergence directly, which is sufficient for our purposes.

Definition 4.2 Let $\boldsymbol{\sigma} = (\sigma_{ij})$ and $\boldsymbol{w} = (w_i)$ be such that $\sigma_{ij} = \sigma_{ji} \in L^1_{\rm loc}(\Omega)$, $w_i \in L^1_{\rm loc}(\Omega)$, for all $1 \le i, j \le d$. Then $\boldsymbol{w}$ is called a *weak divergence* of $\boldsymbol{\sigma}$ if

$$\int_\Omega \sigma_{ij}\varphi_{i,j}\,dx = -\int_\Omega w_i\,\varphi_i\,dx \quad \forall\, \boldsymbol{\varphi} = (\varphi_i) \in C_0^\infty(\Omega)^d.$$

The weak divergence is uniquely determined up to a set of measure zero. From the definition of weak divergence, we see that if $\boldsymbol{\sigma} \in C^1(\overline{\Omega})^{d\times d}_s$ then the classical divergence of $\boldsymbol{\sigma}$ is also its weak divergence. For this reason, we use the notation $\mathrm{Div}\,\boldsymbol{\sigma}$ or $\sigma_{ij,j}$ for the weak divergence of $\boldsymbol{\sigma}$ and, for the sake of simplicity, we omit the adjective "weak".

We now define the space

$$Q_1 = \{\, \boldsymbol{\tau} \in Q : \mathrm{Div}\,\boldsymbol{\tau} \in L^2(\Omega)^d \,\}, \tag{4.20}$$

where, recall, $L^2(\Omega)^d$ is the space (4.1), together with its canonical inner product and the associated norm. The space Q_1 is a Hilbert space endowed with the inner product

$$(\boldsymbol{\sigma}, \boldsymbol{\tau})_{Q_1} = (\boldsymbol{\sigma}, \boldsymbol{\tau})_Q + (\mathrm{Div}\,\boldsymbol{\sigma}, \mathrm{Div}\,\boldsymbol{\tau})_{L^2(\Omega)^d}$$

and the associated norm $\|\cdot\|_{Q_1}$. Also, note that the *divergence* operator $\mathrm{Div} : Q_1 \to L^2(\Omega)^d$ is a linear continuous operator.

If $\boldsymbol{\sigma}$ is a regular function, say $\boldsymbol{\sigma} \in C^1(\overline{\Omega})_s^{d\times d}$, then the *normal* component and the *tangential* part of the stress field $\boldsymbol{\sigma}$ on the boundary are defined by

$$\sigma_\nu = (\boldsymbol{\sigma}\boldsymbol{\nu})\cdot\boldsymbol{\nu}, \quad \boldsymbol{\sigma}_\tau = \boldsymbol{\sigma}\boldsymbol{\nu} - \sigma_\nu\boldsymbol{\nu}. \tag{4.21}$$

Moreover, the following Green's formula holds,

$$\int_\Omega \boldsymbol{\sigma}\cdot\boldsymbol{\varepsilon}(\boldsymbol{v})\,\mathrm{d}x + \int_\Omega \mathrm{Div}\,\boldsymbol{\sigma}\cdot\boldsymbol{v}\,\mathrm{d}x = \int_\Gamma \boldsymbol{\sigma}\boldsymbol{\nu}\cdot\boldsymbol{v}\,\mathrm{d}a \quad \forall\,\boldsymbol{v}\in H^1(\Omega)^d. \tag{4.22}$$

A proof of this formula is based on a standard density argument. First, it follows from the classical Green–Gauss formula that (4.22) is valid for all $\boldsymbol{v}\in C^\infty(\overline{\Omega})^d$; then we use the density of the space $C^\infty(\overline{\Omega})^d$ in $H^1(\Omega)^d$ to obtain that the equality in (4.22) is valid for all $\boldsymbol{v}\in H^1(\Omega)^d$.

4.1.4 Spaces for piezoelectric contact problems

In the study of piezoelectric contact problems, besides the spaces presented above, we need specific spaces associated with the electric variables i.e. the electric potential field and the electric displacement field. To introduce these spaces, we consider a measurable subset $\Gamma_{\mathrm{a}} \subset \Gamma$ such that meas $(\Gamma_{\mathrm{a}}) > 0$.

The electric potential field is to be found in the space

$$W = \{\ \psi\in H^1(\Omega) :\ \psi = 0\ \text{ a.e. on } \Gamma_{\mathrm{a}}\ \} \tag{4.23}$$

where, again, the condition "$\psi = 0$ a.e. on Γ_{a}" is understood in the sense of trace. Note that W is a closed subspace of the space $H^1(\Omega)$. Moreover, since meas $(\Gamma_{\mathrm{a}}) > 0$, the Friedrichs–Poincaré inequality holds, thus,

$$\|\nabla\psi\|_{L^2(\Omega)^d} \ge c_F\,\|\psi\|_{H^1(\Omega)} \quad \forall\,\psi\in W. \tag{4.24}$$

Here $c_F > 0$ is a constant which depends only on Ω and Γ_{a} and $\nabla : H^1(\Omega)\to L^2(\Omega)^d$ represents the *gradient operator*, i.e. $\nabla\psi = (\psi_{,i})$. On W we use the inner product

$$(\varphi,\psi)_W = (\nabla\varphi, \nabla\psi)_{L^2(\Omega)^d}, \tag{4.25}$$

and let $\|\cdot\|_W$ be the associated norm. Also, recall that the norm on the space $H^1(\Omega)$ is given by

$$\|\psi\|_{H^1(\Omega)} = \Big(\|\psi\|^2_{L^2(\Omega)} + \|\nabla\psi\|^2_{L^2(\Omega)^d}\Big)^{1/2}. \tag{4.26}$$

It follows from (4.24)–(4.26) that $\|\cdot\|_{H^1(\Omega)}$ and $\|\cdot\|_W$ are equivalent norms on W. Therefore, $(W, \|\cdot\|_W)$ is a real Hilbert space. Moreover, given a measurable set $\Gamma_3\subset\Gamma$, by the Sobolev trace theorem it follows that there exists a positive constant $\widetilde{c}_0$ which depends only on Ω, $\Gamma_{\rm a}$ and Γ_3, such that

$$\|\psi\|_{L^2(\Gamma_3)} \le \widetilde{c}_0 \|\psi\|_W \quad \forall\, \psi \in W. \tag{4.27}$$

Next, we recall the following useful result.

Theorem 4.3 *Assume that* meas$\,(\Gamma_{\rm a}) > 0$ *and denote by* $\nabla(W)$ *the range of the gradient operator* $\nabla : W \to L^2(\Omega)^d$, *i.e.*

$$\nabla(W) = \{\, \nabla\psi \,:\, \psi \in W \,\}. \tag{4.28}$$

Then $\nabla(W)$ *is a closed subspace of* $L^2(\Omega)^d$.

The proof of Theorem 4.3 is similar to the proof of Theorem 4.1. The difference arises in the fact that now we replace the space Q with the space $L^2(\Omega)^d$, the space V with the space W, the deformation operator with the gradient operator and Korn's inequality with the Friedrichs–Poincaré inequality.

To introduce a function space for the electric displacement fields we need to define the weak divergence of a vector field. This definition is similar to Definition 4.2 and is as follows.

Definition 4.4 Let $\boldsymbol{D} = (D_i)$ and w be such that $D_i \in L^1_{\rm loc}(\Omega)$ for all $1 \le i \le d$ and $w \in L^1_{\rm loc}(\Omega)$. Then w is called a *weak divergence* of $\boldsymbol{D}$ if

$$\int_\Omega D_i \varphi_{,i}\, {\rm d}x = -\int_\Omega w\,\varphi\, {\rm d}x \quad \forall\, \varphi \in C_0^\infty(\Omega).$$

The weak divergence is uniquely determined up to a set of measure zero. From the definition of the weak divergence, we see that if $\boldsymbol{D} \in C^1(\overline{\Omega})^d$ then the classical divergence of $\boldsymbol{D}$ is also its weak divergence. For this reason, we use the notation ${\rm div}\,\boldsymbol{D}$ or $D_{i,i}$ for the weak divergence of $\boldsymbol{D}$ and, for the sake of simplicity, we omit the adjective "weak".

For the electric displacement field, besides the Hilbert space $L^2(\Omega)^d$ we introduce the space

$$\mathcal{W}_1 = \{\, \boldsymbol{D} \in L^2(\Omega)^d :\ {\rm div}\,\boldsymbol{D} \in L^2(\Omega) \,\}. \tag{4.29}$$

This is a Hilbert space endowed with the inner product

$$(\boldsymbol{D}, \boldsymbol{E})_{\mathcal{W}_1} = (\boldsymbol{D}, \boldsymbol{E})_{L^2(\Omega)^d} + ({\rm div}\,\boldsymbol{D}, {\rm div}\,\boldsymbol{E})_{L^2(\Omega)}$$

and the associated norm $\|\cdot\|_{\mathcal{W}_1}$. Note that the divergence operator div : $\mathcal{W}_1 \to L^2(\Omega)$ is a linear continuous operator. Also, recall that when $\boldsymbol{D} \in \mathcal{W}_1$ is a sufficiently regular function, the Green-type formula holds:

$$(\boldsymbol{D}, \nabla\psi)_{L^2(\Omega)^d} + ({\rm div}\,\boldsymbol{D}, \psi)_{L^2(\Omega)} = \int_\Gamma \boldsymbol{D}\cdot\boldsymbol{\nu}\,\psi\, {\rm d}a \quad \forall\, \psi \in H^1(\Omega). \tag{4.30}$$

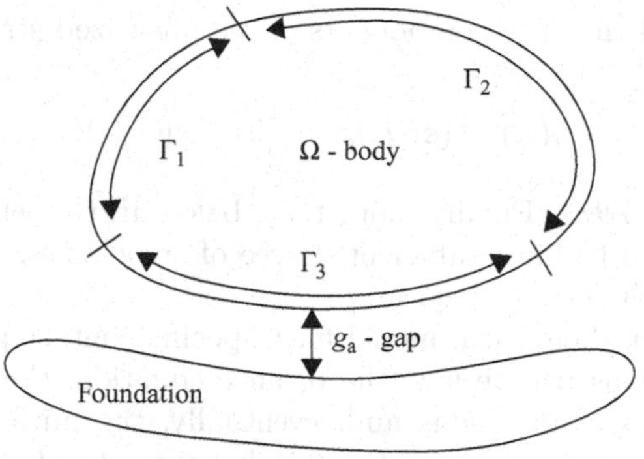

Figure 4.1 The physical setting; Γ_3 is the contact surface

A proof of this formula can be obtained by using similar arguments as those used to prove the Green-type formula (4.22).

4.2 Physical setting and constitutive laws

In this section we present a general physical setting in which most of the contact problems arising in industry and everyday life can be cast. Then we introduce the various constitutive laws considered in this book.

4.2.1 Physical setting

We consider the general physical setting shown in Figure 4.1 and described in what follows. A deformable body occupies, in the reference configuration, an open bounded connected set $\Omega \subset \mathbb{R}^d$ with boundary Γ, composed of three sets $\overline{\Gamma}_1$, $\overline{\Gamma}_2$ and $\overline{\Gamma}_3$, with the mutually disjoint relatively open sets Γ_1, Γ_2 and Γ_3. The body is clamped on Γ_1. Surface tractions of density $\boldsymbol{f}_2$ act on Γ_2 and body forces of density (per unit volume) $\boldsymbol{f}_0$ act in Ω. The body is, or can arrive, in contact on Γ_3 with an obstacle, the so-called foundation. We assume that in the reference configuration there may be a gap, denoted by $g_a = g_a(\boldsymbol{x})$, between Γ_3 and the foundation, which is measured along the normal $\boldsymbol{\nu}$.

We are interested in mathematical models which describe the equilibrium of the mechanical state of the body, in the physical setting above, in the framework of small strain theory. To this end we denote by $\boldsymbol{u}$, $\boldsymbol{\sigma}$ and $\boldsymbol{\varepsilon} = \boldsymbol{\varepsilon}(\boldsymbol{u})$ the displacement vector, the stress tensor, and the linearized strain tensor, respectively. These are functions which depend on the spatial variable $\boldsymbol{x}$ and, eventually, on the time variable t. Nevertheless, in what follows we do not indicate explicitly the dependence of these quantities on $\boldsymbol{x}$ and t, i.e., for instance, we write $\boldsymbol{\sigma}$ instead of $\boldsymbol{\sigma}(\boldsymbol{x})$ or $\boldsymbol{\sigma}(\boldsymbol{x}, t)$. Everywhere below, T is a positive constant, $[0, T]$ denotes the time interval of interest and a dot above a variable will represent the derivative with respect to time. Also, recall that the components of the linearized strain tensor $\boldsymbol{\varepsilon}(\boldsymbol{u})$ are given by

$$\varepsilon_{ij}(\boldsymbol{u}) = (\boldsymbol{\varepsilon}(\boldsymbol{u}))_{ij} = \frac{1}{2}\,(u_{i,j} + u_{j,i}), \tag{4.31}$$

where $u_{i,j} = \partial u_i/\partial x_j$. Finally, note that, below in this chapter, all variables are assumed to have sufficient degree of smoothness consistent with developments they are involved in.

To present a mathematical model for a specific contact process we need to describe the constitutive law, the balance equation, the boundary conditions, the contact conditions and, eventually, the initial conditions. A constitutive law represents a relationship between the stress $\boldsymbol{\sigma}$, the strain $\boldsymbol{\varepsilon}$ and their derivatives, eventually, which characterizes a specific material. It describes the deformations of the body resulting from the action of forces and tractions. Though the constitutive laws must satisfy some basic axioms and invariance principles, they originate mostly from experiments. We refer the reader to [31, 55] for a general description of several diagnostic experiments which provide information needed in constructing constitutive laws for specific materials. And, we note that in this book we restrict ourselves to considering elastic, viscoelastic and viscoplastic materials.

4.2.2 Elastic constitutive laws

A general elastic constitutive law is given by

$$\boldsymbol{\sigma} = \mathcal{F}\boldsymbol{\varepsilon}(\boldsymbol{u}), \tag{4.32}$$

where $\mathcal{F}$ is the *elasticity operator*, assumed to be nonlinear. We allow $\mathcal{F}$ to depend on the location of the point; consequently, all that follows is valid for nonhomogeneous materials. We use the short-hand notation $\mathcal{F}\boldsymbol{\varepsilon}(\boldsymbol{u})$ for $\mathcal{F}(\boldsymbol{x}, \boldsymbol{\varepsilon}(\boldsymbol{u}))$.

In the study of contact problems involving elastic materials, we assume that the operator $\mathcal{F}$ satisfies the following conditions.

$$\left.\begin{array}{l}
\text{(a) } \mathcal{F} : \Omega \times \mathbb{S}^d \to \mathbb{S}^d.\\
\text{(b) There exists } L_{\mathcal{F}} > 0 \text{ such that}\\
\qquad \|\mathcal{F}(\boldsymbol{x}, \boldsymbol{\varepsilon}_1) - \mathcal{F}(\boldsymbol{x}, \boldsymbol{\varepsilon}_2)\| \le L_{\mathcal{F}} \|\boldsymbol{\varepsilon}_1 - \boldsymbol{\varepsilon}_2\|\\
\qquad\quad \forall\, \boldsymbol{\varepsilon}_1, \boldsymbol{\varepsilon}_2 \in \mathbb{S}^d, \text{ a.e. } \boldsymbol{x} \in \Omega.\\
\text{(c) There exists } m_{\mathcal{F}} > 0 \text{ such that}\\
\qquad (\mathcal{F}(\boldsymbol{x}, \boldsymbol{\varepsilon}_1) - \mathcal{F}(\boldsymbol{x}, \boldsymbol{\varepsilon}_2)) \cdot (\boldsymbol{\varepsilon}_1 - \boldsymbol{\varepsilon}_2) \ge m_{\mathcal{F}}\, \|\boldsymbol{\varepsilon}_1 - \boldsymbol{\varepsilon}_2\|^2\\
\qquad\quad \forall\, \boldsymbol{\varepsilon}_1, \boldsymbol{\varepsilon}_2 \in \mathbb{S}^d, \text{ a.e. } \boldsymbol{x} \in \Omega.\\
\text{(d) The mapping } \boldsymbol{x} \mapsto \mathcal{F}(\boldsymbol{x}, \boldsymbol{\varepsilon}) \text{ is measurable on } \Omega,\\
\qquad \text{for any } \boldsymbol{\varepsilon} \in \mathbb{S}^d.\\
\text{(e) The mapping } \boldsymbol{x} \mapsto \mathcal{F}(\boldsymbol{x}, \boldsymbol{0}_{\mathbb{S}^d}) \text{ belongs to } Q.
\end{array}\right\} \tag{4.33}$$

Recall that, here and below, Q represents the space given by (4.2). We observe that assumptions (4.33)(b) and (c) show that the map $\mathcal{F}(\boldsymbol{x},\cdot)$ is Lipschitz continuous and strongly monotone on $\mathbb{S}^d$, for almost every $\boldsymbol{x} \in \Omega$, respectively. Conditions (4.33)(d) and (e) are introduced for mathematical reasons since, combined with condition (4.33)(b), they guarantee that $\boldsymbol{x} \mapsto \mathcal{F}(\boldsymbol{x}, \boldsymbol{\varepsilon}(\boldsymbol{x}))$ belongs to Q whenever $\boldsymbol{\varepsilon} \in Q$. Moreover, it is easy to see that if (4.33) holds, then $\boldsymbol{\varepsilon} \mapsto \mathcal{F}(\boldsymbol{\varepsilon})$ is a strongly monotone Lipschitz continuous operator on the space Q.

In particular, if $\mathcal{F}$ is a linear operator, (4.32) leads to the constitutive law of linearly elastic materials,

$$\sigma_{ij} = f_{ijkl}\varepsilon_{kl}(\boldsymbol{u}),$$

where σ_{ij} are the components of the stress tensor $\boldsymbol{\sigma}$ and f_{ijkl} are the components of the elasticity tensor $\mathcal{F}$. Clearly, assumption (4.33) is satisfied in this particular case, if all the components f_{ijkl} belong to $L^\infty(\Omega)$ and satisfy the usual properties of symmetry and ellipticity, i.e.

$$f_{ijkl} = f_{jikl} = f_{klij},$$

and there exists $m_{\mathcal{F}} > 0$ such that

$$f_{ijkl}\varepsilon_{ij}\varepsilon_{kl} \geq m_{\mathcal{F}}\|\boldsymbol{\varepsilon}\|^2 \qquad \forall\, \boldsymbol{\varepsilon} = (\varepsilon_{ij}) \in \mathbb{S}^d.$$

Due to the symmetry, when $d = 3$ there are only 21 independent components, when $d = 2$ there are only four independent components, and when $d = 1$ there is only one component in the elasticity tensor.

Let $d = 3$. When the material is linear and isotropic, the elasticity tensor is characterized by only two coefficients. Thus, the constitutive law of a linearly elastic isotropic material is given by

$$\boldsymbol{\sigma} = 2\,\mu\,\boldsymbol{\varepsilon}(\boldsymbol{u}) + \lambda\,(\operatorname{tr}\boldsymbol{\varepsilon}(\boldsymbol{u}))\,\boldsymbol{I}_3$$

so that the elasticity operator is

$$\mathcal{F}(\boldsymbol{\varepsilon}) = 2\,\mu\,\boldsymbol{\varepsilon} + \lambda\,(\operatorname{tr}\boldsymbol{\varepsilon})\,\boldsymbol{I}_3.$$

Here λ and μ are the *Lamé coefficients* and satisfy $\lambda > 0$, $\mu > 0$, $\operatorname{tr}\boldsymbol{\varepsilon}(\boldsymbol{u})$ denotes the trace of the tensor $\boldsymbol{\varepsilon}(\boldsymbol{u})$,

$$\operatorname{tr}\boldsymbol{\varepsilon}(\boldsymbol{u}) = \varepsilon_{ii}(\boldsymbol{u}),$$

and, recall, $\boldsymbol{I}_3$ denotes the identity tensor on $\mathbb{R}^3$. In components, we have

$$\sigma_{ij} = 2\,\mu\,\varepsilon_{ij}(\boldsymbol{u}) + \lambda\,\varepsilon_{kk}(\boldsymbol{u})\,\delta_{ij},$$

where δ_{ij} is the Kronecker symbol, i.e., δ_{ij} are the components of the unit 3×3 matrix.

Besides the linear case described above, a second example of elastic constitutive law of the form (4.32) is provided by

$$\boldsymbol{\sigma} = \mathcal{E}\boldsymbol{\varepsilon}(\boldsymbol{u}) + \beta\,(\boldsymbol{\varepsilon}(\boldsymbol{u}) - \mathcal{P}_K\boldsymbol{\varepsilon}(\boldsymbol{u})). \tag{4.34}$$

Here $\mathcal{E}$ is a linear or nonlinear operator which satisfies (4.33), $\beta > 0$, K is a closed convex subset of $\mathbb{S}^d$ such that $\mathbf{0}_{\mathbb{S}^d} \in K$ and $\mathcal{P}_K : \mathbb{S}^d \to K$ denotes the projection operator. The corresponding elasticity operator is nonlinear and is given by

$$\mathcal{F}\boldsymbol{\varepsilon} = \mathcal{E}\boldsymbol{\varepsilon} + \beta\,(\boldsymbol{\varepsilon} - \mathcal{P}_K\boldsymbol{\varepsilon}). \tag{4.35}$$

Usually the set K is defined by

$$K = \{\,\boldsymbol{\varepsilon} \in \mathbb{S}^d : \mathcal{F}(\boldsymbol{\varepsilon}) \le 0\,\}, \tag{4.36}$$

where $\mathcal{F} : \mathbb{S}^d \to \mathbb{R}$ is a convex continuous function such that $\mathcal{F}(\mathbf{0}_{\mathbb{S}^d}) < 0$. Since the projection operator is nonexpansive, the elasticity operator (4.35) satisfies condition (4.33). Recall that $\boldsymbol{\varepsilon} = \mathcal{P}_K\boldsymbol{\varepsilon}$ iff $\boldsymbol{\varepsilon} \in K$ and, therefore, from (4.34) we see that $\boldsymbol{\sigma} = \mathcal{E}\boldsymbol{\varepsilon}(\boldsymbol{u})$ iff $\boldsymbol{\varepsilon}(\boldsymbol{u}) \in K$. Assume that $\mathcal{E}$ is linear. Then, it follows from the above that the material behaves linearly as far as the strain tensor $\boldsymbol{\varepsilon}$ belongs to K. Thus, in this case, the set K represents the domain of linearly elastic behavior of the material (4.34).

A third family of elasticity operators satisfying the conditions (4.33) is provided by nonlinear *Hencky materials*, see [157] for details. For a Hencky material, the constituive law is

$$\boldsymbol{\sigma} = k_0\,(\operatorname{tr}\boldsymbol{\varepsilon}(\boldsymbol{u}))\,\boldsymbol{I}_d + \psi(\|\boldsymbol{\varepsilon}^{\mathrm{D}}(\boldsymbol{u})\|^2)\,\boldsymbol{\varepsilon}^{\mathrm{D}}(\boldsymbol{u}),$$

so that the elasticity operator is

$$\mathcal{F}(\boldsymbol{\varepsilon}) = k_0\,(\operatorname{tr}\boldsymbol{\varepsilon})\,\boldsymbol{I}_d + \psi(\|\boldsymbol{\varepsilon}^{\mathrm{D}}\|^2)\,\boldsymbol{\varepsilon}^{\mathrm{D}}. \tag{4.37}$$

Here, $k_0 > 0$ is a material coefficient, $\boldsymbol{I}_d$ is the identity matrix of order d, $\psi : \mathbb{R} \to \mathbb{R}$ is a constitutive function and $\boldsymbol{\varepsilon}^{\mathrm{D}} = \boldsymbol{\varepsilon}^{\mathrm{D}}(\boldsymbol{u})$ denotes the *deviatoric* part of $\boldsymbol{\varepsilon} = \boldsymbol{\varepsilon}(\boldsymbol{u})$, defined by

$$\boldsymbol{\varepsilon}^{\mathrm{D}} = \boldsymbol{\varepsilon} - \frac{1}{d}\,(\operatorname{tr}\boldsymbol{\varepsilon})\,\boldsymbol{I}_d. \tag{4.38}$$

The function ψ is assumed to be piecewise continuously differentiable, and there exist positive constants c_1, c_2, d_1 and d_2, such that

$$\psi(\xi) \le d_1, \quad -c_1 \le \psi'(\xi) \le 0, \quad c_2 \le \psi(\xi) + 2\,\psi'(\xi)\,\xi \le d_2$$

for all $\xi \ge 0$. The conditions (4.33) are satisfied for the elasticity operator defined in (4.37), see for instance [55, p. 125].

4.2.3 Viscoelastic constitutive laws

A general viscoelastic constitutive law with short memory is given by

$$\boldsymbol{\sigma} = \mathcal{A}\boldsymbol{\varepsilon}(\dot{\boldsymbol{u}}) + \mathcal{B}\boldsymbol{\varepsilon}(\boldsymbol{u}). \tag{4.39}$$

We allow the *viscosity operator* $\mathcal{A}$ and the *elasticity operator* $\mathcal{B}$ to depend on the location of the point; consequently, all that follows is valid for non-homogeneous materials. We use the short-hand notation $\mathcal{A}\boldsymbol{\varepsilon}(\dot{\boldsymbol{u}})$ and $\mathcal{B}\boldsymbol{\varepsilon}(\boldsymbol{u})$ for $\mathcal{A}(\boldsymbol{x}, \boldsymbol{\varepsilon}(\dot{\boldsymbol{u}}))$ and $\mathcal{B}(\boldsymbol{x}, \boldsymbol{\varepsilon}(\boldsymbol{u}))$, respectively.

In the study of mechanical problems involving viscoelastic materials with short memory we assume that the operators $\mathcal{A}$ and $\mathcal{B}$ satisfy the following conditions.

$$\left.\begin{array}{l}
\text{(a) } \mathcal{A} : \Omega \times \mathbb{S}^d \to \mathbb{S}^d. \\
\text{(b) There exists } L_{\mathcal{A}} > 0 \text{ such that} \\
\quad \|\mathcal{A}(\boldsymbol{x}, \boldsymbol{\varepsilon}_1) - \mathcal{A}(\boldsymbol{x}, \boldsymbol{\varepsilon}_2)\| \le L_{\mathcal{A}} \|\boldsymbol{\varepsilon}_1 - \boldsymbol{\varepsilon}_2\| \\
\qquad \forall\, \boldsymbol{\varepsilon}_1, \boldsymbol{\varepsilon}_2 \in \mathbb{S}^d, \ \text{a.e. } \boldsymbol{x} \in \Omega. \\
\text{(c) There exists } m_{\mathcal{A}} > 0 \text{ such that} \\
\quad (\mathcal{A}(\boldsymbol{x}, \boldsymbol{\varepsilon}_1) - \mathcal{A}(\boldsymbol{x}, \boldsymbol{\varepsilon}_2)) \cdot (\boldsymbol{\varepsilon}_1 - \boldsymbol{\varepsilon}_2) \ge m_{\mathcal{A}} \|\boldsymbol{\varepsilon}_1 - \boldsymbol{\varepsilon}_2\|^2 \\
\qquad \forall\, \boldsymbol{\varepsilon}_1, \boldsymbol{\varepsilon}_2 \in \mathbb{S}^d, \ \text{a.e. } \boldsymbol{x} \in \Omega. \\
\text{(d) The mapping } \boldsymbol{x} \mapsto \mathcal{A}(\boldsymbol{x}, \boldsymbol{\varepsilon}) \text{ is measurable on } \Omega, \\
\quad \text{for any } \boldsymbol{\varepsilon} \in \mathbb{S}^d. \\
\text{(e) The mapping } \boldsymbol{x} \mapsto \mathcal{A}(\boldsymbol{x}, \mathbf{0}_{\mathbb{S}^d}) \text{ belongs to } Q.
\end{array}\right\} \tag{4.40}$$

$$\left.\begin{array}{l}
\text{(a) } \mathcal{B} : \Omega \times \mathbb{S}^d \to \mathbb{S}^d. \\
\text{(b) There exists } L_{\mathcal{B}} > 0 \text{ such that} \\
\quad \|\mathcal{B}(\boldsymbol{x}, \boldsymbol{\varepsilon}_1) - \mathcal{B}(\boldsymbol{x}, \boldsymbol{\varepsilon}_2)\| \le L_{\mathcal{B}} \|\boldsymbol{\varepsilon}_1 - \boldsymbol{\varepsilon}_2\| \\
\qquad \forall\, \boldsymbol{\varepsilon}_1, \boldsymbol{\varepsilon}_2 \in \mathbb{S}^d, \ \text{a.e. } \boldsymbol{x} \in \Omega. \\
\text{(c) The mapping } \boldsymbol{x} \mapsto \mathcal{B}(\boldsymbol{x}, \boldsymbol{\varepsilon}) \text{ is measurable on } \Omega, \\
\quad \text{for any } \boldsymbol{\varepsilon} \in \mathbb{S}^d. \\
\text{(d) The mapping } \boldsymbol{x} \mapsto \mathcal{B}(\boldsymbol{x}, \mathbf{0}_{\mathbb{S}^d}) \text{ belongs to } Q.
\end{array}\right\} \tag{4.41}$$

On assumptions (4.40) and (4.41) we have similar comments to those made on assumption (4.33). In particular, we note that these assumptions allow us to consider $\mathcal{A}$ and $\mathcal{B}$ as operators defined on the space Q with values on Q. Moreover, $\mathcal{A}$ is a strongly monotone Lipschitz continuous operator and $\mathcal{B}$ is a Lipschitz continuous operator.

In particular, if $\mathcal{A}$ and $\mathcal{B}$ are linear operators, then (4.39) leads to the Kelvin–Voigt constitutive law

$$\sigma_{ij} = a_{ijkl}\varepsilon_{kl}(\dot{\boldsymbol{u}}) + b_{ijkl}\varepsilon_{kl}(\boldsymbol{u}),$$

where σ_{ij}, a_{ijkl} and b_{ijkl} denote the components of the stress tensor $\boldsymbol{\sigma}$, the viscosity tensor $\mathcal{A}$ and the elasticity tensor $\mathcal{B}$, respectively. Clearly, assumption (4.40) is satisfied if all the components a_{ijkl} belong to $L^\infty(\Omega)$ and satisfy the usual properties of symmetry and ellipticity, i.e.

$$a_{ijkl} = a_{jikl} = a_{klij},$$

and there exists $m_{\mathcal{A}} > 0$ such that

$$a_{ijkl}\varepsilon_{ij}\varepsilon_{kl} \geq m_{\mathcal{A}} \|\boldsymbol{\varepsilon}\|^2 \qquad \forall \boldsymbol{\varepsilon} = (\varepsilon_{ij}) \in \mathbb{S}^d.$$

Assumption (4.41) is satisfied if b_{ijkl} belong to $L^\infty(\Omega)$ and satisfy the same symmetry properties. Due to the symmetry, when $d = 3$ there are only 21 independent components, when $d = 2$ there are only four independent components, and when $d = 1$ there is only one component in each tensor.

Let $d = 3$. The constitutive law for a linearly isotropic Kelvin–Voigt material is given by

$$\boldsymbol{\sigma} = 2\,\theta\,\boldsymbol{\varepsilon}(\dot{\boldsymbol{u}}) + \zeta\,\mathrm{tr}(\boldsymbol{\varepsilon}(\dot{\boldsymbol{u}}))\,\boldsymbol{I}_3 + 2\,\mu\,\boldsymbol{\varepsilon}(\boldsymbol{u}) + \lambda\,\mathrm{tr}(\boldsymbol{\varepsilon}(\boldsymbol{u}))\,\boldsymbol{I}_3$$

or, in components,

$$\sigma_{ij} = 2\,\theta\,\varepsilon_{ij}(\dot{\boldsymbol{u}}) + \zeta\,\varepsilon_{kk}(\dot{\boldsymbol{u}})\,\delta_{ij} + 2\,\mu\,\varepsilon_{ij}(\boldsymbol{u}) + \lambda\,\varepsilon_{kk}(\boldsymbol{u})\,\delta_{ij}.$$

Here, again, λ and μ are the Lamé coefficients whereas θ and ζ represent *viscosity coefficients* which satisfy $\theta > 0$, $\zeta \geq 0$.

A second example of a viscoelastic constitutive law of the form (4.39) is provided by the nonlinear viscoelastic constitutive law

$$\boldsymbol{\sigma} = \mathcal{A}\boldsymbol{\varepsilon}(\dot{\boldsymbol{u}}) + \beta\,(\boldsymbol{\varepsilon}(\boldsymbol{u}) - \mathcal{P}_K\boldsymbol{\varepsilon}(\boldsymbol{u})). \tag{4.42}$$

Here $\mathcal{A}$ is a linear or nonlinear operator which satisfies (4.40), β is a positive coefficient, K is a closed convex subset of $\mathbb{S}^d$ such that $\boldsymbol{0}_{\mathbb{S}^d} \in K$, and $\mathcal{P}_K : \mathbb{S}^d \to K$ denotes the projection operator. Since the projection operator is nonexpansive, the elasticity operator $\mathcal{B}(\boldsymbol{x}, \boldsymbol{\varepsilon}) = \beta\,(\boldsymbol{\varepsilon} - \mathcal{P}_K\boldsymbol{\varepsilon})$ satisfies condition (4.41). And, again, the convex K is usually defined by (4.36), where $\mathcal{F} : \mathbb{S}^d \to \mathbb{R}$ is a convex continuous function such that $\mathcal{F}(\boldsymbol{0}_{\mathbb{S}^d}) < 0$. Recall that $\boldsymbol{\varepsilon} = \mathcal{P}_K\boldsymbol{\varepsilon}$ iff $\boldsymbol{\varepsilon} \in K$ and, therefore, from (4.42) we see that $\boldsymbol{\sigma} = \mathcal{A}\boldsymbol{\varepsilon}(\dot{\boldsymbol{u}})$ iff $\boldsymbol{\varepsilon}(\boldsymbol{u}) \in K$. This shows that the material has a purely viscous behavior if and only if the strain tensor $\boldsymbol{\varepsilon}$ belongs to K. The elastic response of the material appears only for strain tensors $\boldsymbol{\varepsilon}$ which satisfy $\boldsymbol{\varepsilon} \notin K$. We conclude that, in this case, the set K could be interpreted as the domain of purely viscous behavior of the material.

Note that, since the derivative of the displacement field appears in the constitutive law (4.39), in solving contact problems with viscoelastic materials we have to prescribe the displacement field at the initial moment $t = 0$. Therefore, we supplement (4.39) with the initial condition

$$\boldsymbol{u}(0) = \boldsymbol{u}_0 \quad \text{in} \quad \Omega, \tag{4.43}$$

in which $\boldsymbol{u}_0$ is a given function, the initial displacement.

A general viscoelastic constitutive law with long memory is given by

$$\boldsymbol{\sigma}(t) = \mathcal{F}\boldsymbol{\varepsilon}(\boldsymbol{u}(t)) + \int_0^t \mathcal{R}(t-s, \boldsymbol{\varepsilon}(\boldsymbol{u}(s)))\,\mathrm{ds}. \tag{4.44}$$

We allow the *elasticity operator* $\mathcal{F}$ and the *relaxation operator* $\mathcal{R}$ to depend on the location of the point. Also, as shown in (4.44), the operator $\mathcal{R}$ is time-dependent.

In the study of mechanical problems involving viscoelastic materials with long memory we assume that the operator $\mathcal{F}$ satisfies condition (4.33). Also, the relaxation operator $\mathcal{R}$ is such that

$$\left.\begin{array}{l}
\text{(a) } \mathcal{R}\colon \Omega \times [0,T] \times \mathbb{S}^d \to \mathbb{S}^d;\\
\text{(b) } \mathcal{R}(\boldsymbol{x},t,\boldsymbol{\varepsilon}) = (r_{ijkl}(\boldsymbol{x},t)\varepsilon_{kl}) \text{ for all } \boldsymbol{\varepsilon} = (\varepsilon_{ij}) \in \mathbb{S}^d,\\
\qquad t \in [0,T], \text{ a.e. } \boldsymbol{x} \in \Omega;\\
\text{(c) } r_{ijkl} = r_{jikl} = r_{klij} \in C([0,T]; L^\infty(\Omega)),\ 1 \le i,j,k,l \le d.
\end{array}\right\} \tag{4.45}$$

It follows from assumptions (4.45) that $\mathcal{R}$ is a fourth order symmetric tensor, for almost every $\boldsymbol{x} \in \Omega$ and all $t \in [0,T]$.

Let $\mathbf{Q}_\infty$ be the space of fourth order tensor fields given by

$$\mathbf{Q}_\infty = \{\, \mathcal{E} = (e_{ijkl}) \;:\; e_{ijkl} = e_{jikl} = e_{klij} \in L^\infty(\Omega),\ 1 \le i,j,k,l \le d \,\}.$$

It is easy to see that $\mathbf{Q}_\infty$ is a real Banach space with the norm

$$\|\mathcal{E}\|_{\mathbf{Q}_\infty} = \max_{0 \le i,j,k,l \le d} \|e_{ijkl}\|_{L^\infty(\Omega)}$$

and, moreover,

$$\|\mathcal{E}\boldsymbol{\tau}\|_Q \le d\,\|\mathcal{E}\|_{\mathbf{Q}_\infty}\|\boldsymbol{\tau}\|_Q \qquad \forall\, \mathcal{E} \in \mathbf{Q}_\infty,\ \boldsymbol{\tau} \in Q. \tag{4.46}$$

In addition, we note that condition (4.45) is equivalent to the condition

$$\mathcal{R} \in C([0,T]; \mathbf{Q}_\infty). \tag{4.47}$$

For this reason and for the sake of simplicity, in the rest of the book we write the constitutive law (4.44) in the form

$$\boldsymbol{\sigma}(t) = \mathcal{F}\boldsymbol{\varepsilon}(\boldsymbol{u}(t)) + \int_0^t \mathcal{R}(t-s)\boldsymbol{\varepsilon}(\boldsymbol{u}(s))\,\mathrm{ds}. \tag{4.48}$$

In the linear case, the stress tensor $\boldsymbol{\sigma} = (\sigma_{ij})$ which satisfies (4.48) is given by

$$\sigma_{ij}(t) = f_{ijkl}\,\varepsilon_{kl}(\boldsymbol{u}(t)) + \int_0^t r_{ijkl}(t-s)\,\varepsilon_{kl}(\boldsymbol{u}(s))\,\mathrm{ds},$$

where $\mathcal{F} = (f_{ijkl})$ is the elasticity tensor and $\mathcal{R} = (r_{ijkl})$ is the relaxation tensor. Here we allow the coefficients f_{ijkl} and r_{ijkl} to depend on the location of the point in the body.

Let $d = 3$. The constitutive law for a linearly viscoelastic isotropic material with long memory is given by

$$\begin{aligned}\boldsymbol{\sigma}(t) &= 2\,\mu\,\boldsymbol{\varepsilon}(\boldsymbol{u}(t)) + \lambda\,\mathrm{tr}(\boldsymbol{\varepsilon}(\boldsymbol{u}(t)))\,\boldsymbol{I}_3\\ &\quad + 2\int_0^t \theta(t-s)\,\boldsymbol{\varepsilon}(\boldsymbol{u}(s))\,\mathrm{d}s + \int_0^t \zeta(t-s)\,\mathrm{tr}(\boldsymbol{\varepsilon}(\boldsymbol{u}(s)))\,\boldsymbol{I}_3\,\mathrm{d}s\end{aligned}$$

or, in components,

$$\begin{aligned}\sigma_{ij}(t) &= 2\,\mu\,\varepsilon_{ij}(\boldsymbol{u}(t)) + \lambda\,\varepsilon_{kk}(\boldsymbol{u}(t))\,\delta_{ij}\\ &\quad + 2\int_0^t \theta(t-s)\,\varepsilon_{ij}(\boldsymbol{u}(s))\,\mathrm{d}s + \int_0^t \zeta(t-s)\,\varepsilon_{kk}(\boldsymbol{u}(s))\,\delta_{ij}\,\mathrm{d}s.\end{aligned}$$

Here λ and μ are the Lamé coefficients while θ and ζ represent *relaxation coefficients*, which are time-dependent.

4.2.4 Viscoplastic constitutive laws

A general viscoplastic constitutive law may be written as

$$\dot{\boldsymbol{\sigma}} = \mathcal{E}\boldsymbol{\varepsilon}(\dot{\boldsymbol{u}}) + \mathcal{G}(\boldsymbol{\sigma}, \boldsymbol{\varepsilon}(\boldsymbol{u})), \tag{4.49}$$

where $\mathcal{E}$ and $\mathcal{G}$ are material constitutive functions. Here and below, as usual, we do not specify explicitly the dependence of various functions on $\boldsymbol{x} \in \Omega$. We assume in this book that $\mathcal{E}$ and $\mathcal{G}$ satisfy the following conditions.

$$\left.\begin{array}{l}\text{(a) } \mathcal{E} : \Omega \times \mathbb{S}^d \to \mathbb{S}^d.\\ \text{(b) } \mathcal{E}(\boldsymbol{x}, \boldsymbol{\varepsilon}) = (e_{ijkl}(\boldsymbol{x})\varepsilon_{kl}) \text{ for all } \boldsymbol{\varepsilon} = (\varepsilon_{ij}) \in \mathbb{S}^d,\\ \qquad \text{a.e. } \boldsymbol{x} \in \Omega.\\ \text{(c) } e_{ijkl} = e_{jikl} = e_{klij} \in L^\infty(\Omega),\ 1 \le i,j,k,l \le d.\\ \text{(d) There exists } m_{\mathcal{E}} > 0 \text{ such that}\\ \qquad \mathcal{E}(\boldsymbol{x}, \boldsymbol{\tau}) \cdot \boldsymbol{\tau} \ge m_{\mathcal{E}} \|\boldsymbol{\tau}\|^2\ \forall \boldsymbol{\tau} \in \mathbb{S}^d, \text{ a.e. } \boldsymbol{x} \text{ in } \Omega.\end{array}\right\} \tag{4.50}$$

$$\left.\begin{array}{l}\text{(a) } \mathcal{G} : \Omega \times \mathbb{S}^d \times \mathbb{S}^d \to \mathbb{S}^d.\\ \text{(b) There exists } L_{\mathcal{G}} > 0 \text{ such that}\\ \qquad \|\mathcal{G}(\boldsymbol{x}, \boldsymbol{\sigma}_1, \boldsymbol{\varepsilon}_1) - \mathcal{G}(\boldsymbol{x}, \boldsymbol{\sigma}_2, \boldsymbol{\varepsilon}_2)\|\\ \qquad\quad \le L_{\mathcal{G}} (\|\boldsymbol{\sigma}_1 - \boldsymbol{\sigma}_2\| + \|\boldsymbol{\varepsilon}_1 - \boldsymbol{\varepsilon}_2\|)\\ \qquad\qquad \forall \boldsymbol{\sigma}_1, \boldsymbol{\sigma}_2, \boldsymbol{\varepsilon}_1, \boldsymbol{\varepsilon}_2 \in \mathbb{S}^d, \text{ a.e. in } \Omega.\\ \text{(c) The mapping } \boldsymbol{x} \mapsto \mathcal{G}(\boldsymbol{x}, \boldsymbol{\sigma}, \boldsymbol{\varepsilon}) \text{ is measurable,}\\ \qquad \text{for any } \boldsymbol{\sigma}, \boldsymbol{\varepsilon} \in \mathbb{S}^d.\\ \text{(d) The mapping } \boldsymbol{x} \mapsto \mathcal{G}(\boldsymbol{x}, \boldsymbol{0}_{\mathbb{S}^d}, \boldsymbol{0}_{\mathbb{S}^d}) \text{ belongs to } Q.\end{array}\right\} \tag{4.51}$$

It follows now from assumptions (4.50) that $\mathcal{E} \in \mathbf{Q}_\infty$ and, moreover, it is a fourth order invertible tensor, for almost every $\boldsymbol{x} \in \Omega$. Therefore the role of $\varepsilon(\dot{\boldsymbol{u}})$ can be interchanged with that of $\dot{\boldsymbol{\sigma}}$ in (4.49) to obtain

$$\varepsilon(\dot{\boldsymbol{u}}) = \mathcal{E}^{-1}\dot{\boldsymbol{\sigma}} + \widetilde{\mathcal{G}}(\boldsymbol{\sigma}, \varepsilon(\boldsymbol{u})),$$

where $\mathcal{E}^{-1}$ denotes the inverse of $\mathcal{E}$ and $\widetilde{\mathcal{G}}(\boldsymbol{\sigma}, \varepsilon(\boldsymbol{u})) = -\mathcal{E}^{-1}\mathcal{G}(\boldsymbol{\sigma}, \varepsilon(\boldsymbol{u}))$. This equality shows that the rate of deformation $\varepsilon(\dot{\boldsymbol{u}})$, also denoted $\dot{\boldsymbol{\varepsilon}}$, can be decomposed into two parts: the elastic part $\dot{\boldsymbol{\varepsilon}}^e = \mathcal{E}^{-1}\dot{\boldsymbol{\sigma}}$ and the viscoplastic one given by $\dot{\boldsymbol{\varepsilon}}^p = \widetilde{\mathcal{G}}(\boldsymbol{\sigma}, \varepsilon(\boldsymbol{u}))$. The set of points $(\boldsymbol{\sigma}, \boldsymbol{\varepsilon}) \in \mathbb{S}^d \times \mathbb{S}^d$ for which the viscoplastic part $\dot{\boldsymbol{\varepsilon}}^p$ vanishes is called the *domain of elastic behavior* of the material at $\boldsymbol{x} \in \Omega$. It is characterized by the equality $\widetilde{\mathcal{G}}(\boldsymbol{x}, \boldsymbol{\sigma}, \boldsymbol{\varepsilon}) = \mathbf{0}_{\mathbb{S}^d}$ or, equivalently, $\mathcal{G}(\boldsymbol{x}, \boldsymbol{\sigma}, \boldsymbol{\varepsilon}) = \mathbf{0}_{\mathbb{S}^d}$.

A concrete example of an elastic-viscoplastic constitutive law of this form is *Perzyna's law* given by

$$\varepsilon(\dot{\boldsymbol{u}}) = \mathcal{E}^{-1}\dot{\boldsymbol{\sigma}} + \frac{1}{\mu}(\boldsymbol{\sigma} - \mathcal{P}_K\boldsymbol{\sigma}). \tag{4.52}$$

Here $\mathcal{E}$ is a fourth order tensor which satisfies (4.50), $\mu > 0$ is a viscosity constant, K is a closed convex subset of $\mathbb{S}^d$ such that $\mathbf{0}_{\mathbb{S}^d} \in K$ and, as usual, $\mathcal{P}_K$ represents the projection operator. Note that in this case the function $\mathcal{G}$ does not depend on $\boldsymbol{\varepsilon}$ and is given by

$$\mathcal{G}(\boldsymbol{\sigma}, \boldsymbol{\varepsilon}) = -\frac{1}{\mu}\mathcal{E}(\boldsymbol{\sigma} - \mathcal{P}_K\boldsymbol{\sigma}).$$

The conditions (4.51) are satisfied. Since $\boldsymbol{\sigma} = \mathcal{P}_K\boldsymbol{\sigma}$ iff $\boldsymbol{\sigma} \in K$, from (4.52) we see that viscoplastic deformations occur only for the stress tensors $\boldsymbol{\sigma}$ which do not belong to K. We conclude that, in this case, the set K represents the domain of elastic behavior of the material.

A relatively simple one-dimensional example of constitutive law of the form (4.49) in which a full coupling of the stress and the strain is involved in the function $\mathcal{G}$ is given by

$$\dot{\sigma} = E\varepsilon(\dot{u}) + G(\sigma, \varepsilon(u)), \tag{4.53}$$

where $E > 0$ is the Young modulus, $\varepsilon(u) = \frac{du}{dx}$ and $G : \mathbb{R} \times \mathbb{R} \to \mathbb{R}$ is the function

$$G(\sigma, \varepsilon) = \begin{cases} -k_1F_1(\sigma - f(\varepsilon)) & \text{if} \quad \sigma > f(\varepsilon), \\ 0 & \text{if} \quad g(\varepsilon) \le \sigma \le f(\varepsilon), \\ k_2F_2(g(\varepsilon) - \sigma) & \text{if} \quad \sigma < g(\varepsilon). \end{cases} \tag{4.54}$$

Here k_1, $k_2 > 0$ are viscosity constants, f, $g : \mathbb{R} \to \mathbb{R}$ are Lipschitz continuous functions such that $g(\varepsilon) \le f(\varepsilon)$ for all $\varepsilon \in \mathbb{R}$, and F_1, $F_2 : \mathbb{R}_+ \to \mathbb{R}$ are increasing functions which satisfy $F_1(0) = F_2(0) = 0$. More details on the constitutive law (4.53), (4.54) can be found in [31, p. 35]. Note that the domain of elastic behavior is characterized by the inequalities $g(\varepsilon) \le \sigma \le f(\varepsilon)$.

Assume now that $g(\varepsilon) < 0 < f(\varepsilon)$ for all $\varepsilon \in \mathbb{R}$. In this case viscoplastic deformations occur only for $\sigma > f(\varepsilon)$, in traction, and for $\sigma < g(\varepsilon)$, in compression. Therefore, since the yield limit (in traction and in compression) depends on the deformation, we say that the viscoplastic constitutive law (4.53), (4.54) describes a hardening property of the material.

Note that, since the derivatives of the displacement and the stress field appear in the constitutive law (4.49), in solving contact problems with viscoplastic materials we have to prescribe the displacement field and the stress field at the initial moment $t = 0$. Therefore, we supplement (4.49) with the initial conditions

$$\boldsymbol{u}(0) = \boldsymbol{u}_0, \quad \boldsymbol{\sigma}(0) = \boldsymbol{\sigma}_0 \quad \text{in} \quad \Omega, \tag{4.55}$$

in which $\boldsymbol{u}_0$ and $\boldsymbol{\sigma}_0$ represent the initial displacement and the initial stress, respectively.

4.2.5 The von Mises convex

We end this section with some details on the properties of the von Mises convex, which present interest on their own. The von Mises convex is a convex set in the space of symmetric tensors of the second order, $\mathbb{S}^d$. To define it, we consider the *von Mises function*

$$\mathcal{F}(\boldsymbol{\varepsilon}) = \frac{1}{2}\,\|\boldsymbol{\varepsilon}^{\mathrm{D}}\|^2 - g^2, \tag{4.56}$$

where $\boldsymbol{\varepsilon}^{\mathrm{D}}$ represents the deviatoric part of the tensor $\boldsymbol{\varepsilon}$, (4.38), and g is a positive constant. The corresponding set (4.36) is given by

$$K = \{\, \boldsymbol{\varepsilon} \in \mathbb{S}^d \ :\ \|\boldsymbol{\varepsilon}^{\mathrm{D}}\| \le g\sqrt{2}\,\}. \tag{4.57}$$

Using the convexity of the norm it is easy to see that K is a convex subset of $\mathbb{S}^d$. Moreover, since $g > 0$ it follows that $\boldsymbol{0}_{\mathbb{S}^d} \in K$ and, therefore, K is not empty. Finally, since $\boldsymbol{\varepsilon} \mapsto \boldsymbol{\varepsilon}^{\mathrm{D}}$ is a continuous mapping, it follows that K is a closed subset of $\mathbb{S}^d$.

The convex set defined by (4.57) is called the *von Mises convex.* It can be used in (4.34), (4.42) and (4.52) to provide examples of nonlinear elastic, viscoelastic and viscoplastic constitutive laws, respectively. It is also intensively used in the theory of plasticity, see [31, 52, 55, 66], for instance. In this case the constant $g\sqrt{2}$ represents the minimum magnitude of the deviator strain tensor for which the plastic deformations appear and, for this reason, it is called the *yield limit.*

In what follows, we present two properties of the projection operator on the von Mises convex, denoted $\mathcal{P}_K$, which will be used in Section 6.3. We start with the following explicit formula.

Proposition 4.5 *Let K denote the convex set defined by (4.57). Then for every $\varepsilon \in \mathbb{S}^d$ we have*

$$\mathcal{P}_K \varepsilon = \begin{cases} \varepsilon & \text{if} \quad \varepsilon \in K, \\ \dfrac{1}{d}\,(\operatorname{tr} \varepsilon)\, \boldsymbol{I}_d + \dfrac{g\sqrt{2}}{\|\varepsilon^{\mathrm{D}}\|}\,\varepsilon^{\mathrm{D}} & \text{if} \quad \varepsilon \notin K. \end{cases} \tag{4.58}$$

Proof Let $\varepsilon \in \mathbb{S}^d$ and denote by $\widetilde{\varepsilon}$ the element of $\mathbb{S}^d$ defined by

$$\widetilde{\varepsilon} = \begin{cases} \varepsilon & \text{if} \quad \varepsilon \in K, \\ \dfrac{1}{d}\,(\operatorname{tr} \varepsilon)\, \boldsymbol{I}_d + \dfrac{g\sqrt{2}}{\|\varepsilon^{\mathrm{D}}\|}\,\varepsilon^{\mathrm{D}} & \text{if} \quad \varepsilon \notin K. \end{cases} \tag{4.59}$$

Note that the element $\widetilde{\varepsilon}$ is well defined since for elements ε which do not belong to K we have $\|\varepsilon^{\mathrm{D}}\| > g\sqrt{2}$ and, therefore, $\|\varepsilon^{\mathrm{D}}\| \neq 0$. We shall prove that

$$\widetilde{\varepsilon} \in K \quad \text{and} \quad (\widetilde{\varepsilon} - \varepsilon) \cdot (\boldsymbol{\tau} - \widetilde{\varepsilon}) \geq 0 \quad \forall\, \boldsymbol{\tau} \in K. \tag{4.60}$$

To this end, assume first that $\varepsilon \in K$. Then, using (4.59) it follows that $\widetilde{\varepsilon} = \varepsilon$ and, therefore, it is easy to see that (4.60) holds.

Next, assume that $\varepsilon \notin K$. Then, using (4.59) it follows that

$$\widetilde{\varepsilon} = \frac{1}{d}\,(\operatorname{tr} \varepsilon)\, \boldsymbol{I}_d + \frac{g\sqrt{2}}{\|\varepsilon^{\mathrm{D}}\|}\,\varepsilon^{\mathrm{D}} \tag{4.61}$$

and, since $\operatorname{tr} \varepsilon^{\mathrm{D}} = 0$, we deduce that $\operatorname{tr} \widetilde{\varepsilon} = \operatorname{tr} \varepsilon$ which implies that

$$\widetilde{\varepsilon}^{\mathrm{D}} = \widetilde{\varepsilon} - \frac{1}{d}\,(\operatorname{tr} \widetilde{\varepsilon})\, \boldsymbol{I}_d = \widetilde{\varepsilon} - \frac{1}{d}\,(\operatorname{tr} \varepsilon)\, \boldsymbol{I}_d = \frac{g\sqrt{2}}{\|\varepsilon^{\mathrm{D}}\|}\,\varepsilon^{\mathrm{D}}.$$

Therefore,

$$\|\widetilde{\varepsilon}^{\mathrm{D}}\| = g\sqrt{2}. \tag{4.62}$$

Note that equality (4.62) combined with the definition (4.57) implies that

$$\widetilde{\varepsilon} \in K. \tag{4.63}$$

Let $\boldsymbol{\tau} \in K$. Then, using (4.61) and (4.38) it follows that

$$\begin{aligned}
&(\varepsilon - \widetilde{\varepsilon}) \cdot (\boldsymbol{\tau} - \widetilde{\varepsilon}) \\
&= \Big(\varepsilon - \frac{1}{d}\,(\operatorname{tr} \varepsilon)\, \boldsymbol{I}_d - \frac{g\sqrt{2}}{\|\varepsilon^{\mathrm{D}}\|}\,\varepsilon^{\mathrm{D}}\Big) \cdot \Big(\boldsymbol{\tau} - \frac{1}{d}\,(\operatorname{tr} \varepsilon)\, \boldsymbol{I}_d - \frac{g\sqrt{2}}{\|\varepsilon^{\mathrm{D}}\|}\,\varepsilon^{\mathrm{D}}\Big) \\
&= \varepsilon^{\mathrm{D}}\Big(1 - \frac{g\sqrt{2}}{\|\varepsilon^{\mathrm{D}}\|}\Big) \cdot \Big(\boldsymbol{\tau}^{\mathrm{D}} + \frac{1}{d}\,(\operatorname{tr} \boldsymbol{\tau})\, \boldsymbol{I}_d - \frac{1}{d}\,(\operatorname{tr} \varepsilon)\, \boldsymbol{I}_d - \frac{g\sqrt{2}}{\|\varepsilon^{\mathrm{D}}\|}\,\varepsilon^{\mathrm{D}}\Big) \\
&= \Big(1 - \frac{g\sqrt{2}}{\|\varepsilon^{\mathrm{D}}\|}\Big)\Big(\varepsilon^{\mathrm{D}} \cdot \boldsymbol{\tau}^{\mathrm{D}} + \frac{1}{d}\,(\operatorname{tr} \boldsymbol{\tau} - \operatorname{tr} \varepsilon)\, \varepsilon^{\mathrm{D}} \cdot \boldsymbol{I}_d - g\sqrt{2}\,\|\varepsilon^{\mathrm{D}}\|\Big).
\end{aligned}$$

Next, we use the identity $\boldsymbol{\varepsilon}^{\mathrm{D}} \cdot \boldsymbol{I}_d = 0$ to find that

$$(\boldsymbol{\varepsilon} - \widetilde{\boldsymbol{\varepsilon}}) \cdot (\boldsymbol{\tau} - \widetilde{\boldsymbol{\varepsilon}}) = \Big(1 - \frac{g\sqrt{2}}{\|\boldsymbol{\varepsilon}^{\mathrm{D}}\|}\Big)\Big(\boldsymbol{\varepsilon}^{\mathrm{D}} \cdot \boldsymbol{\tau}^{\mathrm{D}} - g\sqrt{2}\,\|\boldsymbol{\varepsilon}^{\mathrm{D}}\|\Big). \tag{4.64}$$

And, since $\boldsymbol{\varepsilon} \notin K$, it follows that $\|\boldsymbol{\varepsilon}^{\mathrm{D}}\| > g\sqrt{2}$ which implies that

$$1 - \frac{g\sqrt{2}}{\|\boldsymbol{\varepsilon}^{\mathrm{D}}\|} > 0. \tag{4.65}$$

Thus, using (4.64), (4.65) and the Cauchy–Schwarz inequality yields

$$(\boldsymbol{\varepsilon} - \widetilde{\boldsymbol{\varepsilon}}) \cdot (\boldsymbol{\tau} - \widetilde{\boldsymbol{\varepsilon}}) \le \Big(1 - \frac{g\sqrt{2}}{\|\boldsymbol{\varepsilon}^{\mathrm{D}}\|}\Big)\Big(\|\boldsymbol{\tau}^{\mathrm{D}}\| - g\sqrt{2}\Big)\,\|\boldsymbol{\varepsilon}^{\mathrm{D}}\|. \tag{4.66}$$

On the other hand, recall that $\boldsymbol{\tau} \in K$ and, therefore,

$$\|\boldsymbol{\tau}^{\mathrm{D}}\| \le g\sqrt{2}. \tag{4.67}$$

We use (4.65)–(4.67) to see that

$$(\boldsymbol{\varepsilon} - \widetilde{\boldsymbol{\varepsilon}}) \cdot (\boldsymbol{\tau} - \widetilde{\boldsymbol{\varepsilon}}) \le 0. \tag{4.68}$$

We now combine (4.63) and (4.68) to conclude that (4.60) holds in this case, too.

Using (4.60) and Proposition 1.12 it follows that

$$\widetilde{\boldsymbol{\varepsilon}} = \mathcal{P}_K \boldsymbol{\varepsilon}. \tag{4.69}$$

Proposition 4.5 is now a consequence of equalities (4.59) and (4.69). ■

We consider in what follows two positive constants g_1 and g_2 and we denote by K_1 and K_2 the von Mises convex sets defined by

$$K_1 = \{\, \boldsymbol{\varepsilon} \in \mathbb{S}^d \ :\ \|\boldsymbol{\varepsilon}^{\mathrm{D}}\| \le g_1\sqrt{2}\,\}, \tag{4.70}$$

$$K_2 = \{\, \boldsymbol{\varepsilon} \in \mathbb{S}^d \ :\ \|\boldsymbol{\varepsilon}^{\mathrm{D}}\| \le g_2\sqrt{2}\,\}. \tag{4.71}$$

We have the following result.

Proposition 4.6 *Let $g_1, g_2 > 0$ and denote by $\mathcal{P}_{K_1}$ and $\mathcal{P}_{K_2}$ the projection operators on the convex (4.70) and (4.71), respectively. Then, the following inequality holds:*

$$\|\mathcal{P}_{K_1}\boldsymbol{\varepsilon} - \mathcal{P}_{K_2}\boldsymbol{\varepsilon}\| \le \sqrt{2}\,|g_1 - g_2| \quad \forall\, \boldsymbol{\varepsilon} \in \mathbb{S}^d. \tag{4.72}$$

Proof It is easy to see that inequality (4.72) holds in the case $g_1 = g_2$. Therefore, we assume in what follows that $g_1 \ne g_2$ and, to make a choice, we assume that $g_1 < g_2$. Let $\boldsymbol{\varepsilon} \in \mathbb{S}^d$. We have three possibilities, as follows.

(i) $\|\varepsilon^{\mathrm{D}}\| \le g_1\sqrt{2} < g_2\sqrt{2}$. In this case (4.70) and (4.71) imply that $\varepsilon \in K_1$ and $\varepsilon \in K_2$, respectively. Therefore, using (4.58) we obtain that $\mathcal{P}_{K_1}\varepsilon = \mathcal{P}_{K_2}\varepsilon = \varepsilon$, which implies (4.72).

(ii) $g_1\sqrt{2} < \|\varepsilon^{\mathrm{D}}\| \le g_2\sqrt{2}$. In this case (4.70) and (4.71) imply that $\varepsilon \notin K_1$ and $\varepsilon \in K_2$, respectively. Therefore, using (4.58) again, we obtain that

$$\mathcal{P}_{K_1}\varepsilon = \frac{1}{d}\,(\mathrm{tr}\,\varepsilon)\,\boldsymbol{I}_d + \frac{g_1\sqrt{2}}{\|\varepsilon^{\mathrm{D}}\|}\,\varepsilon^{\mathrm{D}}, \quad \mathcal{P}_{K_2}\varepsilon = \varepsilon.$$

We use these equalities and (4.38) to see that

$$\|\mathcal{P}_{K_1}\varepsilon - \mathcal{P}_{K_2}\varepsilon\| = \left\| \frac{g_1\sqrt{2}}{\|\varepsilon^{\mathrm{D}}\|}\,\varepsilon^{\mathrm{D}} - \varepsilon^{\mathrm{D}} \right\|$$

$$= \left| \frac{g_1\sqrt{2}}{\|\varepsilon^{\mathrm{D}}\|} - 1 \right| \|\varepsilon^{\mathrm{D}}\| = \|\varepsilon^{\mathrm{D}}\| - g_1\sqrt{2} \le g_2\sqrt{2} - g_1\sqrt{2},$$

which implies (4.72).

(iii) $g_1\sqrt{2} < g_2\sqrt{2} < \|\varepsilon^{\mathrm{D}}\|$. In this case (4.70) and (4.71) imply that $\varepsilon \notin K_1$ and $\varepsilon \notin K_2$, respectively. Therefore, from (4.58) we obtain that

$$\mathcal{P}_{K_1}\varepsilon = \frac{1}{d}\,(\mathrm{tr}\,\varepsilon)\,\boldsymbol{I}_d + \frac{g_1\sqrt{2}}{\|\varepsilon^{\mathrm{D}}\|}\,\varepsilon^{\mathrm{D}}, \quad \mathcal{P}_{K_2}\varepsilon = \frac{1}{d}\,(\mathrm{tr}\,\varepsilon)\,\boldsymbol{I}_d + \frac{g_2\sqrt{2}}{\|\varepsilon^{\mathrm{D}}\|}\,\varepsilon^{\mathrm{D}}.$$

These equalities imply that

$$\|\mathcal{P}_{K_1}\varepsilon - \mathcal{P}_{K_2}\varepsilon\| = \left\| \frac{g_1\sqrt{2}}{\|\varepsilon^{\mathrm{D}}\|}\,\varepsilon^{\mathrm{D}} - \frac{g_2\sqrt{2}}{\|\varepsilon^{\mathrm{D}}\|}\,\varepsilon^{\mathrm{D}} \right\|$$

$$= |g_1\sqrt{2} - g_2\sqrt{2}| = \sqrt{2}\,|g_1 - g_2|,$$

which shows that (4.72) holds.

So, the inequality (4.72) is satisfied in each of the cases (i)–(iii) above, which concludes the proof. ■

4.3 Modelling of elastic contact problems

In this section we describe the basic equations and boundary conditions which are used to derive mathematical models in the study of contact problems with elastic materials. We consider only the case when the data and the unknowns are time-independent and, therefore, the contact process is static.

4.3.1 Preliminaries

We assume that the body is elastic and, therefore, we model the material's behavior with a constitutive law of the form (4.32). We refer to the physical setting described in Section 4.2 and we denote in what follows by $\boldsymbol{u} = \boldsymbol{u}(\boldsymbol{x})$ and $\boldsymbol{\sigma} = \boldsymbol{\sigma}(\boldsymbol{x})$ the displacement and the stress field, respectively. The state of the system is completely determined by $(\boldsymbol{u}, \boldsymbol{\sigma})$, in other words, $\boldsymbol{u} : \Omega \to \mathbb{R}^d$ and $\boldsymbol{\sigma} : \Omega \to \mathbb{S}^d$ will play the role of the unknowns in elastic contact problems. To present a mathematical model for a specific elastic contact process, besides the constitutive law (4.32) presented in Section 4.2, we need to describe the balance equation, the boundary conditions and the contact conditions.

We assume that the body forces and surface tractions do not depend on time. Therefore, the balance equation for the stress field is

$$\operatorname{Div} \boldsymbol{\sigma} + \boldsymbol{f}_0 = \boldsymbol{0} \quad \text{in} \quad \Omega. \tag{4.73}$$

Equation (4.73) is called the *equation of equilibrium.* Here Div is the divergence operator, that is $\operatorname{Div} \boldsymbol{\sigma} = (\sigma_{ij,j})$, and recall that $\sigma_{ij,j} = \frac{\partial \sigma_{ij}}{\partial x_j}$. Equation (4.73) shows that the applied external force of density $\boldsymbol{f}_0$ is fully balanced by the internal forces that are represented by $-\operatorname{Div} \boldsymbol{\sigma}$ and is derived from the principle of momentum conservation.

Also, we assume that the elastic body is held fixed on Γ_1 and, therefore,

$$\boldsymbol{u} = \boldsymbol{0} \quad \text{on} \quad \Gamma_1, \tag{4.74}$$

which represents the *displacement boundary condition.* Known tractions of density $\boldsymbol{f}_2$ act on the portion Γ_2 thus,

$$\boldsymbol{\sigma\nu} = \boldsymbol{f}_2 \quad \text{on} \quad \Gamma_2. \tag{4.75}$$

This condition is called the *traction boundary condition.* We remark that all the results in this book hold true when $\Gamma_2 = \emptyset$. Also, replacing condition (4.74) with a more general one, $\boldsymbol{u} = \boldsymbol{u}_{\mathrm{D}}$ on Γ_1, introduces no further difficulties in analysis of related contact problems, for a given $\boldsymbol{u}_{\mathrm{D}}$ lying in an appropriate function space.

On the other hand, the assumption $\operatorname{meas}(\Gamma_1) > 0$ is essential in the study of the contact problems presented in this book. Without this assumption, mathematically, the problem becomes noncoercive and many of the estimates and results below do not hold. This accurately reflects the physical situation, since when $\Gamma_1 = \emptyset$ the body is not held in place, and it may move freely in space as a rigid body.

4.3.2 Contact conditions

We describe now the various boundary conditions on the contact surface Γ_3, which is where our main interest lies. These are divided naturally into the conditions in the normal direction and those in the tangential directions. To describe them we use the indices ν and τ to represent the normal component and the tangential part of vectors and tensors, defined by (4.8) and (4.21), respectively. We start with the presentation of the conditions in the normal direction, called also *contact conditions* or *contact laws*, and we present the conditions in the tangential directions in the next subsection.

To begin we assume that in the reference configuration the body is in contact with the foundation on Γ_3. Therefore, in this case there is no gap, that is $g_{\rm a} = 0$.

The so-called *bilateral contact* describes the situation when contact between the body and the foundation is maintained during the process. It can be found in many machines and in moving parts and components of mechanical equipment. The corresponding contact condition is given by

$$u_\nu = 0 \quad \text{on} \quad \Gamma_3. \tag{4.76}$$

Condition (4.76) was considered by several authors, see for instance [55, 133] and the references therein. In the engineering literature the bilateral contact is often described in terms of stress, with an equality of the form

$$-\sigma_\nu = F \quad \text{on} \quad \Gamma_3, \tag{4.77}$$

where F is a given positive function. It results from (4.77) that σ_ν is negative and, therefore, the reaction of the foundation is towards the body. Condition (4.77) was considered in the mathematical literature, too, see e.g. [40, 119]. It also arises in geophysics, in the study of earthquake models, see for instance [23, 61, 62, 63] and the references therein.

The so-called *normal compliance contact condition* describes a deformable foundation. It assigns a reactive normal pressure that depends on the interpenetration of the asperities on the body's surface and those of the foundation. A general expression for the normal reactive pressure is

$$-\sigma_\nu = p_\nu(u_\nu) \quad \text{on} \quad \Gamma_3, \tag{4.78}$$

where p_ν is a nonnegative prescribed function which vanishes for negative argument. Indeed, when $u_\nu < 0$ there is no contact and the normal pressure vanishes. When there is contact then u_ν is positive and represents a measure of the interpenetration of the asperities. Then, condition (4.78) shows that the foundation exerts a pressure on the body which depends on the penetration.

A commonly used example of the normal compliance function p_ν is

$$p_\nu(r) = c_\nu r_+. \tag{4.79}$$

Here the constant $c_\nu > 0$ is the surface *stiffness coefficient* and $r_+ = \max\{r, 0\}$ denotes the positive part of r. We can also consider the truncated normal compliance function

$$p_\nu(r) = \begin{cases} c_\nu r_+ & \text{if } r \le \alpha, \\ c_\nu \alpha & \text{if } r > \alpha, \end{cases} \tag{4.80}$$

where α is a positive coefficient related to the wear and hardness of the surface. In this case the contact condition (4.78) means that when the penetration is too large, i.e., when it exceeds α, the obstacle offers no additional resistance to penetration.

The normal compliance contact condition was first introduced in [118] and since then used in many publications, see e.g. [75, 77, 78, 93, 125] and the references therein. The term *normal compliance* was first used in [77, 78].

An idealization of the normal compliance, which is used often in engineering literature and can also be found in mathematical publications, is the *Signorini contact condition*, in which the foundation is assumed to be perfectly rigid. It is obtained, formally, from the normal compliance condition (4.78), (4.79), in the limit when the surface stiffness coefficient becomes infinite, i.e., $c_\nu \to \infty$, and thus interpenetration is not allowed. This leads to the idea of regarding contact with a rigid support as a limiting case of contact with deformable support, whose resistance to compression increases. The Signorini contact condition can be stated in the following complementarity form:

$$u_\nu \le 0, \quad \sigma_\nu \le 0, \quad \sigma_\nu u_\nu = 0 \quad \text{on} \quad \Gamma_3. \tag{4.81}$$

This condition was first introduced in [134] and then used in many papers, see e.g. [133] for further details and references.

The main criticism of the normal compliance contact condition (4.78) arises from the fact that, in principle, it allows an infinite penetration of the foundation, which is not realistic from a physical point of view. For this reason, various versions of the normal compliance condition have been introduced, recently, in the literature. One of them is the so called *normal compliance condition with unilateral constraint* which can be formulated as follows:

$$\left.\begin{array}{l} u_\nu \le g, \quad \sigma_\nu + p(u_\nu) \le 0, \\ (u_\nu - g)(\sigma_\nu + p(u_\nu)) = 0 \end{array}\right\} \quad \text{on} \quad \Gamma_3. \tag{4.82}$$

Here $g > 0$ is given and, again, p is a nonnegative function which vanishes for a negative argument.

Condition (4.82) was first introduced in [70]. It shows that when there is separation between the body and the obstacle (i.e. when $u_\nu < 0$), then the reaction of the foundation vanishes (since $\sigma_\nu = 0$); moreover, the penetration is limited (since $u_\nu \le g$) and g represents its maximum value. When $0 \le u_\nu < g$ then the reaction of the foundation is uniquely determined by the normal displacement (since $-\sigma_\nu = p(u_\nu)$) and, when $u_\nu = g$, the normal stress is not uniquely determined but is subject to the restriction $-\sigma_\nu \geq p(g)$. Such a condition shows that the contact follows a normal compliance condition up to the limit g and then, when this limit is reached, the contact follows a so-called unilateral constraint condition.

The contact condition (4.82) can be interpreted physically in the following way: the foundation is assumed to be made of hard material covered by a thin layer of soft material with thickness g; the soft material is deformable and allows penetration, which is modelled with normal compliance; the hard material is rigid and, therefore, it does not allow penetration. It follows from the above that the foundation has an elastic–rigid behavior; the elastic behavior is given by the layer of the soft material, and the rigid behavior is given by the hard material.

Also, note that when $g = 0$ condition (4.82) becomes the Signorini contact condition. Moreover, when $g > 0$ and $p = 0$, condition (4.82) becomes the Signorini contact condition in a form with a gap function, see (4.84) below. And, finally, in the limit when $g \to \infty$ we recover the normal compliance contact condition with a zero gap function.

Assume now that there is an initial gap $g_a > 0$ between the body and the foundation. Then the normal compliance contact condition becomes

$$-\sigma_\nu = p_\nu(u_\nu - g_a) \quad \text{on} \quad \Gamma_3, \tag{4.83}$$

where p_ν is a nonnegative prescribed function which vanishes for negative argument. Indeed, when $u_\nu < g_a$ there is no contact and the normal pressure vanishes. When there is contact then $u_\nu - g_a$ is positive and represents a measure of the interpenetration of the asperities. Then, condition (4.83) shows that the foundation exerts a pressure on the body which depends on the penetration. Finally, the Signorini contact condition in a form with a gap function is given by

$$u_\nu \leq g_a, \quad \sigma_\nu \leq 0, \quad \sigma_\nu(u_\nu - g_a) = 0 \quad \text{on} \quad \Gamma_3. \tag{4.84}$$

The graphical representation of the contact conditions described above is presented in Figure 4.2.

4.3.3 Friction laws

We turn now to the conditions in the tangential directions, called also *frictional conditions* or *friction laws*. The simplest one is the so-called *frictionless* condition in which the tangential part of the stress (also called the *friction force*) vanishes, i.e.

$$\boldsymbol{\sigma}_\tau = \mathbf{0} \quad \text{on} \quad \Gamma_3. \tag{4.85}$$

This is an idealization of the process, since even completely lubricated surfaces generate shear resistance to tangential motion. However, the frictionless condition (4.85) is a sufficiently good approximation of the reality

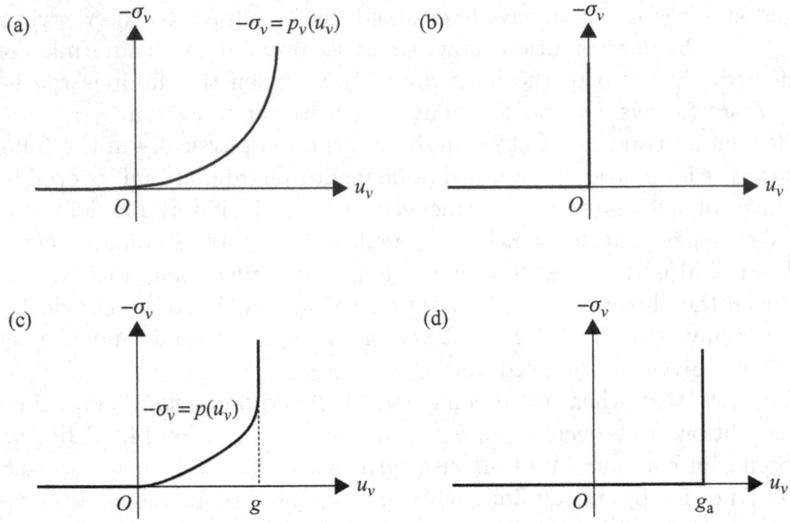

Figure 4.2 Representation of various contact conditions: (a) the normal compliance condition; (b) the Signorini condition without gap; (c) the normal compliance condition with unilateral constraint; (d) the Signorini condition with gap.

in some situations and, for this reason, it was used in several publications, see [55, 133] and the references therein.

In the case when the friction force $\boldsymbol{\sigma}_\tau$ does not vanish on the contact surface, the contact is *frictional.* Frictional contact is usually modelled with the *Coulomb law of dry friction* or its variants. Below, we present the static version of this law, commonly used in frictional contact problems describing the equilibrium of elastic bodies. It states that the magnitude of the friction force is bounded by a function, the so-called *friction bound,* which is the maximal frictional resistance that the surface can generate; also, once slip starts, the friction force opposes the direction of the motion and its magnitude reaches the friction bound. Thus,

$$\|\boldsymbol{\sigma}_\tau\| \le F_{\rm b}, \quad \boldsymbol{\sigma}_\tau = -F_{\rm b}\frac{\boldsymbol{u}_\tau}{\|\boldsymbol{u}_\tau\|} \quad \text{if} \quad \boldsymbol{u}_\tau \ne \boldsymbol{0} \quad \text{on} \quad \Gamma_3, \tag{4.86}$$

where $\boldsymbol{u}_\tau$ is the tangential displacement or slip and $F_{\rm b}$ represents the friction bound. On a nonhomogeneous surface $F_{\rm b}$ depends explicitly on the position $\boldsymbol{x}$ on the surface; it also depends on the process variables and we describe this dependence below in this section.

Note that the Coulomb law (4.86) is characterized by the existence of stick-slip zones on the contact boundary. Indeed, it follows from (4.86) that, if at a point $\boldsymbol{x} \in \Gamma_3$ the inequality $\|\boldsymbol{\sigma}_\tau(\boldsymbol{x})\| < F_b(\boldsymbol{x})$ holds, then $\boldsymbol{u}_\tau(\boldsymbol{x}) = \boldsymbol{0}$ and the material point $\boldsymbol{x}$ is in the so-called *stick zone*; if $\|\boldsymbol{\sigma}_\tau(\boldsymbol{x})\| = F_b(\boldsymbol{x})$ then the point $\boldsymbol{x}$ is in the so-called *slip zone*. We conclude that Coulomb's friction law (4.86) models the phenomenon that slip may occur only when the friction force reaches a critical value, the friction bound F_b.

In certain applications, especially where the bodies are light or the friction is very large, the function F_b in (4.86) does not depend on the process variables and behaves like a function which depends only on the position $\boldsymbol{x}$ on the contact surface. Considering

$$F_b = F_b(\boldsymbol{x}) \tag{4.87}$$

in (4.86) leads to the static version of *Tresca's friction law*, and considerably simplifies the analysis of the corresponding contact problem.

Often, especially in engineering literature, the friction bound F_b is chosen as

$$F_b = F_b(\sigma_\nu) = \mu\,|\sigma_\nu|, \tag{4.88}$$

where $\mu > 0$ is the *coefficient of friction*. The choice (4.88) in (4.86) leads to the *classical version of Coulomb's law* which was intensively studied in the literature, see for instance the references in [133].

When the wear of the contact surface is taken into account, a *modified version of Coulomb's law* is more appropriate. This law has been derived in [144, 145, 146] from thermodynamic considerations, and it is given by (4.86) with the choice

$$F_b = \mu\,|\sigma_\nu|\,(1 - \delta|\sigma_\nu|)_+. \tag{4.89}$$

Here δ is a small positive parameter related to the wear constant of the surface and, again, μ is the coefficient of friction.

The choice (4.77) in (4.88) leads to the friction bound

$$F_b = \mu\,F \tag{4.90}$$

and the choice (4.77) in (4.89) leads to the friction bound

$$F_b = \mu\,F(1 - \delta F)_+. \tag{4.91}$$

It follows from (4.90) and (4.91) that, in the case when μ and δ are given, the friction bound F_b is a given function defined on the contact surface Γ_3 and, therefore, we recover the Tresca friction law.

Various models of frictional contact may be obtained by combining one of the contact conditions presented above with the frictionless condition or with the Coulomb law (4.86), in which the choice of the friction bound is given by (4.87), (4.88) or (4.89). Nevertheless, for physical reasons, this choice is subject to some restrictions. Indeed, the resulting model has to reflect the fact that, in general, when there is no contact between the body and the foundation, then the friction force vanishes. This restriction is satisfied if, for instance, we combine the normal compliance condition (4.83) with (4.88) or (4.89). Indeed, combining (4.83) with (4.88) leads to the friction bound

$$F_{\rm b} = \mu\, p_\nu(u_\nu - g_{\rm a}) \tag{4.92}$$

and combining (4.83) with (4.89) leads to the friction bound

$$F_{\rm b} = \mu\, p_\nu(u_\nu - g_{\rm a})(1 - \delta p_\nu(u_\nu - g_{\rm a}))_+. \tag{4.93}$$

In both these cases it is easy to see that

$$u_\nu < g_{\rm a} \implies p_\nu(u_\nu - g_{\rm a}) = 0 \implies F_{\rm b} = 0 \implies \boldsymbol{\sigma}_\tau = \mathbf{0},$$

which shows that the friction force vanishes when there is no contact. Additional comments on this property will be presented in page 190.

We can unify the expression of the friction bound in (4.92) and (4.93) by taking

$$F_{\rm b} = p_\tau(u_\nu - g_{\rm a}), \tag{4.94}$$

where p_τ is a nonnegative function which vanishes when the argument is negative, i.e., when there is no contact. Indeed, (4.92) and (4.93) can be recovered from (4.94) by taking

$$p_\tau = \mu p_\nu \tag{4.95}$$

and

$$p_\tau = \mu p_\nu (1 - \delta p_\nu)_+, \tag{4.96}$$

respectively.

In variational formulation, frictional contact problems with Coulomb's law lead to variational inequalities involving nondifferentiable functionals and, for this reason, their numerical analysis could present some difficulties. To avoid these difficulties, several regularizations of Coulomb's law (4.86) are used in the literature. A simple example is given by

$$\boldsymbol{\sigma}_\tau = -F_{\rm b}\,\frac{\boldsymbol{u}_\tau}{\sqrt{\|\boldsymbol{u}_\tau\|^2 + \rho^2}} \quad \text{on} \quad \Gamma_3, \tag{4.97}$$

where $\rho > 0$ is a regularization parameter. Note that the friction law (4.97) describes a situation when slip appears even for small tangential shears, which is the case when the surfaces are lubricated by a thin layer of non-Newtonian fluid. We remark that the Coulomb law (4.86) is obtained, formally, from the friction law (4.97) in the limit as $\rho \to 0$.

Conclusion

To conclude, a mathematical model which describes the static process of contact of an elastic body in the physical setting described above consists of finding the unknown functions $\boldsymbol{u} : \Omega \to \mathbb{R}^d$ and $\boldsymbol{\sigma} : \Omega \to \mathbb{S}^d$ which satisfy the constitutive law (4.32) in Ω, the equilibrium equation (4.73), the displacement boundary condition (4.74), the traction boundary condition (4.75), one of the contact conditions (4.76), (4.77), (4.78), (4.81), (4.82), (4.83) or (4.84), and one of the friction laws (4.85), (4.86) or (4.97). Such a mathematical model is represented by a system of nonlinear partial differential equations associated with linear or nonlinear boundary conditions. As results from the above, combining the various contact conditions and friction laws, with various choices of the friction bound, we obtain several mathematical models in the study of contact problems with elastic materials. The analysis of part of these models, selected by taking into account their relevance from a physical or a mathematical point of view, is presented in Chapter 5. We also note that, combining the Signorini contact condition (4.81) or (4.84) with Coulomb's law of dry friction (4.86) leads to important mathematical difficulties, as shown in [133].

4.4 Modelling of elastic-viscoplastic contact problems

In this section we describe the basic equations and boundary conditions which are used to derive mathematical models in the study of quasistatic contact problems with viscoelastic or viscoplastic materials.

4.4.1 Preliminaries

We consider the contact of a viscoelastic or viscoplastic body with a foundation, in the physical setting presented in Section 4.2. Now, the process is evolutionary and, therefore, besides their dependence on the spatial variable $\boldsymbol{x}$, the unknowns and the data of the problem depend also on the time variable t. Hence, we denote in what follows by $\boldsymbol{u} = \boldsymbol{u}(\boldsymbol{x}, t)$ and $\boldsymbol{\sigma} = \boldsymbol{\sigma}(\boldsymbol{x}, t)$ the displacement and the stress field, respectively. Often, we will not indicate explicitly the dependence of the quantities on the spatial variable $\boldsymbol{x}$, or both $\boldsymbol{x}$ and t; i.e., when it is convenient to do so, we write $\boldsymbol{\sigma}(t)$ and $\boldsymbol{u}(t)$, or even $\boldsymbol{\sigma}$ and $\boldsymbol{u}$. Also, recall that a dot above a variable will represent the derivative with respect to time and, therefore, $\dot{\boldsymbol{u}}$ denotes the velocity field.

Let $[0, T]$ be the time interval of interest with $T > 0$. The state of the system is completely determined by $(\boldsymbol{u}, \boldsymbol{\sigma})$; in other words, the functions $\boldsymbol{u} : \Omega \times [0, T] \to \mathbb{R}^d$ and $\boldsymbol{\sigma} : \Omega \times [0, T] \to \mathbb{S}^d$ will play the role of the unknowns in viscoelastic and viscoplastic contact problems. We assume that the behavior of the body's material is modelled by one of: the viscoelastic constitutive law with short memory (4.39), the viscoelastic constitutive law with long memory (4.48), or the viscoplastic constitutive law (4.49). To complete a model of the contact processes, besides the constitutive laws above, we need to describe the balance equation, the boundary conditions, the contact conditions and the initial conditions.

We assume that the body forces and surface tractions change slowly in time so that the inertia of the mechanical system is negligible and, therefore, the process is quasistatic. In other words, the balance equation for the stress field is the equation of equilibrium, that is

$$\mathrm{Div}\,\boldsymbol{\sigma} + \boldsymbol{f}_0 = \boldsymbol{0} \quad \text{in} \quad \Omega \times (0, T). \tag{4.98}$$

Moreover, taking into account the dependence of the variables on time, the displacement boundary condition is given by

$$\boldsymbol{u} = \boldsymbol{0} \quad \text{on} \quad \Gamma_1 \times (0, T), \tag{4.99}$$

and the traction boundary condition reads

$$\boldsymbol{\sigma\nu} = \boldsymbol{f}_2 \quad \text{on} \quad \Gamma_2 \times (0, T). \tag{4.100}$$

Note that the equilibrium equation (4.98) is similar to equation (4.73), the only difference arising in the fact that now the functions depend on time and, therefore, it is satisfied in $\Omega \times (0, T)$ instead of Ω. A similar comment can be made on the boundary conditions (4.99) and (4.100), when compared with the boundary conditions (4.74) and (4.75), respectively.

4.4.2 Contact conditions and friction laws

The contact conditions we use in the mathematical models with viscoelastic and viscoplastic materials are similar to those used for elastic materials described in Section 4.3. Again, the difference arises from the fact that now the variables depend on time; moreover, sometimes we formulate the frictional contact conditions in terms of the normal component and tangential part of the velocity field, denoted by $\dot{u}_\nu$ and $\dot{\boldsymbol{u}}_\tau$, respectively. Therefore, in the viscoelastic and viscoplastic problems the bilateral contact conditions (4.76) and (4.77) read

$$u_\nu = 0 \quad \text{on} \quad \Gamma_3 \times (0, T) \tag{4.101}$$

and

$$-\sigma_\nu = F \quad \text{on} \quad \Gamma_3 \times (0, T), \tag{4.102}$$

respectively, where F is now a given time-dependent positive function. The normal compliance contact condition (4.83) is replaced by

$$-\sigma_\nu = p_\nu(u_\nu - g_{\mathrm{a}}) \quad \text{on} \quad \Gamma_3 \times (0, T), \tag{4.103}$$

where, again, p_ν is a nonnegative prescribed function which vanishes for negative argument and g_{a} represents the gap function. The *Signorini contact condition* (4.84) takes the form

$$u_\nu \le g_{\mathrm{a}}, \quad \sigma_\nu \le 0, \quad \sigma_\nu(u_\nu - g_{\mathrm{a}}) = 0 \quad \text{on} \quad \Gamma_3 \times (0, T) \tag{4.104}$$

and, finally, the normal compliance contact condition with unilateral constraint, (4.82), reads

$$\left.\begin{array}{r} u_\nu \le g, \quad \sigma_\nu + p(u_\nu) \le 0 \\ (u_\nu - g)(\sigma_\nu + p(u_\nu)) = 0 \end{array}\right\} \quad \text{on} \quad \Gamma_3 \times (0,T). \tag{4.105}$$

The so-called *normal damped response* contact condition was also used in the study of various contact problems with viscoelastic materials, mainly to describe a lubricated contact. References can be found in [55, 133]. This condition assumes that the normal stress on the contact surface is related to the normal velocity, that is

$$-\sigma_\nu = p_\nu(\dot{u}_\nu) \quad \text{on} \quad \Gamma_3 \times (0,T). \tag{4.106}$$

Here p_ν is a prescribed function. An example is given by taking

$$p_\nu(r) = d_\nu\, r_+ + p_0. \tag{4.107}$$

The contact condition (4.106), (4.107) was used in [126] in order to model the setting when the foundation is covered with a thin lubricant layer, say oil. Here $d_\nu > 0$ is the damping resistance coefficient and $p_0 > 0$ is the given oil pressure. In this case the lubricant layer presents resistance, or damping, only when the surface moves towards the foundation, but does nothing when it recedes. Another choice of p_ν is

$$p_\nu(r) = d_\nu\, |r|^{q-1} r. \tag{4.108}$$

Here $d_\nu > 0$, $0 < q \le 1$ and the normal contact stress depends on a power of the normal velocity, which mimics the behavior of a nonlinear viscous dashpot. For this reason we say that the contact condition (4.106) associated with the function (4.108) describes a *viscous contact*. This condition was considered in [55, 132], for instance.

We turn now to the description of the friction laws for contact processes with viscoelastic and viscoplastic materials. First, we note that the frictionless condition (4.85) takes the form

$$\boldsymbol{\sigma}_\tau = \mathbf{0} \quad \text{on} \quad \Gamma_3 \times (0,T). \tag{4.109}$$

Next, the static Coulomb's law of dry friction (4.86) is replaced now by its quasistatic version, that is

$$\|\boldsymbol{\sigma}_\tau\| \le F_\mathrm{b}, \quad \boldsymbol{\sigma}_\tau = -\,F_\mathrm{b}\frac{\dot{\boldsymbol{u}}_\tau}{\|\dot{\boldsymbol{u}}_\tau\|} \quad \text{if} \quad \dot{\boldsymbol{u}}_\tau \ne \mathbf{0} \quad \text{on} \quad \Gamma_3 \times (0,T), \tag{4.110}$$

where $\dot{\boldsymbol{u}}_\tau$ is the tangential velocity or slip rate and, again, F_b represents the friction bound. On a nonhomogeneous surface, F_b depends explicitly on the position $\boldsymbol{x}$ on the surface; it also depends on the process variables, as described on pages 109–110. For instance, the choice (4.87) in (4.110) leads to the quasistatic version of Tresca's friction law and the choice (4.88) and (4.89) in (4.110) leads to the classical and modified versions of Coulomb's law, respectively, both in their quasistatic variant.

Note that the Coulomb law (4.110) is characterized by the existence of stick-slip zones on the contact boundary at each time moment $t \in [0,T]$. Indeed, it follows from (4.110) that if in a point $\boldsymbol{x} \in \Gamma_3$ the inequality $\|\boldsymbol{\sigma}_\tau(\boldsymbol{x},t)\| < F_b(\boldsymbol{x})$ holds, then $\dot{\boldsymbol{u}}_\tau(\boldsymbol{x},t) = \boldsymbol{0}$ and the material point $\boldsymbol{x}$ is in the so-called *stick zone*; if $\|\boldsymbol{\sigma}_\tau(\boldsymbol{x},t)\| = F_b(\boldsymbol{x})$ then the point $\boldsymbol{x}$ is in the so-called *slip zone*. It follows from the above that Coulomb's friction law (4.110) models the phenomenon that slip may occur only when the friction force reaches a critical value, the friction bound F_b.

Conclusion

To conclude, a mathematical model which describes the quasistatic contact of a viscoelastic or viscoplastic body in the physical setting described above consists of finding the unknown functions $\boldsymbol{u} : \Omega \times [0,T] \to \mathbb{R}^d$ and $\boldsymbol{\sigma} : \Omega \times [0,T] \to \mathbb{S}^d$ which satisfy one of the constitutive laws (4.39), (4.48) or (4.49) in $\Omega \times (0,T)$, the equilibrium equation (4.98), the displacement boundary condition (4.99), the traction boundary condition (4.100), one of the contact conditions (4.101), (4.102), (4.103), (4.104), (4.105) or (4.106), and one of the friction laws (4.109) or (4.110). And, if the derivatives of the unknowns appear in these equations or conditions, we supplement them with the initial conditions (4.43) or (4.55). Such a mathematical model is represented by an evolutionary system of nonlinear partial differential equations associated with linear or nonlinear boundary conditions and with an initial condition for the displacement or for both the displacement and the stress field. As results from the above, several combinations of the contact and friction boundary conditions are possible and, therefore, several mathematical models in the study of contact problems with viscoelastic and viscoplastic materials can be considered. The analysis of part of these models, selected by taking into account their relevance from physical or from mathematical point of view, is presented in Chapter 6.

4.5 Modelling of piezoelectric contact problems

In this section we describe the physical setting of piezoelectric contact problems together with the constitutive laws and the basic equations and boundary conditions which are used in their study.

4.5.1 Physical setting and preliminaries

The piezoelectric effect is characterized by the coupling between the mechanical and electrical properties of the materials. This coupling leads to

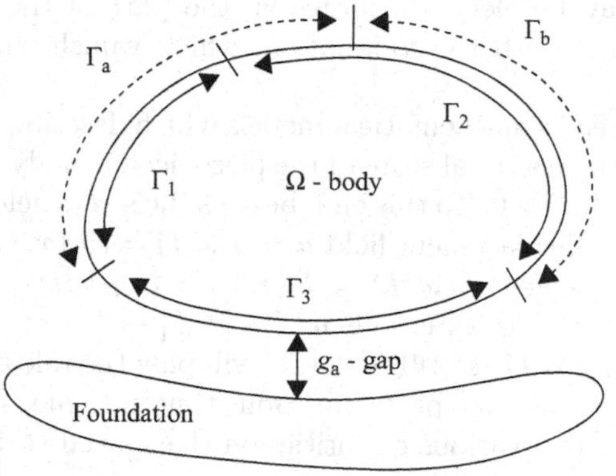

Figure 4.3 The physical setting for piezoelectric material; Γ_3 is the contact surface.

the appearance of electric potential when mechanical stress is present and, conversely, mechanical stress is generated when electric potential is applied. A deformable material which exhibits such a behavior is called a *piezoelectric* material. Piezoelectric materials are used as switches and actuators in many engineering systems, in radio-electronics, electro-acoustics and measuring equipments. Piezoelectric materials for which the mechanical properties are elastic are also called *electro-elastic* materials and piezoelectric materials for which the mechanical properties are viscoelastic are also called *electro-viscoelastic* materials. Finally, piezoelectric materials for which the mechanical properties are viscoplastic are also called *electro-viscoplastic* materials.

To model contact problems with piezoelectric materials we refer to the physical setting shown in Figure 4.3 and described in what follows. There, Ω represents the reference configuration of a piezoelectric body and, besides the partition of Γ into three sets $\overline{\Gamma}_1$, $\overline{\Gamma}_2$ and $\overline{\Gamma}_3$, which corresponds to the mechanical boundary conditions described in Sections 4.3 and 4.4, we consider a partition of $\overline{\Gamma}_1 \cup \overline{\Gamma}_2$ into two sets $\overline{\Gamma}_a$ and $\overline{\Gamma}_b$ with mutually disjoint relatively open sets Γ_a and Γ_b, which corresponds to the electrical boundary conditions.

Everywhere below we assume that meas $(\Gamma_1) > 0$ and meas $(\Gamma_a) > 0$. The body is clamped on Γ_1 and surface tractions of density $\boldsymbol{f}_2$ act on Γ_2. We also assume that the electric potential vanishes on Γ_a and a surface electric charge of density q_b is prescribed on Γ_b. The body is, or can arrive, in contact on Γ_3 with an obstacle, the so-called foundation. As in Section 4.3 we assume that in the reference configuration there may be a gap, denoted by g_a, between Γ_3 and the foundation, which is measured along the outer normal $\boldsymbol{\nu}$. Also, we assume that the foundation is either an insulator or electrically conductive. In this last case its potential is maintained at

φ_F and there may be electric charges on the part of the body's surface which is in contact with the foundation, which vanish when the contact is lost.

We are interested in mathematical models which describe the evolution of the mechanical and electrical state of the piezoelectric body during the time interval $[0, T]$, with $T > 0$. To this end, besides the stress field $\boldsymbol{\sigma} = \boldsymbol{\sigma}(\boldsymbol{x}, t) = (\sigma_{ij}(\boldsymbol{x}, t))$ and the displacement field $\boldsymbol{u} = \boldsymbol{u}(\boldsymbol{x}, t) = (u_i(\boldsymbol{x}, t))$, we introduce the *electric displacement field* $\boldsymbol{D} = \boldsymbol{D}(\boldsymbol{x}, t) = (D_i(\boldsymbol{x}, t))$ and the *electric potential* $\varphi = \varphi(\boldsymbol{x}, t)$. The functions $\boldsymbol{u}\colon \Omega\times[0,T] \to \mathbb{R}^d$, $\boldsymbol{\sigma}\colon \Omega\times[0,T] \to \mathbb{S}^d$, $\varphi\colon \Omega\times[0,T] \to \mathbb{R}$ and $\boldsymbol{D}\colon \Omega\times[0,T] \to \mathbb{R}^d$ will play the role of the unknowns of the piezoelectric contact problem. From time to time, we suppress the explicit dependence of various quantities on the spatial variable $\boldsymbol{x}$, or both $\boldsymbol{x}$ and t.

Assume that the process is quasistatic. Then, the equilibrium equation for the stress field is equation (4.98) in which, recall, $\boldsymbol{f}_0$ represents the density (per unit volume) of body forces. The balance equation for the electric displacement field is

$$\operatorname{div} \boldsymbol{D} - q_0 = 0 \quad \text{in} \quad \Omega \times (0, T). \tag{4.111}$$

Here and below q_0 is the density of electric charges and "div" is the divergence operator, that is $\operatorname{div} \boldsymbol{D} = D_{i,i}$.

We turn now to the boundary conditions on Γ_1, Γ_2, Γ_a and Γ_b. First, since the piezoelectric body is clamped on Γ_1, we impose the displacement boundary condition (4.99). Moreover, the traction boundary condition on the boundary Γ_2 is given by (4.100). Next, since the electric potential vanishes on Γ_a during the process, we impose the boundary condition

$$\varphi = 0 \quad \text{on} \quad \Gamma_a \times (0, T). \tag{4.112}$$

Also, we recall that a surface electric charge of density q_b is prescribed on Γ_b and, therefore,

$$\boldsymbol{D} \cdot \boldsymbol{\nu} = q_b \quad \text{on} \quad \Gamma_b \times (0, T). \tag{4.113}$$

Assume now that the unknowns and the data on the model do not depend on time and, therefore, the process is static. Then we use the equilibrium equation (4.73) and the balance equation for the electric displacement field becomes

$$\operatorname{div} \boldsymbol{D} - q_0 = 0 \quad \text{in} \quad \Omega. \tag{4.114}$$

The displacement and traction boundary conditions are (4.74) and (4.75) and, moreover, the boundary conditions (4.112) and (4.113) hold on Γ_a and Γ_b, respectively.

To complete a mathematical model for a specific piezoelectric contact process, besides the balance equations and the boundary conditions above, we need to introduce the constitutive laws and the contact boundary conditions.

4.5.2 Constitutive laws

As in the purely mechanical case, the constitutive laws describe the particular behavior of the material the piezoelectric body is made of. To introduce them we use the *electric field* $\boldsymbol{E}(\varphi)$ defined by

$$\boldsymbol{E}(\varphi) = -\nabla\varphi = -(\varphi_{,i}) \tag{4.115}$$

where, recall, $\varphi_{,i} = \frac{\partial \varphi}{\partial x_i}$.

A general electro-elastic constitutive law is given by

$$\boldsymbol{\sigma} = \mathcal{F}\boldsymbol{\varepsilon}(\boldsymbol{u}) - \mathcal{P}^\top \boldsymbol{E}(\varphi), \tag{4.116}$$

where $\mathcal{F}$ is the *elasticity operator*, assumed to be nonlinear, $\mathcal{P} = (p_{ijk})$ represents the third-order *piezoelectric tensor* and $\mathcal{P}^\top$ denotes its transpose. The components p_{ijk} are such that $p_{ijk} = p_{ikj}$ for all $1 \le i,\, j,\, k \le d$ and are assumed to belong to the space $L^\infty(\Omega)$, usually. Recall also that the tensors $\mathcal{P}$ and $\mathcal{P}^\top$ satisfy the equality

$$\mathcal{P}\boldsymbol{\sigma}\cdot\boldsymbol{v} = \boldsymbol{\sigma}\cdot\mathcal{P}^\top\boldsymbol{v} \quad \text{for all } \boldsymbol{\sigma}\in\mathbb{S}^d,\ \boldsymbol{v}\in\mathbb{R}^d \tag{4.117}$$

and the components of the tensor $\mathcal{P}^\top$ are given by $p^\top_{ijk} = p_{kij}$.

We complete (4.116) with a constitutive law for the electric displacement field. This law is of the form

$$\boldsymbol{D} = \mathcal{P}\boldsymbol{\varepsilon}(\boldsymbol{u}) + \boldsymbol{\beta}\boldsymbol{E}(\varphi), \tag{4.118}$$

where $\boldsymbol{\beta} = (\beta_{ij})$ denotes the *electric permittivity tensor*. Usually, the components β_{ij} belong to $L^\infty(\Omega)$ and satisfy the usual properties of symmetry and ellipticity, i.e.

$$\beta_{ij} = \beta_{ji},$$

and there exists $m_\beta > 0$ such that

$$\beta_{ij}\xi_i\xi_j \ge m_\beta\|\boldsymbol{\xi}\|^2 \quad \text{for all } \boldsymbol{\xi} = (\xi_i) \in \mathbb{R}^d.$$

Equation (4.116) indicates that the mechanical properties of the material are described by a nonlinear elastic constitutive relation which takes into account the dependence of the stress field on the electric field. Note that in the case when $\mathcal{P}$ vanishes, equation (4.116) reduces to the purely elastic constitutive law (4.32). Equation (4.118) describes a linear dependence of the electric displacement field $\boldsymbol{D}$ on the strain and electric fields. A nonlinear dependence of $\boldsymbol{D}$ with respect to $\boldsymbol{\varepsilon}(\boldsymbol{u})$ and $\boldsymbol{E}(\varphi)$ could also be considered but, for simplicity, in this book we restrict ourselves to the linear case.

Constitutive laws of the form (4.116) and (4.118) have been frequently employed in the literature in order to model the behavior of piezoelectric materials, see e.g. [15, 17, 121] and the references therein. In the linear case, these constitutive laws become

$$\begin{aligned} \sigma_{ij} &= f_{ijkl}\, \varepsilon_{kl}(\boldsymbol{u}) + p_{kij}\varphi_{,k}\,, \\ D_i &= p_{ijk}\, \varepsilon_{jk}(\boldsymbol{u}) - \beta_{ij}\varphi_{,j}\,, \end{aligned}$$

where f_{ijkl} are the components of the elasticity tensor $\mathcal{F}$.

A general electro-viscoelastic constitutive law with short memory is given by

$$\boldsymbol{\sigma} = \mathcal{A}\varepsilon(\dot{\boldsymbol{u}}) + \mathcal{B}\varepsilon(\boldsymbol{u}) - \mathcal{P}^\top \boldsymbol{E}(\varphi) \tag{4.119}$$

in which $\mathcal{A}$ is the *viscosity operator*, assumed to be nonlinear, $\mathcal{B}$ is the *elasticity operator* and, again, $\mathcal{P}^\top$ is the transpose of the piezoelectric tensor $\mathcal{P}$. We complete (4.119) with the linear constitutive law (4.118) for the electric displacement field.

Equation (4.119) indicates that the mechanical properties of the material are described by a nonlinear viscoelastic constitutive law with short memory which takes into account the dependence of the stress field on the electric field. It has been intensively employed in the literature, see for instance [10, 11, 84, 106] and the references therein. Note that in the case when $\mathcal{P}$ vanishes, equation (4.119) reduces to the purely viscoelastic constitutive law (4.39).

A general electro-viscoplastic constitutive law can be obtained as follows. First, extending the arguments in [31] to the electro-mechanic case, we assume that the stress tensor $\boldsymbol{\sigma}$ has an additive decomposition of the form

$$\boldsymbol{\sigma} = \boldsymbol{\sigma}^{\mathrm{EE}} + \boldsymbol{\sigma}^{\mathrm{EVP}}, \tag{4.120}$$

where $\boldsymbol{\sigma}^{\mathrm{EE}}$ represents the electro-elastic part of the stress and $\boldsymbol{\sigma}^{\mathrm{EVP}}$ is the electro-viscoplastic part. Next, for the electro-elastic stress we use a linear version of the constitutive law (4.116),

$$\boldsymbol{\sigma}^{\mathrm{EE}} = \mathcal{E}\varepsilon(\boldsymbol{u}) - \mathcal{P}^\top \boldsymbol{E}(\varphi), \tag{4.121}$$

in which $\mathcal{E}$ denotes the elasticity tensor. Also, for the electro-viscoplastic stress, we use an evolutionary equation,

$$\dot{\boldsymbol{\sigma}}^{\mathrm{EVP}} = \mathcal{G}(\boldsymbol{\sigma}, \varepsilon(\boldsymbol{u}), \boldsymbol{D}, \boldsymbol{E}(\varphi)), \tag{4.122}$$

where $\mathcal{G}$ is a constitutive function, generally nonlinear. We now combine (4.120)–(4.122) to obtain

$$\dot{\boldsymbol{\sigma}} = \mathcal{E}\varepsilon(\dot{\boldsymbol{u}}) - \mathcal{P}^\top \boldsymbol{E}(\dot{\varphi}) + \mathcal{G}(\boldsymbol{\sigma}, \varepsilon(\boldsymbol{u}), \boldsymbol{D}, \boldsymbol{E}(\varphi)). \tag{4.123}$$

We use a similar argument for the electric displacement field $\boldsymbol{D}$ and equation (4.118) for its linear part to obtain

$$\dot{\boldsymbol{D}} = \mathcal{B}\boldsymbol{E}(\dot{\varphi}) + \mathcal{P}\varepsilon(\dot{\boldsymbol{u}}) + G(\boldsymbol{D}, \boldsymbol{E}(\varphi), \boldsymbol{\sigma}, \varepsilon(\boldsymbol{u})). \tag{4.124}$$

Here, again, G is a nonlinear constitutive function.

Note that in the case when $\mathcal{P}$ vanishes and $\mathcal{G}$ does not depend on $\boldsymbol{D}$ and $\boldsymbol{E}(\varphi)$, equation (4.123) reduces to the rate-type viscoplastic constitutive law (4.49) introduced on page 98; also, in the case when $\mathcal{P}$ vanishes and G does not depend on $\boldsymbol{\sigma}$ and $\boldsymbol{\varepsilon}(\boldsymbol{u})$, constitutive laws of the form (4.124) were considered in [143]. And, finally, constitutive laws of the form (4.123)–(4.124) in which $\mathcal{G}$ does not depend on $\boldsymbol{D}$ and G does not depend on $\boldsymbol{\sigma}$ were considered in [56]. There, the unique solvability of a piezoelectric contact problem was proved, error estimates for fully discrete schemes were obtained and numerical simulations were performed, under the assumption that the foundation is an insulator.

4.5.3 Contact conditions

In the study of contact problems for piezoelectric materials we consider both the case when the foundation is an insulator and the case when the foundation is electrically conductive.

Assume that the foundation is an insulator. Then, the piezoelectric effect does not influence the frictional contact conditions and, therefore, we can use these conditions in the form presented in Sections 4.3 and 4.4 of this chapter. More precisely, if the body is electro-elastic, then we use the contact conditions and the friction laws in Section 4.3 and, if the body is electro-viscoelastic or electro-viscoplastic, then we use the contact conditions and the friction laws in Section 4.4. We add to these conditions the electrical contact condition

$$\boldsymbol{D} \cdot \boldsymbol{\nu} = 0. \tag{4.125}$$

This condition is valid on Γ_3 if the process is static and on $\Gamma_3 \times (0, T)$ if the process is quasistatic; it shows that there are no electric charges on the contact surface and was used in several papers, including [17, 90, 97, 136].

Now, we move to the more difficult case in which the foundation is assumed to be conductive. In this case we have to take into account the electrical conductivity of the foundation into the frictional contact conditions. This could be done by assuming that the various functions which appear in these conditions depend on the difference of the electric potential of the body's surface and the foundation.

For instance, a version of the normal compliance contact condition (4.83) or (4.103) is given by

$$-\sigma_\nu = h_\nu(\varphi - \varphi_{\mathrm{F}})\, p_\nu(u_\nu - g_{\mathrm{a}}). \tag{4.126}$$

This condition is valid on Γ_3 if the process is static and on $\Gamma_3 \times (0, T)$ if the process is quasistatic. Here, φ_{F} represents the electric potential of the foundation, g_{a} denotes the initial gap and $h_\nu : \Gamma_3 \times \mathbb{R} \to \mathbb{R}_+$ and $p_\nu : \mathbb{R} \to \mathbb{R}_+$ are prescribed nonnegative functions which represent the stiffness coefficient and the normal compliance function, respectively. Note that (4.126) shows that the stiffness coefficient depends on the difference of the electric potential of the body's surface and the foundation.

A version of the normal compliance condition with unilateral constraint (4.82) or (4.105) is given by

$$\left.\begin{aligned} &u_\nu \le g, \quad \sigma_\nu + h_\nu(\varphi - \varphi_{\rm F})p_\nu(u_\nu) \le 0 \\ &(u_\nu - g)(\sigma_\nu + h_\nu(\varphi - \varphi_{\rm F})p_\nu(u_\nu)) = 0 \end{aligned}\right\} \tag{4.127}$$

where, again, h_ν is a stiffness coefficient and g represents the maximum interpenetration of body's and foundations' asperities. This condition is valid on Γ_3 if the process is static and on $\Gamma_3 \times (0,T)$ if the process is quasistatic.

In a similar way, assuming the dependence of the friction bound on the difference of the electric potential of the body's surface and the foundation, the friction law (4.86) leads to the following static version of Coulomb's law of dry friction

$$\left.\begin{aligned} &\|\boldsymbol{\sigma}_\tau\| \le F_{\rm b}(\varphi - \varphi_{\rm F}), \\ &-\boldsymbol{\sigma}_\tau = F_{\rm b}(\varphi - \varphi_{\rm F})\,\frac{\boldsymbol{u}_\tau}{\|\boldsymbol{u}_\tau\|} \quad \text{if} \quad \boldsymbol{u}_\tau \ne \boldsymbol{0} \end{aligned}\right\} \quad \text{on } \Gamma_3. \tag{4.128}$$

In particular, if the function $F_{\rm b}$ does not depend on $\varphi - \varphi_{\rm F}$, nor on the rest of the process variables, from (4.128) we obtain Tresca's friction law. Finally, if $F_{\rm b}$ vanishes, (4.128) reduces to the frictionless condition $\boldsymbol{\sigma}_\tau = \boldsymbol{0}$.

Next, the quasistatic version of the static Coulomb's law of dry friction (4.128) is given by

$$\left.\begin{aligned} &\|\boldsymbol{\sigma}_\tau\| \le F_{\rm b}(\varphi - \varphi_{\rm F}), \\ &-\boldsymbol{\sigma}_\tau = F_{\rm b}(\varphi - \varphi_{\rm F})\,\frac{\dot{\boldsymbol{u}}_\tau}{\|\dot{\boldsymbol{u}}_\tau\|} \quad \text{if} \quad \dot{\boldsymbol{u}}_\tau \ne \boldsymbol{0} \end{aligned}\right\} \quad \text{on } \Gamma_3 \times (0,T) \tag{4.129}$$

where $\dot{\boldsymbol{u}}_\tau$ is the tangential velocity or slip rate and, again, $F_{\rm b}$ represents the friction bound.

We turn to the electrical condition on the contact surface. We assume that

$$\boldsymbol{D}\cdot\boldsymbol{\nu} = p_{\rm e}(u_\nu - g_{\rm a})\,h_{\rm e}(\varphi - \varphi_{\rm F}), \tag{4.130}$$

where $p_{\rm e}$ and $h_{\rm e}$ are given real-valued functions. Again, this condition is valid on Γ_3 if the process is static and on $\Gamma_3 \times (0,T)$ if the process is quasistatic. It represents a regularized condition which may be obtained as follows.

First, recall that the foundation is assumed to be electrically conductive and its potential is maintained at $\varphi_{\rm F}$. The gap is assumed to be an insulator (say, it is filled with air) and, therefore, when there is no contact at a point on the surface (i.e. when $u_\nu < g_{\rm a}$), the normal component of the electric displacement field vanishes, so that there are no free electric charges on the surface. Thus,

$$u_\nu < g_{\mathrm{a}} \implies \boldsymbol{D}\cdot\boldsymbol{\nu} = 0. \tag{4.131}$$

During the process of contact (i.e. when $u_\nu \geq g_{\mathrm{a}}$) the normal component of the electric displacement field or the free charge is assumed to depend on the difference of the electric potential of the body's surface and the foundation. Thus,

$$u_\nu \geq g_{\mathrm{a}} \implies \boldsymbol{D}\cdot\boldsymbol{\nu} = k_{\mathrm{e}}\, h_{\mathrm{e}}(\varphi - \varphi_{\mathrm{F}}), \tag{4.132}$$

where h_{e} is a prescribed real-valued function and $k_{\mathrm{e}} > 0$ represents the *electric conductivity coefficient.* A possible choice of the function h_{e} is $h_{\mathrm{e}}(r) = r$. We combine (4.131), (4.132) to obtain

$$\boldsymbol{D}\cdot\boldsymbol{\nu} = k_{\mathrm{e}}\, \chi_{[0,\infty)}(u_\nu - g_{\mathrm{a}})\, h_{\mathrm{e}}(\varphi - \varphi_{\mathrm{F}}), \tag{4.133}$$

where $\chi_{[0,\infty)}$ is the characteristic function of the interval $[0,\infty)$, that is

$$\chi_{[0,\infty)}(r) = \begin{cases} 0 & \text{if } r < 0, \\ 1 & \text{if } r \geq 0. \end{cases}$$

Condition (4.133) describes a perfect electrical contact and is somewhat similar to the Signorini contact condition introduced on page 106. Both conditions may be over-idealizations in many applications. To make it more realistic, we regularize condition (4.133) with condition (4.130) in which p_{e} is a nonnegative function which describes the electric conductivity of the foundation. The reason for this regularization is mathematical, since we need to avoid the discontinuity in the free electric charge when contact is established. Nevertheless, we note that this regularization seems to be reasonable from physical point of view as shown in the two examples below, which provide possible choices for the function p_{e}.

A first choice is given by

$$p_{\mathrm{e}}(r) = \begin{cases} 0 & \text{if } r < 0, \\ k_{\mathrm{e}}\, \delta^{-1} r & \text{if } 0 \leq r \leq \delta, \\ k_{\mathrm{e}} & \text{if } r > \delta, \end{cases} \tag{4.134}$$

where $\delta > 0$ is a small parameter. This choice means that during the process of contact the electric conductivity increases up to k_{e} as the contact among the surface asperities improves, and stabilizes when the penetration $u_\nu - g_{\mathrm{a}}$ reaches the value δ. A second choice is given by

$$p_{\mathrm{e}}(r) = \begin{cases} 0 & \text{if } r < -\delta, \\ k_{\mathrm{e}}\, \dfrac{r+\delta}{\delta} & \text{if } -\delta \leq r \leq 0, \\ k_{\mathrm{e}} & \text{if } r > 0, \end{cases} \tag{4.135}$$

where, again, $\delta > 0$ is a small parameter. This choice means that the air is electrically conductive under the critical thickness δ and behaves like an insulator only above this critical thickness, which justifies the use of the electric conductivity coefficient $p_e(u_\nu - g_a)$ instead of $k_e\,\chi_{[0,\infty)}(u_\nu - g_a)$. Note also that when $p_e \equiv 0$ condition (4.130) leads to the electric contact condition (4.125) used in the case when the obstacle is a perfect insulator.

Conclusion

To conclude, a mathematical model which describes a contact process of a piezoelectric body is obtained by gathering the balance equation for the stress and electric displacement fields, the constitutive equations, the boundary conditions on Γ_1, Γ_2, Γ_a, Γ_b for the displacement field, the stress vector, the electric potential and the electric displacement field, respectively, together with the frictional contact conditions and the electrical contact conditions. And, in the case of electro-viscoelastic or electro-viscoplastic materials, it is completed with appropriate initial conditions. Based on the equations and boundary conditions presented in this section, in Chapter 7 we construct and analyze three mathematical models which describe frictional contact processes with piezoelectric materials.

5

Analysis of elastic contact problems

In this chapter we illustrate the use of the abstract results obtained in Chapter 2, in the study of representative frictionless and frictional contact problems with elastic materials. The contact is either bilateral or it is modelled with the Signorini condition or normal compliance. The friction is modelled with Coulomb's law and its versions. For each problem we provide a variational formulation which is in a form of a variational inequality for the displacement field. Then we use the abstract existence and uniqueness results presented in Chapter 2 to prove the unique weak solvability of the corresponding contact problem. For some of the problems we also provide dual variational formulations and convergence results. Everywhere in this chapter we use the function spaces in Section 4.1 and we construct the mathematical models by combining the various contact and friction laws described in Section 4.3.

5.1 The Signorini contact problem

In this section we provide various results in the analysis of the Signorini contact problem for elastic materials. This includes existence, uniqueness and convergence results, among others.

5.1.1 Problem statement

We assume that the contact is frictionless and we model it with the Signorini condition in a form with a gap function. Therefore, the classical model of the process is the following.

Problem 5.1 *Find a displacement field* $\boldsymbol{u} : \Omega \to \mathbb{R}^d$ *and a stress field* $\boldsymbol{\sigma} : \Omega \to \mathbb{S}^d$ *such that*

$$\boldsymbol{\sigma} = \mathcal{F}\boldsymbol{\varepsilon}(\boldsymbol{u}) \qquad \text{in } \Omega, \tag{5.1}$$

$$\operatorname{Div} \boldsymbol{\sigma} + \boldsymbol{f}_0 = \boldsymbol{0} \qquad \text{in } \Omega, \tag{5.2}$$

$$\boldsymbol{u} = \boldsymbol{0} \qquad \text{on } \Gamma_1, \tag{5.3}$$

$$\boldsymbol{\sigma\nu} = \boldsymbol{f}_2 \qquad \text{on } \Gamma_2, \tag{5.4}$$

$$u_\nu \le g_{\mathrm{a}}, \quad \sigma_\nu \le 0, \quad \sigma_\nu(u_\nu - g_{\mathrm{a}}) = 0 \qquad \text{on } \Gamma_3, \tag{5.5}$$

$$\boldsymbol{\sigma}_\tau = \boldsymbol{0} \qquad \text{on } \Gamma_3. \tag{5.6}$$

Note that (5.1) represents the elastic constitutive law (4.32). Also, (5.2) is the equation of equilibrium (4.73), (5.3) represents the displacement boundary condition (4.74), and (5.4) is the traction boundary condition (4.75). Finally, (5.5) represents the Signorini condition (4.84) in which g_{a} represents the gap function, and (5.6) represents the frictionless condition (4.85).

This is the *classical formulation* of the problem, and by this we mean that the unknowns and the data are smooth functions such that all the derivatives and all the conditions are satisfied in the usual sense at each point. However, in general, the contact problems do not have any *classical solutions*, i.e., solutions which have the degree of smoothness consistent with developments they are involved in, and some of the conditions will be satisfied in a weaker sense that has to be made precise. Moreover, in many contact problems, the frictional contact conditions impose a ceiling on the regularity or smoothness of the solutions, even if all the problem data are very smooth. To allow for the conditions to be satisfied in a weaker sense, we need to reformulate the contact problems in the so-called *variational formulation.* Variational formulations of contact problems may have solutions which do not have the necessary regularity or smoothness of the classical solutions, but we still call them solutions and now we write that such solutions are *weak solutions* to the corresponding contact problems. We note that the variational formulations not only represent a mathematical necessity, but also are very useful practically, since they can often be employed directly in the finite element method for the problem's numerical approximation.

We turn to the variational formulation of Problem 5.1 and, to this end, we assume that the elasticity operator satisfies condition (4.33). The densities of body forces and surface tractions have the regularity

$$\boldsymbol{f}_0 \in L^2(\Omega)^d, \quad \boldsymbol{f}_2 \in L^2(\Gamma_2)^d, \tag{5.7}$$

and, finally, the gap function is such that

$$g_{\mathrm{a}} \in L^2(\Gamma_3) \text{ and } g_{\mathrm{a}}(\boldsymbol{x}) \ge 0 \text{ a.e. } \boldsymbol{x} \in \Gamma_3. \tag{5.8}$$

We use the space V, (4.9), and we note that

$$\boldsymbol{v} \mapsto \int_{\Omega} \boldsymbol{f}_0 \cdot \boldsymbol{v} \, \mathrm{d}x + \int_{\Gamma_2} \boldsymbol{f}_2 \cdot \boldsymbol{v} \, \mathrm{d}a \quad \forall \, \boldsymbol{v} \in V$$

is a linear continuous functional on the space V. Therefore, we may apply the Riesz representation theorem (page 17) to define the element $\boldsymbol{f} \in V$ by the equality

$$(\boldsymbol{f}, \boldsymbol{v})_V = \int_{\Omega} \boldsymbol{f}_0 \cdot \boldsymbol{v} \, \mathrm{d}x + \int_{\Gamma_2} \boldsymbol{f}_2 \cdot \boldsymbol{v} \, \mathrm{d}a \quad \forall \, \boldsymbol{v} \in V. \tag{5.9}$$

We also introduce the set of admissible displacements fields, defined by

$$U = \{ \boldsymbol{v} \in V \ : \ v_\nu \leq g_{\mathrm{a}} \ \text{ a.e. on } \Gamma_3 \}, \tag{5.10}$$

and we note that U is a closed convex subset of V such that $\boldsymbol{0}_V \in U$.

Assume now that the mechanical problem 5.1 has a classical solution $(\boldsymbol{u}, \boldsymbol{\sigma})$, i.e. $\boldsymbol{u}$ and $\boldsymbol{\sigma}$ are regular functions which satisfy (5.1)–(5.6) at each point. Let $\boldsymbol{v} \in U$. We use Green's formula (4.22) and equation (5.2) to find that

$$\int_{\Omega} \boldsymbol{\sigma} \cdot (\boldsymbol{\varepsilon}(\boldsymbol{v}) - \boldsymbol{\varepsilon}(\boldsymbol{u})) \, \mathrm{d}x = \int_{\Omega} \boldsymbol{f}_0 \cdot (\boldsymbol{v} - \boldsymbol{u}) \, \mathrm{d}x + \int_{\Gamma} \boldsymbol{\sigma\nu} \cdot (\boldsymbol{v} - \boldsymbol{u}) \, \mathrm{d}a.$$

Then we split the boundary integral over Γ_1, Γ_2 and Γ_3 and, since $\boldsymbol{v} - \boldsymbol{u} = \boldsymbol{0}$ a.e. on Γ_1, $\boldsymbol{\sigma\nu} = \boldsymbol{f}_2$ on Γ_2, and

$$\boldsymbol{\sigma\nu} \cdot (\boldsymbol{v} - \boldsymbol{u}) = \sigma_\nu (v_\nu - u_\nu) + \boldsymbol{\sigma}_\tau \cdot (\boldsymbol{v}_\tau - \boldsymbol{u}_\tau) \quad \text{on } \Gamma_3,$$

we deduce that

$$\int_{\Omega} \boldsymbol{\sigma} \cdot (\boldsymbol{\varepsilon}(\boldsymbol{v}) - \boldsymbol{\varepsilon}(\boldsymbol{u})) \, \mathrm{d}x = \int_{\Omega} \boldsymbol{f}_0 \cdot (\boldsymbol{v} - \boldsymbol{u}) \, \mathrm{d}x + \int_{\Gamma_2} \boldsymbol{f}_2 \cdot (\boldsymbol{v} - \boldsymbol{u}) \, \mathrm{d}a$$
$$+ \int_{\Gamma_3} \sigma_\nu (v_\nu - u_\nu) \, \mathrm{d}a + \int_{\Gamma_3} \boldsymbol{\sigma}_\tau \cdot (\boldsymbol{v}_\tau - \boldsymbol{u}_\tau) \, \mathrm{d}a. \tag{5.11}$$

We now use (5.11), (5.6) and (5.9) to obtain

$$(\boldsymbol{\sigma}, \boldsymbol{\varepsilon}(\boldsymbol{v}) - \boldsymbol{\varepsilon}(\boldsymbol{u}))_Q = (\boldsymbol{f}, \boldsymbol{v} - \boldsymbol{u})_V + \int_{\Gamma_3} \sigma_\nu (v_\nu - u_\nu) \, \mathrm{d}a. \tag{5.12}$$

Note that the boundary conditions (5.5) and the definition of the set U show that $\boldsymbol{u} \in U$. Moreover,

$$\sigma_\nu(v_\nu - u_\nu) = \sigma_\nu(v_\nu - g_{\rm a}) + \sigma_\nu(g_{\rm a} - u_\nu) = \sigma_\nu(v_\nu - g_{\rm a}) \geq 0 \quad \text{a.e. on } \Gamma_3,$$

which implies that

$$\int_{\Gamma_3} \sigma_\nu(v_\nu - u_\nu)\,{\rm d}a \geq 0. \tag{5.13}$$

Inequality (5.13) and equality (5.12) yield

$$(\boldsymbol{\sigma}, \boldsymbol{\varepsilon}(\boldsymbol{v}) - \boldsymbol{\varepsilon}(\boldsymbol{u}))_Q \geq (\boldsymbol{f}, \boldsymbol{v} - \boldsymbol{u})_V. \tag{5.14}$$

We now use the constitutive law (5.1) and (5.14) to derive the following variational formulation of the frictionless contact problem 5.1.

Problem 5.2 *Find a displacement field* $\boldsymbol{u}$ *such that*

$$\boldsymbol{u} \in U, \quad (\mathcal{F}\boldsymbol{\varepsilon}(\boldsymbol{u}), \boldsymbol{\varepsilon}(\boldsymbol{v}) - \boldsymbol{\varepsilon}(\boldsymbol{u}))_Q \geq (\boldsymbol{f}, \boldsymbol{v} - \boldsymbol{u})_V \quad \forall\, \boldsymbol{v} \in U. \tag{5.15}$$

We note that Problem 5.2 represents an elliptic variational inequality of the first kind for the displacement field. Also, we note that Problem 5.2 is formally equivalent to the mechanical problem (5.1)–(5.6). Indeed, if $\boldsymbol{u}$ represents a regular solution of the variational inequality (5.15) and $\boldsymbol{\sigma} = \mathcal{F}\boldsymbol{\varepsilon}(\boldsymbol{u})$, by using arguments similar to those used in [40, 55, 75, 133] it can be proved that $(\boldsymbol{u}, \boldsymbol{\sigma})$ satisfies (5.1)–(5.6) in a generalized sense.

5.1.2 Existence and uniqueness

In the study of Problem 5.2 we have the following existence and uniqueness result.

Theorem 5.3 *Assume (4.33), (5.7) and (5.8). Then there exists a unique solution to Problem 5.2.*

Proof Given $\boldsymbol{u} \in V$, it is easy to check that $\boldsymbol{v} \mapsto (\mathcal{F}\boldsymbol{\varepsilon}(\boldsymbol{u}), \boldsymbol{\varepsilon}(\boldsymbol{v}))_Q$ is a linear continuous functional on the space V. Therefore, we apply the Riesz representation theorem to define the operator $A : V \to V$ by

$$(A\boldsymbol{u}, \boldsymbol{v})_V = (\mathcal{F}\boldsymbol{\varepsilon}(\boldsymbol{u}), \boldsymbol{\varepsilon}(\boldsymbol{v}))_Q \quad \forall\, \boldsymbol{u},\, \boldsymbol{v} \in V. \tag{5.16}$$

We note that $\boldsymbol{u}$ is a solution of Problem 5.2 if and only if $\boldsymbol{u}$ is a solution to the variational inequality

$$\boldsymbol{u} \in U, \quad (A\boldsymbol{u}, \boldsymbol{v} - \boldsymbol{u})_V \geq (\boldsymbol{f}, \boldsymbol{v} - \boldsymbol{u})_V \quad \forall\, \boldsymbol{v} \in U. \tag{5.17}$$

For this reason, we investigate in what follows the properties of the operator A. First, we use (5.16) to see that

$$(A\boldsymbol{u} - A\boldsymbol{v}, \boldsymbol{u} - \boldsymbol{v})_V = \int_\Omega (\mathcal{F}\boldsymbol{\varepsilon}(\boldsymbol{u}) - \mathcal{F}\boldsymbol{\varepsilon}(\boldsymbol{v})) \cdot (\boldsymbol{\varepsilon}(\boldsymbol{u}) - \boldsymbol{\varepsilon}(\boldsymbol{v}))\, dx$$

for all $\boldsymbol{u}$, $\boldsymbol{v} \in V$ and, using (4.33)(c), it follows that

$$(A\boldsymbol{u} - A\boldsymbol{v}, \boldsymbol{u} - \boldsymbol{v})_V \geq m_\mathcal{F} \|\boldsymbol{u} - \boldsymbol{v}\|_V^2 \qquad \forall\, \boldsymbol{u}, \boldsymbol{v} \in V. \tag{5.18}$$

On the other hand, using (5.16) and (4.33)(b) yields

$$\|A\boldsymbol{u} - A\boldsymbol{v}\|_V \leq L_\mathcal{F} \|\boldsymbol{u} - \boldsymbol{v}\|_V \qquad \forall\, \boldsymbol{u}, \boldsymbol{v} \in V. \tag{5.19}$$

We conclude by (5.18) and (5.19) that A is a strongly monotone Lipschitz continuous operator on the space V. Moreover, recall that U is a nonempty closed convex subset of V. Theorem 5.3 is now a consequence of Theorem 2.1. ■

Let $\boldsymbol{u}$ be the solution of Problem 5.2 obtained in Theorem 5.3 and let $\boldsymbol{\sigma} = \mathcal{F}\boldsymbol{\varepsilon}(\boldsymbol{u})$. Using assumption (4.33) it follows that $\boldsymbol{\sigma} \in Q$ and, moreover, inequality (5.15) implies that

$$(\boldsymbol{\sigma}, \boldsymbol{\varepsilon}(\boldsymbol{v}) - \boldsymbol{\varepsilon}(\boldsymbol{u}))_Q \geq (\boldsymbol{f}, \boldsymbol{v} - \boldsymbol{u})_V \quad \forall\, \boldsymbol{v} \in U.$$

Let $\boldsymbol{\varphi}$ be an arbitrary element of $C_0^\infty(\Omega)^d$ and let $\boldsymbol{v} = \boldsymbol{u} \pm \boldsymbol{\varphi}$. It follows that $\boldsymbol{v} \in U$ and testing with this element in the previous inequality yields

$$(\boldsymbol{\sigma}, \boldsymbol{\varepsilon}(\boldsymbol{\varphi}))_Q = (\boldsymbol{f}, \boldsymbol{\varphi})_V.$$

We now use (5.9) and Definition 4.2 of the divergence operator Div to see that

$$(\mathrm{Div}\,\boldsymbol{\sigma}, \boldsymbol{\varphi})_{L^2(\Omega)^d} + (\boldsymbol{f}_0, \boldsymbol{\varphi})_{L^2(\Omega)^d} = 0 \quad \forall\, \boldsymbol{\varphi} \in C_0^\infty(\Omega)^d$$

and by the density of the space $C_0^\infty(\Omega)^d$ in $L^2(\Omega)^d$ it follows that

$$\mathrm{Div}\,\boldsymbol{\sigma} + \boldsymbol{f}_0 = \boldsymbol{0} \quad \text{in } \Omega. \tag{5.20}$$

Using the regularity of $\boldsymbol{f}_0$ in (5.7) it follows from (5.20) that $\mathrm{Div}\,\boldsymbol{\sigma} \in L^2(\Omega)^d$ and, therefore, recalling (4.20) we find that $\boldsymbol{\sigma} \in Q_1$.

A pair of functions $(\boldsymbol{u}, \boldsymbol{\sigma})$ which satisfies (5.1) and (5.15) is called a *weak solution* of the frictionless contact problem (5.1)–(5.6). We conclude by Theorem 5.3 that the problem (5.1)–(5.6) has a unique weak solution, with regularity $\boldsymbol{u} \in U$ and $\boldsymbol{\sigma} \in Q_1$.

5.1.3 Penalization

In what follows we study the Signorini frictionless contact problem by using the penalization method. To this end we consider the following frictionless contact problem with normal compliance.

Problem 5.4 *Find a displacement field* $\boldsymbol{u}_\mu : \Omega \to \mathbb{R}^d$ *and a stress field* $\boldsymbol{\sigma}_\mu : \Omega \to \mathbb{S}^d$ *such that*

$$\boldsymbol{\sigma}_\mu = \mathcal{F}\boldsymbol{\varepsilon}(\boldsymbol{u}_\mu) \quad \text{in } \Omega, \tag{5.21}$$

$$\text{Div}\,\boldsymbol{\sigma}_\mu + \boldsymbol{f}_0 = \boldsymbol{0} \quad \text{in } \Omega, \tag{5.22}$$

$$\boldsymbol{u}_\mu = \boldsymbol{0} \quad \text{on } \Gamma_1, \tag{5.23}$$

$$\boldsymbol{\sigma}_\mu \boldsymbol{\nu} = \boldsymbol{f}_2 \quad \text{on } \Gamma_2, \tag{5.24}$$

$$-\sigma_{\mu\nu} = \frac{1}{\mu}\, p_\nu(u_{\mu\nu} - g_{\text{a}}) \quad \text{on } \Gamma_3, \tag{5.25}$$

$$\boldsymbol{\sigma}_{\mu\tau} = \boldsymbol{0} \quad \text{on } \Gamma_3. \tag{5.26}$$

Here and below $u_{\mu\nu}$ and $\sigma_{\mu\nu}$ denote the normal components of $\boldsymbol{u}_\mu$ and $\boldsymbol{\sigma}_\mu$, and $\boldsymbol{\sigma}_{\mu\tau}$ represents the tangential part of the tensor $\boldsymbol{\sigma}_\mu$ on the boundary Γ, see equalities (4.8) and (4.21) on pages 86 and 89, respectively. Also, μ is a positive penalization parameter and p_ν represents the normal compliance function. Note that the coefficient μ may be interpreted as a *deformability coefficient* of the foundation, and then $\frac{1}{\mu}$ is the surface *stiffness coefficient*. Indeed, when μ is smaller the reaction force of the foundation to penetration is larger and so the same force will result in a smaller penetration, which means that the foundation is less deformable. When μ is larger the reaction force of the foundation to penetration is smaller, and so the foundation is less stiff and more deformable.

We assume that (4.33), (5.7) and (5.8) hold and, moreover, we assume that the normal compliance function satisfies

$$\left.\begin{array}{l}
\text{(a) } p_\nu : \Gamma_3 \times \mathbb{R} \to \mathbb{R}_+;\\
\text{(b) there exists } L_\nu > 0 \text{ such that}\\
\qquad |p_\nu(\boldsymbol{x}, r_1) - p_\nu(\boldsymbol{x}, r_2)| \le L_\nu |r_1 - r_2|\\
\qquad \forall\, r_1,\, r_2 \in \mathbb{R}, \ \text{a.e. } \boldsymbol{x} \in \Gamma_3;\\
\text{(c) } (p_\nu(\boldsymbol{x}, r_1) - p_\nu(\boldsymbol{x}, r_2))\,(r_1 - r_2) \ge 0\\
\qquad \forall\, r_1,\, r_2 \in \mathbb{R}, \ \text{a.e. } \boldsymbol{x} \in \Gamma_3;\\
\text{(d) } p_\nu(\boldsymbol{x}, r) = 0 \quad \text{if and only if} \quad r \le 0, \ \text{a.e. } \boldsymbol{x} \in \Gamma_3;\\
\text{(e) the mapping } \boldsymbol{x} \mapsto p_\nu(\boldsymbol{x}, r) \text{ is measurable on } \Gamma_3,\\
\qquad \text{for any } r \in \mathbb{R}.
\end{array}\right\} \tag{5.27}$$

We observe that the assumptions (5.27) on $p_\nu(\boldsymbol{x}, \cdot)$ are fairly general. The main severe restriction comes from condition (b) which, roughly speaking, requires that the function grows at most linearly for large values of the argument. From a mechanical point of view, conditions (c) and (d) express the fact that at each point of the contact surface the reaction force increases with the penetration and vanishes when there is no penetration into the foundation. Condition (e) is introduced here for mathematical reasons since, combined with conditions (b) and (d), it guarantees that the mapping $\boldsymbol{x} \mapsto p_\nu(\boldsymbol{x}, \zeta(\boldsymbol{x}))$ belongs to $L^2(\Gamma_3)$ whenever $\zeta \in L^2(\Gamma_3)$. Note also that, moreover, $\zeta \mapsto p_\nu(\boldsymbol{x}, \zeta(\boldsymbol{x}))$ is a monotone Lipschitz continuous operator from $L^2(\Gamma_3)$ to $L^2(\Gamma_3)$. We remark that assumptions (5.27) are satisfied for the functions (4.79) and (4.80) and we conclude that our results below are valid for the corresponding contact problems.

Next, we use the Riesz representation theorem to define the operator $G : V \to V$ by the equality

$$(G\boldsymbol{u}, \boldsymbol{v})_V = \int_{\Gamma_3} p_\nu(u_\nu - g_{\rm a}) v_\nu \, da \quad \forall\, \boldsymbol{u}, \boldsymbol{v} \in V. \tag{5.28}$$

Note that by assumptions (5.8) and (5.27) the integral in (5.28) is well defined.

Using arguments similar to those presented on page 125, we derive the following variational formulation of the frictionless contact problem 5.4.

Problem 5.5 *Find a displacement field $\boldsymbol{u}_\mu$ such that*

$$\boldsymbol{u}_\mu \in V, \quad (\mathcal{F}\boldsymbol{\varepsilon}(\boldsymbol{u}_\mu), \boldsymbol{\varepsilon}(\boldsymbol{v}))_Q + \frac{1}{\mu}(G\boldsymbol{u}_\mu, \boldsymbol{v})_V = (\boldsymbol{f}, \boldsymbol{v})_V \quad \forall\, \boldsymbol{v} \in V. \tag{5.29}$$

We have the following existence, uniqueness and convergence result.

Theorem 5.6 *Assume (4.33), (5.7), (5.8) and (5.27). Then:*
(1) *for each $\mu > 0$ there exists a unique solution $\boldsymbol{u}_\mu \in V$ to Problem 5.5;*
(2) *there exists a unique solution $\boldsymbol{u} \in V$ to Problem 5.2;*
(3) *the solution $\boldsymbol{u}_\mu$ of Problem 5.5 converges strongly to the solution $\boldsymbol{u}$ of Problem 5.2, i.e.*

$$\boldsymbol{u}_\mu \to \boldsymbol{u} \quad \text{in } V \quad \text{as} \quad \mu \to 0. \tag{5.30}$$

The proof of Theorem 5.6 is based on the properties of the operator G that we gather in the following result.

Lemma 5.7 *Assume (5.8) and (5.27). Then the operator (5.28) satisfies condition (2.3) on the space $X = V$, with $K = U$.*

Proof Let $\boldsymbol{u}$, $\boldsymbol{v}$, $\boldsymbol{w} \in V$. We use (5.28) and (5.27) to see that

$$(G\boldsymbol{u} - G\boldsymbol{v}, \boldsymbol{w})_V = \int_{\Gamma_3} (p_\nu(u_\nu - g_{\rm a}) - p_\nu(v_\nu - g_{\rm a})) w_\nu \, da$$

$$\leq \int_{\Gamma_3} |p_\nu(u_\nu - g_{\rm a}) - p_\nu(v_\nu - g_{\rm a})|\,|w_\nu| \, da \leq L_\nu \int_{\Gamma_3} |u_\nu - v_\nu|\,|w_\nu| \, da.$$

Note that $|u_\nu - v_\nu| \le \|\boldsymbol{u} - \boldsymbol{v}\|$, $|w_\nu| \le \|\boldsymbol{w}\|$ a.e. on Γ_3 and, therefore, we obtain

$$(G\boldsymbol{u} - G\boldsymbol{v}, \boldsymbol{w})_V \le L_\nu \int_{\Gamma_3} \|\boldsymbol{u} - \boldsymbol{v}\| \, \|\boldsymbol{w}\| \, da.$$

We now use the Cauchy–Schwartz inequality and (4.13) to see that

$$(G\boldsymbol{u} - G\boldsymbol{v}, \boldsymbol{w})_V \le c_0^2 L_\nu \, \|\boldsymbol{u} - \boldsymbol{v}\|_V \|\boldsymbol{w}\|_V$$

which, in turn, implies that

$$\|G\boldsymbol{u} - G\boldsymbol{v}\|_V \le c_0^2 L_\nu \, \|\boldsymbol{u} - \boldsymbol{v}\|_V. \tag{5.31}$$

Moreover, using (5.27)(c) we have

$$\begin{aligned}(G\boldsymbol{u} - &G\boldsymbol{v}, \boldsymbol{u} - \boldsymbol{v})_V \\ &= \int_{\Gamma_3} \big(p_\nu(u_\nu - g_{\mathrm{a}}) - p_\nu(v_\nu - g_{\mathrm{a}})\big)(u_\nu - v_\nu) \, da \ge 0. \end{aligned} \tag{5.32}$$

Inequalities (5.31) and (5.32) show that G is a monotone Lipschitz continuous operator, i.e. it satisfies condition (2.3)(a).

On the other hand, if $\boldsymbol{u} \in V$ and $\boldsymbol{v} \in U$ then, using (5.27) it is easy to see that

$$\begin{aligned} p_\nu(u_\nu - g_{\mathrm{a}})(v_\nu - g_{\mathrm{a}}) \le 0 \quad \text{a.e. on } \Gamma_3, \\ p_\nu(u_\nu - g_{\mathrm{a}})(g_{\mathrm{a}} - u_\nu) \le 0 \quad \text{a.e. on } \Gamma_3 \end{aligned}$$

and, therefore,

$$\begin{aligned}(G\boldsymbol{u}, \boldsymbol{v} - \boldsymbol{u})_V &= \int_{\Gamma_3} p_\nu(u_\nu - g_{\mathrm{a}})(v_\nu - u_\nu) \, da \\ = \int_{\Gamma_3} p_\nu(u_\nu - g_{\mathrm{a}})&(v_\nu - g_{\mathrm{a}}) \, da + \int_{\Gamma_3} p_\nu(u_\nu - g_{\mathrm{a}})(g_{\mathrm{a}} - u_\nu) \, da \le 0.\end{aligned}$$

We conclude from this inequality that G satisfies condition (2.3)(b).

Finally, assume that $G\boldsymbol{u} = \boldsymbol{0}_V$. Then $(G\boldsymbol{u}, \boldsymbol{u})_V = 0$ and, therefore,

$$\int_{\Gamma_3} p_\nu(u_\nu - g_{\mathrm{a}}) u_\nu \, da = 0. \tag{5.33}$$

We use (5.27)(a), (c) and (5.8) to obtain the inequalities

$$p_\nu(u_\nu - g_{\mathrm{a}}) u_\nu \ge p_\nu(u_\nu - g_{\mathrm{a}}) g_{\mathrm{a}} \ge 0 \quad \text{a.e. on } \Gamma_3.$$

Therefore, since the integrand is positive, we deduce from (5.33) that $p_\nu(u_\nu - g_{\mathrm{a}}) u_\nu = 0$ a.e. on Γ_3. This equality combined with assumption

(5.27)(d) on the normal compliance function p_ν implies that $u_\nu \le g_a$ a.e. on Γ_3 and, therefore, we deduce that $\boldsymbol{u} \in U$.

Conversely, if $\boldsymbol{u} \in U$ it follows that $u_\nu \le g_a$ a.e. on Γ_3 and, using assumption (5.27)(d) it follows that $p_\nu(u_\nu - g_a) = 0$ a.e. on Γ_3. Using the definition (5.28) of the operator G we deduce that $(G\boldsymbol{u}, \boldsymbol{v})_V = 0$ for all $\boldsymbol{v} \in V$, which implies that $G\boldsymbol{u} = \boldsymbol{0}_V$.

It follows from the above that G satisfies the equivalence condition (2.3)(c), which concludes the proof. ■

We are now in position to present the proof of Theorem 5.6.

Proof Consider the operator $A : V \to V$ defined by (5.16). Then, it is easy to see that, for each $\mu > 0$, the function $\boldsymbol{u}_\mu$ is a solution of Problem 5.5 if and only if $\boldsymbol{u}_\mu$ is a solution to the nonlinear equation

$$\boldsymbol{u}_\mu \in V, \quad A\boldsymbol{u}_\mu + \frac{1}{\mu} G\boldsymbol{u}_\mu = \boldsymbol{f}.$$

On the other hand, it follows from the proof of Theorem 5.3 that $\boldsymbol{u}$ is a solution of Problem 5.2 if and only if $\boldsymbol{u}$ is a solution to the variational inequality (5.17). Recall that U is a nonempty closed convex subset of V, (5.18) and (5.19) imply that A is a strongly monotone Lipschitz continuous operator and, finally, Lemma 5.7 shows that the operator G satisfies condition (2.3). Theorem 5.3 is now a consequence of Theorem 2.2. ■

The interest in Theorem 5.6 is twofold; first, it provides the existence and uniqueness of the solution to the variational inequality (5.15) and, therefore, it recovers the existence result obtained in Theorem 5.3; second, it shows that the solution of (5.15) represents the strong limit of the sequence of solutions $\boldsymbol{u}_\mu$ of the nonlinear equation (5.29), as $\mu \to 0$.

In addition to the mathematical interest in this convergence result, it is important from the mechanical point of view, since (5.30) shows that the weak solution of the elastic frictionless contact problem with a rigid obstacle may be approached as closely as one wishes by the solution of an elastic frictionless contact problem with a deformable foundation, with a sufficiently small deformability coefficient. It is also important from a numerical point of view since it shows that solving the variational inequality (5.15) leads to solving the nonlinear equation (5.29) for a "small" parameter μ and, to this end, various numerical methods can be used, see e.g. [48].

5.1.4 Dual variational formulation

Next, we consider the Signorini frictionless contact problem 5.1 in the particular case when $g_a = 0$. The classical formulation of this problem is the following.

Problem 5.8 *Find a displacement field* $\boldsymbol{u} : \Omega \to \mathbb{R}^d$ *and a stress field* $\boldsymbol{\sigma} : \Omega \to \mathbb{S}^d$ *such that*

$$\boldsymbol{\sigma} = \mathcal{F}\boldsymbol{\varepsilon}(\boldsymbol{u}) \qquad \text{in } \Omega, \tag{5.34}$$

$$\operatorname{Div} \boldsymbol{\sigma} + \boldsymbol{f}_0 = \boldsymbol{0} \qquad \text{in } \Omega, \tag{5.35}$$

$$\boldsymbol{u} = \boldsymbol{0} \qquad \text{on } \Gamma_1, \tag{5.36}$$

$$\boldsymbol{\sigma\nu} = \boldsymbol{f}_2 \qquad \text{on } \Gamma_2, \tag{5.37}$$

$$u_\nu \le 0, \quad \sigma_\nu \le 0, \quad \sigma_\nu u_\nu = 0 \qquad \text{on } \Gamma_3, \tag{5.38}$$

$$\boldsymbol{\sigma}_\tau = \boldsymbol{0} \qquad \text{on } \Gamma_3. \tag{5.39}$$

In the study of Problem 5.8 we assume that the elasticity operator $\mathcal{F}$ satisfies (4.33) and the densities of body forces and surface tractions have the regularity (5.7). We use the space V, (4.9), and the element $\boldsymbol{f} \in V$ defined by (5.9). Also, we introduce the set of admissible displacements fields U_0 and the set of admissible stress fields Σ_0 defined by

$$U_0 = \{\, \boldsymbol{v} \in V \; : \; v_\nu \le 0 \;\text{ a.e. on } \Gamma_3 \,\}, \tag{5.40}$$

$$\Sigma_0 = \{\, \boldsymbol{\tau} \in Q \; : \; (\boldsymbol{\tau}, \boldsymbol{\varepsilon}(\boldsymbol{v}))_Q \ge (\boldsymbol{f}, \boldsymbol{v})_V \quad \forall\, \boldsymbol{v} \in U_0 \,\}. \tag{5.41}$$

As shown on page 126, the variational formulation of Problem 5.8, in terms of the displacements, is the following.

Problem 5.9 *Find a displacement field* $\boldsymbol{u}$ *such that*

$$\boldsymbol{u} \in U_0, \quad (\mathcal{F}\boldsymbol{\varepsilon}(\boldsymbol{u}), \boldsymbol{\varepsilon}(\boldsymbol{v}) - \boldsymbol{\varepsilon}(\boldsymbol{u}))_Q \ge (\boldsymbol{f}, \boldsymbol{v} - \boldsymbol{u})_V \quad \forall\, \boldsymbol{v} \in U_0. \tag{5.42}$$

Problem 5.9 is also called the *primal variational formulation* of the Signorini contact problem (5.34)–(5.39). Also, from Theorem 5.3 we have the following existence and uniqueness result.

Theorem 5.10 *Assume (4.33) and (5.7). Then there exists a unique solution to Problem 5.9.*

Our aim in this subsection is to study a variational formulation of Problem 5.8 in which the unknown is the stress tensor. To this end we assume in what follows that $(\boldsymbol{u}, \boldsymbol{\sigma})$ represents a regular solution to (5.34)–(5.39) and we consider an arbitrary element $\boldsymbol{v} \in U_0$. Then, Green's formula (4.22) combined with (5.35)–(5.37), (5.39) and (5.9) imply that

$$(\boldsymbol{\sigma}, \boldsymbol{\varepsilon}(\boldsymbol{v}))_Q = (\boldsymbol{f}, \boldsymbol{v})_V + \int_{\Gamma_3} \sigma_\nu v_\nu \, da. \tag{5.43}$$

Next, by (5.38) and the definition of U_0 it follows that $\sigma_\nu v_\nu \ge 0$ a.e. on Γ_3 and, therefore, (5.43) implies that

$$(\boldsymbol{\sigma}, \boldsymbol{\varepsilon}(\boldsymbol{v}))_Q \ge (\boldsymbol{f}, \boldsymbol{v})_V. \tag{5.44}$$

On the other hand, since (5.37) implies that $\boldsymbol{u} \in U_0$, by taking $\boldsymbol{v} = \boldsymbol{u}$ in (5.43) we obtain that

$$(\boldsymbol{\sigma}, \boldsymbol{\varepsilon}(\boldsymbol{u}))_Q = (\boldsymbol{f}, \boldsymbol{u})_V + \int_{\Gamma_3} \sigma_\nu u_\nu \, da.$$

Combining this equality with equality $\sigma_\nu u_\nu = 0$ a.e. on Γ_3, included in (5.38), yields

$$(\boldsymbol{\sigma}, \boldsymbol{\varepsilon}(\boldsymbol{u}))_Q = (\boldsymbol{f}, \boldsymbol{u})_V. \tag{5.45}$$

Inequality (5.44) shows that

$$\boldsymbol{\sigma} \in \Sigma_0. \tag{5.46}$$

Also, equality (5.45) and (5.41) imply that

$$(\boldsymbol{\tau} - \boldsymbol{\sigma}, \boldsymbol{\varepsilon}(\boldsymbol{u}))_Q \geq 0 \quad \forall \boldsymbol{\tau} \in \Sigma_0. \tag{5.47}$$

Finally, assumption (4.33) shows that $\mathcal{F}$ is an invertible operator and (5.34) implies that

$$\boldsymbol{\varepsilon}(\boldsymbol{u}) = \mathcal{F}^{-1}\boldsymbol{\sigma}. \tag{5.48}$$

We combine (5.46)–(5.48) to derive the following variational formulation of Problem 5.8, in terms of the stress field.

Problem 5.11 *Find a stress field $\boldsymbol{\sigma}$ such that*

$$\boldsymbol{\sigma} \in \Sigma_0, \qquad (\mathcal{F}^{-1}\boldsymbol{\sigma}, \boldsymbol{\tau} - \boldsymbol{\sigma})_Q \geq 0 \quad \forall \boldsymbol{\tau} \in \Sigma_0. \tag{5.49}$$

Problem 5.11 is called the *dual variational formulation* of the Signorini frictionless contact problem 5.8. In the study of this problem we have the following existence and uniqueness result.

Theorem 5.12 *Assume (4.33) and (5.7). Then there exists a unique solution to Problem 5.11.*

Proof It follows from (5.41) that Σ_0 is a closed convex subset of Q and, from (4.11) it follows that $\boldsymbol{\varepsilon}(\boldsymbol{f}) \in \Sigma_0$, i.e. Σ_0 is nonempty. Moreover, assumption (4.33) on the elasticity operator and Proposition 1.25 show that $\mathcal{F}^{-1} : Q \to Q$ is a strongly monotone Lipschitz continuous operator. Theorem 5.12 is now a direct consequence of Theorem 2.1. ∎

Note that the solution $\boldsymbol{\sigma}$ obtained above has the regularity $\boldsymbol{\sigma} \in Q_1$. Indeed, by the definition of the set Σ_0 we have $(\boldsymbol{\sigma}, \boldsymbol{\varepsilon}(\boldsymbol{v}))_Q \geq (\boldsymbol{f}, \boldsymbol{v})_V$ for all $\boldsymbol{v} \in V$ and, taking $\boldsymbol{v} = \pm\boldsymbol{\varphi}$ in this inequality, it follows that

$$(\boldsymbol{\sigma}, \boldsymbol{\varepsilon}(\boldsymbol{\varphi}))_Q = (\boldsymbol{f}, \boldsymbol{\varphi})_V \quad \forall \boldsymbol{\varphi} \in C_0^\infty(\Omega)^d.$$

We use now (5.9) and arguments similar to those on page 127 to see that $\mathrm{Div}\,\boldsymbol{\sigma} + \boldsymbol{f}_0 = \mathbf{0}$ in Ω. Using the regularity $\boldsymbol{f}_0 \in L^2(\Omega)^d$ in (5.7) it follows that $\mathrm{Div}\,\boldsymbol{\sigma} \in L^2(\Omega)^d$ and, therefore, $\boldsymbol{\sigma} \in Q_1$.

The link between the solutions $\boldsymbol{\sigma}$ and $\boldsymbol{u}$ of the variational problems 5.11 and 5.9 is provided by the following result.

Theorem 5.13 *Assume (4.33) and (5.7).*

(1) *If $\boldsymbol{u}$ is the solution to Problem 5.9 obtained in Theorem 5.10 and $\boldsymbol{\sigma}$ denotes the function given by*

$$\boldsymbol{\sigma} = \mathcal{F}\boldsymbol{\varepsilon}(\boldsymbol{u}), \tag{5.50}$$

then $\boldsymbol{\sigma}$ is a solution to Problem 5.11.

(2) *Conversely, if $\boldsymbol{\sigma}$ is the solution to Problem 5.11 obtained in Theorem 5.12 then there exists a unique element $\boldsymbol{u} \in V$ such that (5.50) holds and, moreover, $\boldsymbol{u}$ is a solution to Problem 5.9.*

Proof (1) Assume that $\boldsymbol{u}$ satisfies (5.42) and let $\boldsymbol{\sigma}$ be given by (5.50). Then, using the properties (4.33) of the operator $\mathcal{F}$, it follows that $\boldsymbol{\sigma} \in Q$. Moreover, by (5.42) we have

$$(\boldsymbol{\sigma}, \boldsymbol{\varepsilon}(\boldsymbol{v}) - \boldsymbol{\varepsilon}(\boldsymbol{u}))_Q \geq (\boldsymbol{f}, \boldsymbol{v} - \boldsymbol{u})_V \quad \forall\, \boldsymbol{v} \in U_0 \tag{5.51}$$

and, taking in this inequality $\boldsymbol{v} = 2\boldsymbol{u}$ and $\boldsymbol{v} = \mathbf{0}_V$, both in U_0, we obtain

$$(\boldsymbol{\sigma}, \boldsymbol{\varepsilon}(\boldsymbol{u}))_Q = (\boldsymbol{f}, \boldsymbol{u})_V. \tag{5.52}$$

We combine now (5.51) and (5.52) to see that

$$(\boldsymbol{\sigma}, \boldsymbol{\varepsilon}(\boldsymbol{v}))_Q \geq (\boldsymbol{f}, \boldsymbol{v})_V \quad \forall\, \boldsymbol{v} \in U_0,$$

which yields

$$\boldsymbol{\sigma} \in \Sigma_0. \tag{5.53}$$

Next, using (5.50) and (5.52) we have

$$(\mathcal{F}^{-1}\boldsymbol{\sigma}, \boldsymbol{\tau} - \boldsymbol{\sigma})_Q = (\boldsymbol{\tau} - \boldsymbol{\sigma}, \boldsymbol{\varepsilon}(\boldsymbol{u}))_Q = (\boldsymbol{\tau}, \boldsymbol{\varepsilon}(\boldsymbol{u}))_Q - (\boldsymbol{f}, \boldsymbol{u})_V \quad \forall\, \boldsymbol{\tau} \in Q$$

and, by the definition (5.41) of the set Σ_0 we deduce that

$$(\mathcal{F}^{-1}\boldsymbol{\sigma}, \boldsymbol{\tau} - \boldsymbol{\sigma})_Q \geq 0 \qquad \forall\, \boldsymbol{\tau} \in \Sigma_0. \tag{5.54}$$

We now combine (5.53) and (5.54) to conclude that $\boldsymbol{\sigma}$ is a solution to Problem 5.11.

(2) Assume that $\boldsymbol{\sigma}$ satisfies (5.49) and consider an element $\boldsymbol{z} \in Q$ such that

$$(\boldsymbol{z}, \boldsymbol{\varepsilon}(\boldsymbol{v}))_Q = 0 \qquad \forall\, \boldsymbol{v} \in V. \tag{5.55}$$

Using (5.41) and (5.55) we deduce that $\boldsymbol{\sigma} \pm \boldsymbol{z} \in \Sigma_0$ and, testing in (5.49) with $\boldsymbol{\tau} = \boldsymbol{\sigma} \pm \boldsymbol{z}$, we obtain that

$$(\mathcal{F}^{-1}\boldsymbol{\sigma}, \boldsymbol{z})_Q = 0. \tag{5.56}$$

On the other hand, Theorem 4.1 guarantees that the space $\boldsymbol{\varepsilon}(V)$ defined by (4.14) is a closed subspace of Q. Therefore, by Theorem 1.17 we deduce that

$$\boldsymbol{\varepsilon}(V)^{\perp\perp} = \boldsymbol{\varepsilon}(V), \tag{5.57}$$

where, recall, the superscript $\perp$ indicates the orthogonal complement in Q. Using (5.56) and (5.55) we see that $\mathcal{F}^{-1}\boldsymbol{\sigma} \perp \boldsymbol{z}$ for all $\boldsymbol{z} \in \boldsymbol{\varepsilon}(V)^{\perp}$, i.e. $\mathcal{F}^{-1}\boldsymbol{\sigma} \in \boldsymbol{\varepsilon}(V)^{\perp\perp}$ and, by (5.57), it follows that $\mathcal{F}^{-1}\boldsymbol{\sigma} \in \boldsymbol{\varepsilon}(V)$. Therefore, there exists an element $\boldsymbol{u} \in V$ such that $\mathcal{F}^{-1}\boldsymbol{\sigma} = \boldsymbol{\varepsilon}(\boldsymbol{u})$, i.e. (5.50) holds. Moreover, the uniqueness of $\boldsymbol{u}$ which satisfies (5.50) follows from (4.33)(c) and Korn's inequality (4.10).

Next, we prove that $\boldsymbol{u}$ is a solution to Problem 5.9 and, to this end, we start by proving that $\boldsymbol{u} \in U_0$. Arguing by contradiction, we assume in what follows that $\boldsymbol{u} \notin U_0$ and we denote by $\mathcal{P}_0\boldsymbol{u}$ the projection of $\boldsymbol{u}$ on the closed convex subset $U_0 \subset V$. We have $\boldsymbol{u} \neq \mathcal{P}_0\boldsymbol{u}$ and using (1.22) it follows that

$$(\mathcal{P}_0\boldsymbol{u} - \boldsymbol{u}, \boldsymbol{v})_V \geq (\mathcal{P}_0\boldsymbol{u} - \boldsymbol{u}, \mathcal{P}_0\boldsymbol{u})_V > (\mathcal{P}_0\boldsymbol{u} - \boldsymbol{u}, \boldsymbol{u})_V \qquad \forall\, \boldsymbol{v} \in U_0.$$

These inequalities imply that there exists $\alpha \in \mathbb{R}$ such that

$$(\mathcal{P}_0\boldsymbol{u} - \boldsymbol{u}, \boldsymbol{v})_V > \alpha > (\mathcal{P}_0\boldsymbol{u} - \boldsymbol{u}, \boldsymbol{u})_V \qquad \forall\, \boldsymbol{v} \in U_0. \tag{5.58}$$

Let $\widetilde{\boldsymbol{\tau}}$ be the element in Q given by

$$\widetilde{\boldsymbol{\tau}} = \boldsymbol{\varepsilon}(\mathcal{P}_0\boldsymbol{u} - \boldsymbol{u}) \in Q.$$

Then, using (4.11) and (5.58) we have

$$(\widetilde{\boldsymbol{\tau}}, \boldsymbol{\varepsilon}(\boldsymbol{v}))_Q > \alpha > (\widetilde{\boldsymbol{\tau}}, \boldsymbol{\varepsilon}(\boldsymbol{u}))_Q \qquad \forall\, \boldsymbol{v} \in U_0 \tag{5.59}$$

and taking $\boldsymbol{v} = \boldsymbol{0}_V$ in (5.59) yields

$$\alpha < 0. \tag{5.60}$$

Next, assume that there exists $\widetilde{\boldsymbol{v}} \in U_0$ such that

$$(\widetilde{\boldsymbol{\tau}}, \boldsymbol{\varepsilon}(\widetilde{\boldsymbol{v}}))_Q < 0. \tag{5.61}$$

We take $\boldsymbol{v} = \lambda\widetilde{\boldsymbol{v}}$ in (5.59) where $\lambda \geq 0$ to obtain

$$\lambda\,(\widetilde{\boldsymbol{\tau}}, \boldsymbol{\varepsilon}(\widetilde{\boldsymbol{v}}))_Q > \alpha \qquad \forall\, \lambda \geq 0. \tag{5.62}$$

Therefore, passing to the limit in (5.62) as $\lambda \to \infty$ and using (5.61) we deduce that $\alpha \leq -\infty$ which contradicts $\alpha \in \mathbb{R}$. We conclude from the above that

$$(\widetilde{\boldsymbol{\tau}}, \boldsymbol{\varepsilon}(\boldsymbol{v}))_Q \geq 0 \qquad \forall\, \boldsymbol{v} \in U_0. \tag{5.63}$$

Moreover, from (5.49) and (5.41) we deduce that

$$(\boldsymbol{\sigma}, \boldsymbol{\varepsilon}(\boldsymbol{v}))_Q \geq (\boldsymbol{f}, \boldsymbol{v})_V \qquad \forall\, \boldsymbol{v} \in U_0 \tag{5.64}$$

and, combining (5.63) with (5.64), it follows that $\widetilde{\boldsymbol{\tau}} + \boldsymbol{\sigma} \in \Sigma_0$. This allows us to take $\boldsymbol{\tau} = \widetilde{\boldsymbol{\tau}} + \boldsymbol{\sigma}$ in (5.49) and, as a result, we find that

$$(\mathcal{F}^{-1}\boldsymbol{\sigma}, \widetilde{\boldsymbol{\tau}})_Q \geq 0. \tag{5.65}$$

Using (5.50) and (5.65) we deduce that

$$(\widetilde{\boldsymbol{\tau}}, \boldsymbol{\varepsilon}(\boldsymbol{u}))_Q \geq 0 \tag{5.66}$$

and, combining (5.59), (5.60), it follows that

$$(\widetilde{\boldsymbol{\tau}}, \boldsymbol{\varepsilon}(\boldsymbol{u}))_Q < 0. \tag{5.67}$$

The inequalities (5.66) and (5.67) lead to a contradiction. Therefore, we conclude that

$$\boldsymbol{u} \in U_0. \tag{5.68}$$

Next, we note that (4.11) implies that the element $\boldsymbol{\varepsilon}(\boldsymbol{f})$ belongs to the set Σ_0. Therefore, taking $\boldsymbol{\tau} = \boldsymbol{\varepsilon}(\boldsymbol{f})$ in (5.49) and using again (5.50) and (4.11) it follows that

$$(\boldsymbol{f}, \boldsymbol{u})_V \geq (\boldsymbol{\sigma}, \boldsymbol{\varepsilon}(\boldsymbol{u}))_Q. \tag{5.69}$$

On the other hand, since $\boldsymbol{\sigma} \in \Sigma_0$ and $\boldsymbol{u} \in U_0$ we have

$$(\boldsymbol{\sigma}, \boldsymbol{\varepsilon}(\boldsymbol{u}))_Q \geq (\boldsymbol{f}, \boldsymbol{u})_V. \tag{5.70}$$

We now combine (5.69) and (5.70) to obtain

$$(\boldsymbol{\sigma}, \boldsymbol{\varepsilon}(\boldsymbol{u}))_Q = (\boldsymbol{f}, \boldsymbol{u})_V. \tag{5.71}$$

Finally, we use (5.41) and (5.71) to see that

$$(\boldsymbol{\sigma}, \boldsymbol{\varepsilon}(\boldsymbol{v}) - \boldsymbol{\varepsilon}(\boldsymbol{u}))_Q \geq (\boldsymbol{f}, \boldsymbol{v} - \boldsymbol{u})_V \qquad \forall\, \boldsymbol{v} \in U_0$$

and by (5.50) we obtain that

$$(\mathcal{F}\boldsymbol{\varepsilon}(\boldsymbol{u}), \boldsymbol{\varepsilon}(\boldsymbol{v}) - \boldsymbol{\varepsilon}(\boldsymbol{u}))_Q \geq (\boldsymbol{f}, \boldsymbol{v} - \boldsymbol{u})_V \qquad \forall\, \boldsymbol{v} \in U_0. \tag{5.72}$$

We now use (5.68) and (5.72) to conclude that $\boldsymbol{u}$ is a solution to Problem 5.9. ■

The mechanical interpretation of the results in Theorem 5.13, combined with the uniqueness part of Theorems 5.12 and 5.10, is the following.

(1) If the displacement field $\boldsymbol{u}$ is the solution of the primal variational formulation of the Signorini frictionless contact problem (5.34)–(5.39) (i.e. Problem 5.9), then the stress field $\boldsymbol{\sigma}$ associated with $\boldsymbol{u}$ by the elastic constitutive law $\boldsymbol{\sigma} = \mathcal{F}\boldsymbol{\varepsilon}(\boldsymbol{u})$ is the solution of the dual variational formulation of the Signorini frictionless contact problem (5.34)–(5.39) (i.e. Problem 5.11).

(2) If the stress field $\boldsymbol{\sigma}$ is the solution of the dual variational formulation of the Signorini frictionless contact problem (5.34)–(5.39) (i.e. Problem 5.11), then there exists a unique displacement field $\boldsymbol{u} \in V$ associated with $\boldsymbol{\sigma}$ by the elastic constitutive law $\boldsymbol{\sigma} = \mathcal{F}\boldsymbol{\varepsilon}(\boldsymbol{u})$ and, moreover, $\boldsymbol{u}$ is the solution of the primal variational formulation of the Signorini frictionless contact problem (5.34)–(5.39) (i.e. Problem 5.9).

Under the assumptions of Theorem 5.12 it also results that, if the displacement field $\boldsymbol{u}$ is the solution of the primal variational formulation and the stress field $\boldsymbol{\sigma}$ is the solution of the dual variational formulation, then $\boldsymbol{u}$ and $\boldsymbol{\sigma}$ are connected by the elastic constitutive law $\boldsymbol{\sigma} = \mathcal{F}\boldsymbol{\varepsilon}(\boldsymbol{u})$. For this reason, we refer to the pair $(\boldsymbol{u}, \boldsymbol{\sigma})$ given by Theorems 5.10 and 5.12 as a *weak solution* to the Signorini frictionless contact problem (5.34)–(5.39) and we conclude that this problem has a unique weak solution.

We end this subsection with the remark that the assumption $g_a = 0$ was essential in the proof of Theorem 5.13. Indeed, unlike the convex U defined by (5.10), the convex U_0 defined by (5.40) is a positive cone, i.e. it has the property that $\boldsymbol{v} \in U_0$ implies $\lambda\boldsymbol{v} \in U_0$, for all $\lambda \geq 0$. This property was used several times in the proof of Theorem 5.13.

5.1.5 *Minimization*

Everywhere below we assume that the elasticity operator $\mathcal{F}$ is linear, symmetric and positively defined, i.e. it satisfies

$$\left.\begin{array}{l}
\text{(a) } \mathcal{F} : \Omega \times \mathbb{S}^d \to \mathbb{S}^d;\\
\text{(b) } \mathcal{F}(\boldsymbol{x}, \boldsymbol{\varepsilon}) = (f_{ijkl}(\boldsymbol{x})\varepsilon_{kl}) \text{ for all } \boldsymbol{\varepsilon} = (\varepsilon_{ij}) \in \mathbb{S}^d,\\
\qquad \text{a.e. } \boldsymbol{x} \in \Omega;\\
\text{(c) } f_{ijkl} = f_{jikl} = f_{klij} \in L^\infty(\Omega),\ 1 \leq i, j, k, l \leq d;\\
\text{(d) there exists } m_{\mathcal{F}} > 0 \text{ such that}\\
\qquad \mathcal{F}(\boldsymbol{x}, \boldsymbol{\tau}) \cdot \boldsymbol{\tau} \geq m_{\mathcal{F}} \|\boldsymbol{\tau}\|^2\ \forall \boldsymbol{\tau} \in \mathbb{S}^d, \text{ a.e. } \boldsymbol{x} \text{ in } \Omega.
\end{array}\right\} \quad (5.73)$$

Under this assumption we prove that solving the primal and dual variational formulations of the Signorini contact problem (5.34)–(5.39) is equivalent to solving appropriate minimization problems. Then we use Theorem 1.37 to obtain existence and uniqueness results.

We start by considering the bilinear forms $a : V \times V \to \mathbb{R}$ and $a^* : Q \times Q \to \mathbb{R}$ defined by

$$a(\boldsymbol{u}, \boldsymbol{v}) = (\mathcal{F}\boldsymbol{\varepsilon}(\boldsymbol{u}), \boldsymbol{\varepsilon}(\boldsymbol{v}))_Q \qquad \forall\, \boldsymbol{u}, \boldsymbol{v} \in V, \tag{5.74}$$

$$a^*(\boldsymbol{\sigma}, \boldsymbol{\tau}) = (\mathcal{F}^{-1}\boldsymbol{\sigma}, \boldsymbol{\tau})_Q \qquad \forall\, \boldsymbol{\sigma}, \boldsymbol{\tau} \in Q, \tag{5.75}$$

where $\mathcal{F}^{-1}$ denotes the inverse of the operator $\mathcal{F}$, see Proposition 1.25. It follows from (5.73) that a is a symmetric continuous and V-elliptic bilinear form and a^* is a symmetric continuous and Q-elliptic bilinear form. Moreover, a and a^* satisfy

$$a(\boldsymbol{v}, \boldsymbol{v}) \geq m_{\mathcal{F}} \|\boldsymbol{v}\|_V^2 \qquad \forall\, \boldsymbol{v} \in V,$$
$$a^*(\boldsymbol{\tau}, \boldsymbol{\tau}) \geq m_{\mathcal{F}^{-1}} \|\boldsymbol{\tau}\|_Q^2 \qquad \forall\, \boldsymbol{\tau} \in Q,$$

where $m_{\mathcal{F}^{-1}}$ is the strong monotonicity constant of $\mathcal{F}^{-1}$ see, again, Proposition 1.25.

Consider now the Signorini frictionless contact problem (5.34)–(5.39) and assume that (5.73) and (5.7) hold. Also, consider the *energy function* $J_0 : U_0 \to \mathbb{R}$ and the *complementary energy function* $J_0^* : \Sigma_0 \to \mathbb{R}$ defined by

$$J_0(\boldsymbol{v}) = \frac{1}{2}\, a(\boldsymbol{v}, \boldsymbol{v}) - (\boldsymbol{f}, \boldsymbol{v})_V \qquad \forall\, \boldsymbol{v} \in U_0,$$

$$J_0^*(\boldsymbol{\tau}) = \frac{1}{2}\, a^*(\boldsymbol{\tau}, \boldsymbol{\tau}) \qquad \forall\, \boldsymbol{\tau} \in \Sigma_0,$$

where, recall, $\boldsymbol{f}$ is given by (5.9).

We have the following results.

Theorem 5.14 *Assume (5.73) and (5.7). Then:*

(1) *the element* $\boldsymbol{u}$ *is a solution to Problem 5.9 iff*

$$\boldsymbol{u} \in U_0, \qquad J_0(\boldsymbol{v}) \geq J_0(\boldsymbol{u}) \quad \forall\, \boldsymbol{v} \in U_0; \tag{5.76}$$

(2) *there exists a unique solution* $\boldsymbol{u}$ *to the minimization problem (5.76) (and, therefore, there exists a unique solution* $\boldsymbol{u}$ *to Problem 5.9);*

(3) *the element* $\boldsymbol{\sigma}$ *is a solution to Problem 5.11 iff*

$$\boldsymbol{\sigma} \in \Sigma_0, \qquad J_0^*(\boldsymbol{\tau}) \geq J_0^*(\boldsymbol{\sigma}) \quad \forall\, \boldsymbol{\tau} \in \Sigma_0; \tag{5.77}$$

(4) *there exists a unique solution* $\boldsymbol{\sigma}$ *to the minimization problem (5.77) (and, therefore, there exists a unique solution* $\boldsymbol{\sigma}$ *to Problem 5.11).*

Proof Theorem 5.14 is a direct consequence of Theorem 1.37. More precisely, for the proofs of (1)–(2) we apply Theorem 1.37 with $X = V$, $K = U_0$, $j \equiv 0$, $f = \boldsymbol{f}$ and a given by (5.74), which imply that $J = J_0$.

And for the proof of (3)–(4) we apply Theorem 1.37 with $X = Q$, $K = \Sigma_0$, $j \equiv 0$ and $f = \mathbf{0}_Q$ and a given by (5.75), which imply that $J = J_0^*$. ■

The mechanical interpretation of Theorem 5.14 is the following. Assume that the elastic operator $\mathcal{F}$ satisfies (5.73). Then:

(1) a displacement field $\boldsymbol{u}$ is a solution to the primal variational formulation of the Signorini frictionless contact problem (5.34)–(5.39) if and only if $\boldsymbol{u}$ is a minimizer of the energy function J_0 on the set of admissible displacement fields U_0;

(2) the energy function J_0 has a unique minimizer on the set of admissible displacement fields and, therefore, there exists a unique solution to the primal variational formulation of the Signorini frictionless contact problem (5.34)–(5.39);

(3) a stress field $\boldsymbol{\sigma}$ is a solution to the dual variational formulation of the Signorini frictionless contact problem (5.34)–(5.39) if and only if $\boldsymbol{\sigma}$ is a minimizer of the complementary energy function J_0^* on the set of admissible stress fields Σ_0;

(4) the energy function J_0^* has a unique minimizer on the set of admissible stress fields and, therefore, there exists a unique solution to the dual variational formulation of the Signorini frictionless contact problem (5.34)–(5.39).

5.1.6 *One-dimensional example*

Consider now the one dimensional case. Then, the solution of the elastic contact problem with normal compliance can be directly computed, especially when the elasticity operator is linear and the body forces are constant. The calculus is an instructive exercise and, for this reason, we present it in detail. Then we use the convergence result in Theorem 5.6 to obtain the exact solution of the corresponding Signorini contact problem.

Thus, we consider a cantilever elastic rod $\overline{\Omega} = [0, L]$ which is fixed at its left end $x = 0$ and is subjected to the action of a body force of density f_0 in the x-direction as shown in Figure 5.1. An initial gap g_a is assumed between its right end $x = L$ and an obstacle with normal compliance of stiffness $\frac{1}{\mu}$, $\mu > 0$. This problem corresponds to problem (5.21)–(5.26) with $\Omega = (0, L)$, $\Gamma_1 = \{0\}$, $\Gamma_2 = \emptyset$, $\Gamma_3 = \{L\}$.

We use a linearly elastic constitutive law

$$\sigma = E\,\varepsilon(u),$$

where $E > 0$ represents the Young modulus of the material and $\varepsilon(u) = \frac{du}{dx}$. Also, for the normal compliance contact function we choose

$$p_\nu(r) = r_+$$

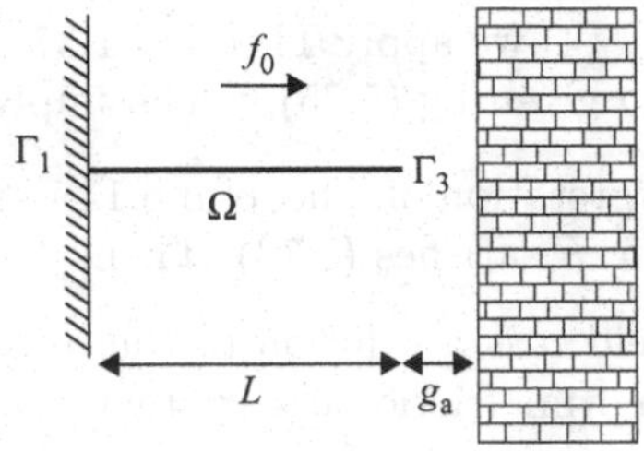

Figure 5.1 Contact of a rod with a foundation

and, for simplicity, we assume in what follows that $L = 1$ and f_0 does not depend on x. With these assumptions, a complete description of the one-dimensional problem we are studying is the following one.

Problem 5.15 *Find a displacement field* $u_\mu : [0,1] \to \mathbb{R}$ *and a stress field* $\sigma_\mu : [0,1] \to \mathbb{R}$ *such that*

$$\sigma_\mu(x) = E\,\frac{\mathrm{d}u_\mu}{\mathrm{d}x}(x) \qquad \forall\, x \in [0,1], \tag{5.78}$$

$$\frac{\mathrm{d}\sigma_\mu}{\mathrm{d}x}(x) + f_0 = 0 \qquad \forall\, x \in [0,1], \tag{5.79}$$

$$u_\mu(0) = 0, \tag{5.80}$$

$$-\sigma_\mu(1) = \frac{1}{\mu}\,(u_\mu(1) - g_a)_+ . \tag{5.81}$$

The exact solution of Problem 5.15 can be obtained through an elementary calculation which we present in what follows. First, we integrate (5.79) to obtain

$$\sigma_\mu(x) = -f_0 x + \alpha \qquad \forall\, x \in [0,1], \tag{5.82}$$

where α is a constant of integration. Then we combine (5.82) and (5.78) to deduce that

$$\frac{\mathrm{d}u_\mu}{\mathrm{d}x}(x) = -\frac{f_0}{E}\,x + \frac{\alpha}{E} \qquad \forall\, x \in [0,1] \tag{5.83}$$

and, finally, we integrate (5.83) with condition (5.80) to see that

$$u_\mu(x) = -\frac{f_0}{2E}\,x^2 + \frac{\alpha}{E}\,x \qquad \forall\, x \in [0,1]. \tag{5.84}$$

We turn now to determinate the values of the constant α and, to this end, we use the normal compliance contact condition (5.81).

First, we assume that $u_\mu(1) < g_a$. Then (5.81) implies that $\sigma_\mu(1) = 0$ and (5.82) yields $\alpha = f_0$. It follows now from (5.84) and (5.82) that

$$u_\mu(x) = -\frac{f_0}{2E}\,x^2 + \frac{f_0}{E}\,x, \quad \sigma_\mu(x) = -f_0 x + f_0 \qquad \forall\, x \in [0,1]. \tag{5.85}$$

Now, since $u_\mu(1) < g_a$, the first equality in (5.85) yields $\frac{f_0}{2E} < g_a$. To conclude, we proved that

$$u_\mu(1) < g_a \Longrightarrow \begin{cases} u_\mu(x) = -\dfrac{f_0}{2E}\,x^2 + \dfrac{f_0}{E}\,x & \forall\, x \in [0,1], \\ \sigma_\mu(x) = -f_0 x + f_0 & \forall\, x \in [0,1], \\ \dfrac{f_0}{2E} < g_a. \end{cases} \tag{5.86}$$

Assume now that $u_\mu(1) \geq g_a$. Then (5.81), (5.82) and (5.84) imply that

$$-f_0 + \alpha = \frac{f_0}{2\mu E} - \frac{\alpha}{\mu E} + \frac{g_a}{\mu}$$

and, therefore,

$$\alpha = \frac{f_0 + 2f_0\mu E + 2g_a E}{2(\mu E + 1)}\,.$$

It follows now from (5.84) and (5.82) that

$$\left.\begin{aligned} u_\mu(x) &= -\frac{f_0}{2E}\,x^2 + \frac{f_0 + 2f_0\mu E + 2g_a E}{2E(\mu E + 1)}\,x \qquad \forall\, x \in [0,1], \\ \sigma_\mu(x) &= -f_0 x + \frac{f_0 + 2f_0\mu E + 2g_a E}{2(\mu E + 1)} \qquad \forall\, x \in [0,1]. \end{aligned}\right\} \tag{5.87}$$

Since $u_\mu(1) \geq g_a$, the first equality in (5.87) yields $\frac{f_0}{2E} \geq g_a$. To conclude, we proved that

$$u_\mu(1) \geq g_a \Longrightarrow \begin{cases} u_\mu(x) = -\dfrac{f_0}{2E}\,x^2 + \dfrac{f_0 + 2f_0\mu E + 2g_a E}{2E(\mu E + 1)}\,x \\ \qquad \forall\, x \in [0,1], \\ \sigma_\mu(x) = -f_0 x + \dfrac{f_0 + 2f_0\mu E + 2g_a E}{2(\mu E + 1)} \\ \qquad \forall\, x \in [0,1], \\ \dfrac{f_0}{2E} \geq g_a. \end{cases} \tag{5.88}$$

Assume now that $\frac{f_0}{2E} < g_a$. Then, arguing by contradiction, (5.88) implies that $u_\mu(1) < g_a$ and, therefore, (5.86) shows that the solution of the contact problem (5.78)–(5.81) is given by (5.85). In this case there is no contact (since $u_\mu(1) < g_a$) and the point $x = 1$ is stress free (since $\sigma_\mu(1) = 0$).

Next, assume that $\frac{f_0}{2E} \geq g_a$. Then, arguing by contradiction, (5.86) implies that $u_\mu(1) \geq g_a$ and, therefore, (5.88) shows that the solution of the contact problem (5.78)–(5.81) is given by (5.87). In this case there is contact (since $u_\mu(1) \geq g_a$) and the reaction of the obstacle is towards the rod (since $\sigma_\mu(1) = \frac{-f_0+2g_a E}{2(\mu E+1)} \leq 0$).

We conclude from the above that the solution of the one-dimensional contact problem with normal compliance is given by (5.85) if $\frac{f_0}{2E} < g_a$, and (5.87) if $\frac{f_0}{2E} \geq g_a$. It is easy to check that if $\frac{f_0}{2E} < g_a$ then the functions (u_μ, σ_μ) given by (5.85) satisfy Problem 5.15 and, if $\frac{f_0}{2E} \geq g_a$, then the functions (u_μ, σ_μ) given by (5.87) satisfy Problem 5.15. The uniqueness of the solution is guaranteed by Theorem 5.6.

Next, we consider the contact problem in the case when the obstacle is rigid and, therefore, we model the contact with the Signorini condition. A complete description of this problem is the following.

Problem 5.16 *Find a displacement field* $u : [0,1] \to \mathbb{R}$ *and a stress field* $\sigma : [0,1] \to \mathbb{R}$ *such that*

$$\begin{aligned}
\sigma(x) &= E\frac{du}{dx}(x) \qquad \forall x \in [0,1],\\
\frac{d\sigma}{dx}(x) + f_0 &= 0 \qquad \forall x \in [0,1],\\
u(0) &= 0,\\
u(1) \leq g_a, \qquad \sigma(1) \leq 0, &\qquad \sigma(1)(u(1) - g_a) = 0.
\end{aligned}$$

The exact solution of Problem 5.16 can be obtained by passing to the limit as $\mu \to 0$ in the solution of Problem 5.15, as suggested by the convergence result (5.30) in Theorem 5.6.

Assume that $\frac{f_0}{2E} < g_a$. Then the solution of (5.78)–(5.81) is given by (5.85) and, therefore, we have

$$\lim_{\mu\to 0} u_\mu(x) = \lim_{\mu\to 0}\Big(-\frac{f_0}{2E}x^2 + \frac{f_0}{E}x\Big) = -\frac{f_0}{2E}x^2 + \frac{f_0}{E}x,$$

$$\lim_{\mu\to 0} \sigma_\mu(x) = \lim_{\mu\to 0}(-f_0 x + f_0) = -f_0 x + f_0$$

for all $x \in [0,1]$. Next, assume that $\frac{f_0}{2E} \geq g_a$. Then the solution of (5.78)–(5.81) is given by (5.87) and, therefore,

$$\lim_{\mu\to 0} u_\mu(x) = \lim_{\mu\to 0} \Big(-\frac{f_0}{2E}\,x^2 + \frac{f_0 + 2f_0\mu E + 2g_a E}{2E(\mu E+1)}\,x \Big)$$
$$= -\frac{f_0}{2E}\,x^2 + \Big(\frac{f_0}{2E} + g_a\Big)\,x,$$

$$\lim_{\mu\to 0} \sigma_\mu(x) = \lim_{\mu\to 0} \Big(-f_0 x + \frac{f_0 + 2f_0\mu E + 2g_a E}{2(\mu E+1)} \Big) = -f_0 x + \frac{f_0}{2} + g_a E,$$

for all $x \in [0,1]$.

Consider now the couple of functions (u,σ) defined by

$$\left.\begin{aligned} u(x) &= -\frac{f_0}{2E}\,x^2 + \frac{f_0}{E}\,x \qquad \forall\, x \in [0,1], \\ \sigma(x) &= -f_0 x + f_0 \qquad \forall\, x \in [0,1] \end{aligned}\right\} \tag{5.89}$$

if $\frac{f_0}{2E} < g_a$ and

$$\left.\begin{aligned} u(x) &= -\frac{f_0}{2E}\,x^2 + \Big(\frac{f_0}{2E} + g_a\Big)\,x \qquad \forall\, x \in [0,1], \\ \sigma(x) &= -f_0 x + \frac{f_0}{2} + g_a E \qquad \forall\, x \in [0,1] \end{aligned}\right\} \tag{5.90}$$

if $\frac{f_0}{2E} \geq g_a$. It is easy to see that if $\frac{f_0}{2E} < g_a$ then the functions (u,σ) given by (5.89) satisfy Problem 5.16 and, if $\frac{f_0}{2E} \geq g_a$ then the functions (u,σ) given by (5.90) satisfy Problem 5.16. The uniqueness of the solution is guaranteed by Theorem 5.6.

A brief examination of the solution of the Signorini problem above shows that if $\frac{f_0}{2E} < g_a$ then there is no contact (since $u(1) < g_a$) and the point $x = 1$ is stress free (since $\sigma(1) = 0$). Also, if $\frac{f_0}{2E} \geq g_a$ then there is contact (since $u(1) = g_a$) and the reaction of the obstacle is towards the rod (since $\sigma(1) = -\frac{f_0}{2} + g_a E \leq 0$). We also note that as far as there is no contact, the solution of the Signorini problem and the solution of the normal compliance problem are the same.

5.2 Frictional contact problems

In this section we provide various results in the analysis of two frictional contact problems for elastic materials. This includes existence, uniqueness and convergence results, among others.

5.2.1 Statement of the problems

In the first problem we assume that the normal stress is prescribed on the contact surface and we use the static version of Coulomb's law of dry friction. Therefore, the classical model of the process is the following.

Problem 5.17 *Find a displacement field* $\boldsymbol{u} : \Omega \to \mathbb{R}^d$ *and a stress field* $\boldsymbol{\sigma} : \Omega \to \mathbb{S}^d$ *such that*

$$\boldsymbol{\sigma} = \mathcal{F}\boldsymbol{\varepsilon}(\boldsymbol{u}) \qquad \text{in } \Omega, \tag{5.91}$$

$$\operatorname{Div}\boldsymbol{\sigma} + \boldsymbol{f}_0 = \boldsymbol{0} \qquad \text{in } \Omega, \tag{5.92}$$

$$\boldsymbol{u} = \boldsymbol{0} \qquad \text{on } \Gamma_1, \tag{5.93}$$

$$\boldsymbol{\sigma\nu} = \boldsymbol{f}_2 \qquad \text{on } \Gamma_2, \tag{5.94}$$

$$-\sigma_\nu = F \qquad \text{on } \Gamma_3, \tag{5.95}$$

$$\|\boldsymbol{\sigma}_\tau\| \le \mu\,|\sigma_\nu|, \quad \boldsymbol{\sigma}_\tau = -\mu\,|\sigma_\nu|\,\frac{\boldsymbol{u}_\tau}{\|\boldsymbol{u}_\tau\|} \quad \text{if} \quad \boldsymbol{u}_\tau \neq \boldsymbol{0} \qquad \text{on } \Gamma_3. \tag{5.96}$$

Recall that (5.91) represents the elastic constitutive law (4.32). Also, (5.92) is the equation of equilibrium (4.73), (5.93) represents the displacement boundary condition (4.74) and (5.94) is the traction boundary condition (4.75). Finally, (5.95) represents the contact condition (4.77) and (5.96) is the friction law (4.86), associated with the friction bound (4.88).

In the second problem we use the contact condition (4.76) associated with Tresca's friction law (4.86), (4.87). Therefore, the classical model of the process is the following.

Problem 5.18 *Find a displacement field* $\boldsymbol{u} : \Omega \to \mathbb{R}^d$ *and a stress field* $\boldsymbol{\sigma} : \Omega \to \mathbb{S}^d$ *such that*

$$\boldsymbol{\sigma} = \mathcal{F}\boldsymbol{\varepsilon}(\boldsymbol{u}) \qquad \text{in } \Omega, \tag{5.97}$$

$$\operatorname{Div}\boldsymbol{\sigma} + \boldsymbol{f}_0 = \boldsymbol{0} \qquad \text{in } \Omega, \tag{5.98}$$

$$\boldsymbol{u} = \boldsymbol{0} \qquad \text{on } \Gamma_1, \tag{5.99}$$

$$\boldsymbol{\sigma\nu} = \boldsymbol{f}_2 \qquad \text{on } \Gamma_2, \tag{5.100}$$

$$u_\nu = 0 \qquad \text{on } \Gamma_3, \tag{5.101}$$

$$\|\boldsymbol{\sigma}_\tau\| \le F_{\rm b}, \quad \boldsymbol{\sigma}_\tau = -\,F_{\rm b}\,\frac{\boldsymbol{u}_\tau}{\|\boldsymbol{u}_\tau\|} \quad \text{if} \quad \boldsymbol{u}_\tau \neq \boldsymbol{0} \qquad \text{on } \Gamma_3. \tag{5.102}$$

We turn to the variational formulation of Problems 5.17 and 5.18. To this end we assume in what follows that the elasticity operator satisfies condition (4.33), the densities of body forces and surface tractions have the regularity (5.7) and the prescribed normal stress satisfies

$$F \in L^2(\Gamma_3) \quad \text{and} \quad F(\boldsymbol{x}) \ge 0 \text{ a.e. } \boldsymbol{x} \in \Gamma_3. \tag{5.103}$$

The coefficient of friction satisfies

$$\mu \in L^\infty(\Gamma_3) \ \text{ and } \ \mu(\boldsymbol{x}) \geq 0 \text{ a.e. } \boldsymbol{x} \in \Gamma_3 \tag{5.104}$$

and, finally, the friction bound is such that

$$F_{\rm b} \in L^2(\Gamma_3) \ \text{ and } \ F_{\rm b}(\boldsymbol{x}) \geq 0 \text{ a.e. } \boldsymbol{x} \in \Gamma_3. \tag{5.105}$$

In the study of Problem 5.17 we use the space V, (4.9), and the element $\boldsymbol{f} \in V$ defined by

$$(\boldsymbol{f}, \boldsymbol{v})_V = \int_\Omega \boldsymbol{f}_0 \cdot \boldsymbol{v}\, {\rm d}x + \int_{\Gamma_2} \boldsymbol{f}_2 \cdot \boldsymbol{v}\, {\rm d}a - \int_{\Gamma_3} F v_\nu\, {\rm d}a \quad \forall\, \boldsymbol{v} \in V. \tag{5.106}$$

Also, we consider the functional $j : V \to \mathbb{R}$ given by

$$j(\boldsymbol{v}) = \int_{\Gamma_3} \mu F \, \|\boldsymbol{v}_\tau\|\, {\rm d}a \quad \forall\, \boldsymbol{v} \in V. \tag{5.107}$$

Note that by assumptions (5.103) and (5.104) the integral in (5.107) is well defined.

In the study of Problem 5.18 we use the space V_1, (4.19), the element $\boldsymbol{f}_1 \in V_1$ defined by

$$(\boldsymbol{f}_1, \boldsymbol{v})_V = \int_\Omega \boldsymbol{f}_0 \cdot \boldsymbol{v}\, {\rm d}x + \int_{\Gamma_2} \boldsymbol{f}_2 \cdot \boldsymbol{v}\, {\rm d}a \quad \forall\, \boldsymbol{v} \in V_1 \tag{5.108}$$

and the functional $j_1 : V_1 \to \mathbb{R}$ given by

$$j_1(\boldsymbol{v}) = \int_{\Gamma_3} F_{\rm b} \, \|\boldsymbol{v}_\tau\|\, {\rm d}a \quad \forall\, \boldsymbol{v} \in V_1. \tag{5.109}$$

Again, note that by assumption (5.105) the integral in (5.109) is well defined.

Assume that the mechanical problem (5.91)–(5.96) has a regular solution $(\boldsymbol{u}, \boldsymbol{\sigma})$ and let $\boldsymbol{v} \in V$. First, we recall equality (5.11) on page 125, i.e.

$$\int_\Omega \boldsymbol{\sigma} \cdot (\boldsymbol{\varepsilon}(\boldsymbol{v}) - \boldsymbol{\varepsilon}(\boldsymbol{u}))\, {\rm d}x = \int_\Omega \boldsymbol{f}_0 \cdot (\boldsymbol{v} - \boldsymbol{u})\, {\rm d}x + \int_{\Gamma_2} \boldsymbol{f}_2 \cdot (\boldsymbol{v} - \boldsymbol{u})\, {\rm d}a$$
$$+ \int_{\Gamma_3} \sigma_\nu (v_\nu - u_\nu)\, {\rm d}a + \int_{\Gamma_3} \boldsymbol{\sigma}_\tau \cdot (\boldsymbol{v}_\tau - \boldsymbol{u}_\tau)\, {\rm d}a. \tag{5.110}$$

Next, we use the friction law (5.96) to see that

$$\boldsymbol{\sigma}_\tau \cdot (\boldsymbol{v}_\tau - \boldsymbol{u}_\tau) \geq \mu\, |\sigma_\nu|\, \|\boldsymbol{u}_\tau\| - \mu\, |\sigma_\nu|\, \|\boldsymbol{v}_\tau\| \qquad \text{a.e. on } \Gamma_3. \tag{5.111}$$

Indeed, in the points of Γ_3 where $\boldsymbol{u}_\tau \neq \boldsymbol{0}$ we have

$$\begin{aligned}\boldsymbol{\sigma}_\tau \cdot (\boldsymbol{v}_\tau - \boldsymbol{u}_\tau) &= -\mu\,|\sigma_\nu|\,\frac{\boldsymbol{u}_\tau}{\|\boldsymbol{u}_\tau\|} \cdot (\boldsymbol{v}_\tau - \boldsymbol{u}_\tau)\\ &\geq \mu\,|\sigma_\nu|\,\|\boldsymbol{u}_\tau\| - \mu\,|\sigma_\nu|\,\|\boldsymbol{v}_\tau\|,\end{aligned}$$

since $\boldsymbol{u}_\tau \cdot \boldsymbol{u}_\tau = \|\boldsymbol{u}_\tau\|^2$ and $\boldsymbol{u}_\tau \cdot \boldsymbol{v}_\tau \leq \|\boldsymbol{u}_\tau\|\,\|\boldsymbol{v}_\tau\|$; at the points of Γ_3 where $\boldsymbol{u}_\tau = \boldsymbol{0}$ we have

$$\begin{aligned}\boldsymbol{\sigma}_\tau \cdot (\boldsymbol{v}_\tau - \boldsymbol{u}_\tau) = \boldsymbol{\sigma}_\tau \cdot \boldsymbol{v}_\tau &\geq -\|\boldsymbol{\sigma}_\tau\|\,\|\boldsymbol{v}_\tau\|\\ &\geq -\mu\,|\sigma_\nu|\,\|\boldsymbol{v}_\tau\| = \mu\,|\sigma_\nu|\,\|\boldsymbol{u}_\tau\| - \mu\,|\sigma_\nu|\,\|\boldsymbol{v}_\tau\|,\end{aligned}$$

since $\|\boldsymbol{\sigma}_\tau\| \leq \mu\,|\sigma_\nu|$ and $\|\boldsymbol{u}_\tau\| = 0$. We integrate (5.111) on Γ_3 to obtain

$$\int_{\Gamma_3} \boldsymbol{\sigma}_\tau \cdot (\boldsymbol{v}_\tau - \boldsymbol{u}_\tau)\,da \geq \int_{\Gamma_3} \mu\,|\sigma_\nu|\,(\|\boldsymbol{u}_\tau\| - \|\boldsymbol{v}_\tau\|)\,da. \tag{5.112}$$

We now combine (5.110) and (5.112), then we use the boundary condition (5.95) to find

$$\begin{aligned}&\int_\Omega \boldsymbol{\sigma} \cdot (\boldsymbol{\varepsilon}(\boldsymbol{v}) - \boldsymbol{\varepsilon}(\boldsymbol{u}))\,dx + \int_{\Gamma_3} \mu\,F(\|\boldsymbol{v}_\tau\| - \|\boldsymbol{u}_\tau\|)\,da\\ &\quad\geq \int_\Omega \boldsymbol{f}_0 \cdot (\boldsymbol{v} - \boldsymbol{u})\,dx + \int_{\Gamma_2} \boldsymbol{f}_2 \cdot (\boldsymbol{v} - \boldsymbol{u})\,da - \int_{\Gamma_3} F(v_v - u_\nu)\,da.\end{aligned}$$

Next, we use the notation (5.106) and (5.107) to deduce that

$$(\boldsymbol{\sigma}, \boldsymbol{\varepsilon}(\boldsymbol{v}) - \boldsymbol{\varepsilon}(\boldsymbol{u}))_Q + j(\boldsymbol{v}) - j(\boldsymbol{u}) \geq (\boldsymbol{f}, \boldsymbol{v} - \boldsymbol{u})_V. \tag{5.113}$$

Finally, we use the constitutive law (5.91) and (5.113) to derive the following variational formulation of the frictionless contact problem 5.17.

Problem 5.19 *Find a displacement field $\boldsymbol{u} \in V$ such that*

$$(\mathcal{F}\boldsymbol{\varepsilon}(\boldsymbol{u}), \boldsymbol{\varepsilon}(\boldsymbol{v}) - \boldsymbol{\varepsilon}(\boldsymbol{u}))_Q + j(\boldsymbol{v}) - j(\boldsymbol{u}) \geq (\boldsymbol{f}, \boldsymbol{v} - \boldsymbol{u})_V \quad \forall\,\boldsymbol{v} \in V. \tag{5.114}$$

We use a similar procedure to derive the following variational formulation of Problem 5.18.

Problem 5.20 *Find a displacement field $\boldsymbol{u} \in V_1$ such that*

$$(\mathcal{F}\boldsymbol{\varepsilon}(\boldsymbol{u}), \boldsymbol{\varepsilon}(\boldsymbol{v}) - \boldsymbol{\varepsilon}(\boldsymbol{u}))_Q + j_1(\boldsymbol{v}) - j_1(\boldsymbol{u}) \geq (\boldsymbol{f}_1, \boldsymbol{v} - \boldsymbol{u})_V \quad \forall\,\boldsymbol{v} \in V_1. \tag{5.115}$$

We note that Problems 5.19 and 5.20 represent elliptic variational inequalities of the second kind for the displacement field. Therefore, we apply the abstract results in Section 2.2 to provide the unique solvability of these problems.

5.2.2 Existence and uniqueness

In the study of the contact problem 5.19 we have the following existence and uniqueness result.

Theorem 5.21 *Assume (4.33), (5.7), (5.103) and (5.104). Then there exists a unique solution to Problem 5.19.*

Proof We use the operator $A : V \to V$ defined by (5.16) and we recall that A is a strongly monotone Lipschitz continuous operator on the space V. Also, we use (5.103) and (5.104) to see that the functional j defined by (5.107) is a seminorm on V and, moreover, it satisfies

$$j(\boldsymbol{v}) \leq \|\mu\|_{L^\infty(\Gamma_3)} \|F\|_{L^2(\Gamma_3)} \|\boldsymbol{v}\|_{L^2(\Gamma_3)^d} \qquad \forall\, \boldsymbol{v} \in V.$$

Therefore, (4.13) yields

$$j(\boldsymbol{v}) \leq c_0 \|\mu\|_{L^\infty(\Gamma_3)} \|F\|_{L^2(\Gamma_3)} \|\boldsymbol{v}\|_V \qquad \forall\, \boldsymbol{v} \in V. \tag{5.116}$$

Inequality (5.116) and Corollary 1.35 on page 28 show that j is a convex lower semicontinuous function on V. Theorem 5.21 is now a consequence of Theorem 2.8. ■

Let $\boldsymbol{u}$ be the solution of Problem 5.19 obtained in Theorem 5.21 and let $\boldsymbol{\sigma} = \mathcal{F}\boldsymbol{\varepsilon}(\boldsymbol{u})$. It follows from properties (4.33) of the operator $\mathcal{F}$ that $\boldsymbol{\sigma} \in Q$. Moreover, taking $\boldsymbol{v} = \boldsymbol{u} \pm \boldsymbol{\varphi}$ in (5.114) implies that

$$(\boldsymbol{\sigma}, \boldsymbol{\varepsilon}(\boldsymbol{\varphi}))_Q = (\boldsymbol{f}, \boldsymbol{\varphi})_V \quad \forall\, \boldsymbol{\varphi} \in C_0^\infty(\Omega)^d$$

which, in turn, yields

$$\operatorname{Div} \boldsymbol{\sigma} + \boldsymbol{f}_0 = \boldsymbol{0} \quad \text{in } \Omega.$$

Using the regularity $\boldsymbol{f}_0 \in L^2(\Omega)^d$ in (5.7) it follows that $\operatorname{Div} \boldsymbol{\sigma} \in L^2(\Omega)^d$ and, therefore, $\boldsymbol{\sigma} \in Q_1$.

A pair of functions $(\boldsymbol{u}, \boldsymbol{\sigma})$ which satisfies (5.91) and (5.114) is called a *weak solution* of the frictionless contact problem (5.91)–(5.96). We conclude by Theorem 5.21 that problem (5.91)–(5.96) has a unique weak solution, with regularity $\boldsymbol{u} \in V$ and $\boldsymbol{\sigma} \in Q_1$.

We now obtain the following existence and uniqueness result in the study of Problem 5.20.

Theorem 5.22 *Assume (4.33), (5.7) and (5.105). Then there exists a unique solution to Problem 5.20.*

Proof We consider the operator A_1 defined by

$$(A_1 \boldsymbol{u}, \boldsymbol{v})_V = (\mathcal{F}\boldsymbol{\varepsilon}(\boldsymbol{u}), \boldsymbol{\varepsilon}(\boldsymbol{v}))_Q \quad \forall\, \boldsymbol{u}, \boldsymbol{v} \in V_1. \tag{5.117}$$

We use (4.33)(c) to see that

$$(A_1 \boldsymbol{u} - A_1 \boldsymbol{v}, \boldsymbol{u} - \boldsymbol{v})_V \geq m_{\mathcal{F}} \, \|\boldsymbol{u} - \boldsymbol{v}\|_V^2 \qquad \forall \, \boldsymbol{u}, \, \boldsymbol{v} \in V_1. \tag{5.118}$$

On the other hand, by (4.33)(b) it follows that

$$\|A_1 \boldsymbol{u} - A_1 \boldsymbol{v}\|_V \leq L_{\mathcal{F}} \, \|\boldsymbol{u} - \boldsymbol{v}\|_V \qquad \forall \, \boldsymbol{u}, \, \boldsymbol{v} \in V_1. \tag{5.119}$$

We conclude by (5.118) and (5.119) that A_1 is a strongly monotone Lipschitz continuous operator on the space V_1. Moreover, it is easy to see that the functional j_1 defined by (5.109) is a continuous seminorm on V_1. Theorem 5.22 is now a consequence of Theorem 2.8. ■

A pair of functions $(\boldsymbol{u}, \boldsymbol{\sigma})$ which satisfies (5.97) and (5.115) is called a *weak solution* of the frictionless contact problem (5.97)–(5.102). We conclude by Theorem 5.22 that problem (5.97)–(5.102) has a unique weak solution. Moreover, using arguments similar to those used above, it follows that the weak solution has the regularity $\boldsymbol{u} \in V_1$ and $\boldsymbol{\sigma} \in Q_1$.

5.2.3 A convergence result

We present in what follows a result concerning the dependence of the solution with respect to the frictional conditions. Such results can be obtained for both Problems 5.19 and 5.20 but, to avoid repetition, we restrict ourselves to the case of Problem 5.20. To this end, we assume in what follows that (4.33), (5.7) and (5.105) hold and we denote by $\boldsymbol{u}$ the solution of Problem 5.20 obtained in Theorem 5.22. Also, for each $\rho > 0$, we assume that $F_{b\rho}$ is a given friction bound which satisfies (5.105). Denote by $j_{1\rho} : V_1 \to \mathbb{R}$ the functional defined by

$$j_{1\rho}(\boldsymbol{v}) = \int_{\Gamma_3} F_{b\rho} \, \|\boldsymbol{v}_\tau\| \, da \quad \forall \, \boldsymbol{v} \in V_1 \tag{5.120}$$

and consider the following variational problem.

Problem 5.23 *Find a displacement field $\boldsymbol{u}_\rho \in V_1$ such that*

$$(\mathcal{F}\boldsymbol{\varepsilon}(\boldsymbol{u}_\rho), \boldsymbol{\varepsilon}(\boldsymbol{v}) - \boldsymbol{\varepsilon}(\boldsymbol{u}_\rho))_Q + j_{1\rho}(\boldsymbol{v}) - j_{1\rho}(\boldsymbol{u}_\rho)$$
$$\geq (\boldsymbol{f}_1, \boldsymbol{v} - \boldsymbol{u}_\rho)_V \quad \forall \, \boldsymbol{v} \in V_1. \tag{5.121}$$

We deduce from Theorem 5.22 that Problem 5.23 has a unique solution $\boldsymbol{u}_\rho \in V_1$, for each $\rho > 0$.

Consider now the assumption

$$F_{b\rho} \to F_b \quad \text{in } L^2(\Gamma_3) \quad \text{as} \quad \rho \to 0. \tag{5.122}$$

Then the behavior of the solution $\boldsymbol{u}_\rho$ as ρ converges to zero is given in the following theorem.

Theorem 5.24 *Under the assumption (5.122), the solution $\boldsymbol{u}_\rho$ of Problem 5.23 converges to the solution $\boldsymbol{u}$ of Problem 5.20, i.e.*

$$u_\rho \to u \quad in\ V_1 \quad as \quad \rho \to 0. \tag{5.123}$$

Proof Let $\rho > 0$ and let $\boldsymbol{u}_1,\ \boldsymbol{u}_2 \in V_1$. We use (5.120) and (5.109) to see that

$$j_{1\rho}(\boldsymbol{u}_1) - j_{1\rho}(\boldsymbol{u}_2) + j_1(\boldsymbol{u}_2) - j_1(\boldsymbol{u}_1) = \int_{\Gamma_3} (F_{b\rho} - F_b)(\|\boldsymbol{u}_{1\tau}\| - \|\boldsymbol{u}_{2\tau}\|)\, da$$

and, since

$$\Big| \|\boldsymbol{u}_{1\tau}\| - \|\boldsymbol{u}_{2\tau}\| \Big| \le \|\boldsymbol{u}_{1\tau} - \boldsymbol{u}_{2\tau}\| \le \|\boldsymbol{u}_1 - \boldsymbol{u}_2\| \quad \text{a.e. on } \Gamma_3, \tag{5.124}$$

it follows that

$$\begin{aligned} j_{1\rho}(\boldsymbol{u}_1) - j_{1\rho}(\boldsymbol{u}_2) + j_1(\boldsymbol{u}_2) - j_1(\boldsymbol{u}_1) &\le \int_{\Gamma_3} |F_{b\rho} - F_b|\, \|\boldsymbol{u}_1 - \boldsymbol{u}_2\|\, da \\ &\le \|F_{b\rho} - F_b\|_{L^2(\Gamma_3)} \|\boldsymbol{u}_1 - \boldsymbol{u}_2\|_{L^2(\Gamma_3)^d}. \end{aligned}$$

We now use (4.13) and the previous inequality to see that

$$\begin{aligned} &j_{1\rho}(\boldsymbol{u}_1) - j_{1\rho}(\boldsymbol{u}_2) + j_1(\boldsymbol{u}_2) - j_1(\boldsymbol{u}_1) \\ &\qquad \le c_0 \|F_{b\rho} - F_b\|_{L^2(\Gamma_3)} \|\boldsymbol{u}_1 - \boldsymbol{u}_2\|_V. \end{aligned} \tag{5.125}$$

Take

$$G(\rho) = c_0\, \|F_{b\rho} - F_b\|_{L^2(\Gamma_3)}. \tag{5.126}$$

Then, by (5.125), (5.126) and (5.122) it follows that the functionals $j_{1\rho}$ and j_1 satisfy condition (2.34) on the space V_1. Theorem 5.24 is now a consequence of Theorem 2.11. ∎

In addition to the mathematical interest in the convergence result (5.123), it is of importance from mechanical point of view, since it states that small errors in the estimation of the friction bound lead to small errors in the solution. We conclude that the weak solution of the contact problem (5.97)–(5.102) depends continuously on the friction bound.

5.2.4 Regularization

In what follows we study Problem 5.20 using the regularization method. To this end, for each $\rho > 0$ we consider the following frictional contact problem.

Problem 5.25 *Find a displacement field* $\boldsymbol{u}_\rho : \Omega \to \mathbb{R}^d$ *and a stress field* $\boldsymbol{\sigma}_\rho : \Omega \to \mathbb{S}^d$ *such that*

$$\boldsymbol{\sigma}_\rho = \mathcal{F}\boldsymbol{\varepsilon}(\boldsymbol{u}_\rho) \quad \text{in } \Omega, \tag{5.127}$$

$$\operatorname{Div}\boldsymbol{\sigma}_\rho + \boldsymbol{f}_0 = \boldsymbol{0} \quad \text{in } \Omega, \tag{5.128}$$

$$\boldsymbol{u}_\rho = \boldsymbol{0} \quad \text{on } \Gamma_1, \tag{5.129}$$

$$\boldsymbol{\sigma}_\rho \boldsymbol{\nu} = \boldsymbol{f}_2 \quad \text{on } \Gamma_2, \tag{5.130}$$

$$u_{\rho\nu} = 0 \quad \text{on } \Gamma_3, \tag{5.131}$$

$$-\boldsymbol{\sigma}_{\rho\tau} = F_{\mathrm{b}}\,\frac{\boldsymbol{u}_{\rho\tau}}{\sqrt{\|\boldsymbol{u}_{\rho\tau}\|^2 + \rho^2}} \quad \text{on } \Gamma_3. \tag{5.132}$$

Here and below $u_{\rho\nu}$ and $\sigma_{\rho\nu}$ denote the normal components of $\boldsymbol{u}_\rho$ and $\boldsymbol{\sigma}_\rho$, and $\boldsymbol{u}_{\rho\tau}$, $\boldsymbol{\sigma}_{\rho\tau}$ represent the tangential part of $\boldsymbol{u}_\rho$, $\boldsymbol{\sigma}_\rho$ on the boundary Γ, see equalities (4.8) and (4.21) on pages 86 and 89, respectively. The equations and boundary conditions involved in Problem 5.25 have the same meaning as those in Problem 5.18. The difference arises from the fact that here we replace the Coulomb friction law (5.102) with its regularization (5.132), introduced on page 110.

We assume in what follows that (4.33) and (5.7) hold and we reinforce condition (5.105) by assuming that the friction bound satisfies

$$F_{\mathrm{b}} \in L^\infty(\Gamma_3) \quad \text{and} \quad F_{\mathrm{b}}(\boldsymbol{x}) \ge 0 \text{ a.e. } \boldsymbol{x} \in \Gamma_3. \tag{5.133}$$

Consider now the operator $P_\rho : V_1 \to V_1$ defined by

$$(P_\rho \boldsymbol{u}, \boldsymbol{v})_V = \int_{\Gamma_3} F_{\mathrm{b}}\,\boldsymbol{\phi}_\rho(\boldsymbol{u}_\tau)\cdot \boldsymbol{v}_\tau\, da, \tag{5.134}$$

for all $\boldsymbol{u}$, $\boldsymbol{v} \in V_1$, where $\boldsymbol{\phi}_\rho : \mathbb{R}^d \to \mathbb{R}^d$ is the function given by

$$\boldsymbol{\phi}_\rho(\boldsymbol{v}) = \frac{\boldsymbol{v}}{\sqrt{\|\boldsymbol{v}\|^2 + \rho^2}} \qquad \forall\, \boldsymbol{v} \in \mathbb{R}^d. \tag{5.135}$$

Note that $\|\boldsymbol{\phi}_\rho(\boldsymbol{v})\| \le 1$ for all $\boldsymbol{v} \in \mathbb{R}^d$ and, therefore, the integral in (5.134) is well defined. In addition, consider the functional $j_{1\rho} : V_1 \to \mathbb{R}$ defined by

$$j_{1\rho}(\boldsymbol{v}) = \int_{\Gamma_3} F_{\mathrm{b}}\,(\sqrt{\|\boldsymbol{v}_\tau\|^2 + \rho^2} - \rho)\, da \tag{5.136}$$

for all $\boldsymbol{v} \in V_1$ and note that, again, the integral in (5.136) is well defined.

Using arguments similar to those used in Section 5.1, based on the Green formula (4.22), it is easy to see that if $(\boldsymbol{u}_\rho, \boldsymbol{\sigma}_\rho)$ are regular functions which satisfy (5.128)–(5.132), then

$$(\boldsymbol{\sigma}_\rho, \boldsymbol{\varepsilon}(\boldsymbol{v}))_Q + (P_\rho \boldsymbol{u}_\rho, \boldsymbol{v})_V = (\boldsymbol{f}_1, \boldsymbol{v})_V \quad \forall\, \boldsymbol{v} \in V_1$$

where, recall, $\boldsymbol{f}_1$ is defined by (5.108). We now use the constitutive law (5.127) and the previous equality to derive the following variational formulation of Problem 5.25.

Problem 5.26 *Find a displacement field* $\boldsymbol{u}_\rho \in V_1$ *such that*

$$(\mathcal{F}\boldsymbol{\varepsilon}(\boldsymbol{u}_\rho), \boldsymbol{\varepsilon}(\boldsymbol{v}))_Q + (P_\rho \boldsymbol{u}_\rho, \boldsymbol{v})_V = (\boldsymbol{f}_1, \boldsymbol{v})_V \quad \forall\, \boldsymbol{v} \in V_1. \tag{5.137}$$

We have the following existence, uniqueness and convergence result.

Theorem 5.27 *Assume (4.33), (5.7) and (5.133). Then:*

(1) *for each* $\rho > 0$ *there exists a unique solution* $\boldsymbol{u}_\rho \in V$ *to Problem 5.26;*

(2) *there exists a unique solution* $\boldsymbol{u} \in V$ *to Problem 5.20;*

(3) *the solution* $\boldsymbol{u}_\rho$ *of Problem 5.26 converges strongly to the solution* $\boldsymbol{u}$ *of Problem 5.20, i.e.*

$$\boldsymbol{u}_\rho \to \boldsymbol{u} \quad \text{in } V \quad \text{as} \quad \rho \to 0. \tag{5.138}$$

The proof of Theorem 5.27 will be obtained in several steps and is based on the following three lemmas.

Lemma 5.28 *Assume (5.133). Then for each* $\rho > 0$ *the functional* $j_{1\rho}$ *is Gâteaux differentiable on* V_1 *and, moreover,* $\nabla j_{1\rho} = P_\rho$.

Proof Let $\rho > 0$ and note that, for all $\boldsymbol{u}$, $\boldsymbol{v} \in V$ and $t \neq 0$, we have

$$\frac{j_{1\rho}(\boldsymbol{u} + t\boldsymbol{v}) - j_{1\rho}(\boldsymbol{u})}{t}$$

$$= \int_{\Gamma_3} F_{\mathrm{b}} \frac{\sqrt{\|\boldsymbol{u}_\tau + t\boldsymbol{v}_\tau\|^2 + \rho^2} - \sqrt{\|\boldsymbol{u}_\tau\|^2 + \rho^2}}{t}\, da. \tag{5.139}$$

Also, for all $\boldsymbol{u}$, $\boldsymbol{v} \in V$ and $t \neq 0$, we can write

$$\left| F_{\mathrm{b}} \frac{\sqrt{\|\boldsymbol{u}_\tau + t\boldsymbol{v}_\tau\|^2 + \rho^2} - \sqrt{\|\boldsymbol{u}_\tau\|^2 + \rho^2}}{t} \right|$$

$$= \left| F_{\mathrm{b}} \frac{2\boldsymbol{u}_\tau \cdot \boldsymbol{v}_\tau + t\,\|\boldsymbol{v}_\tau\|^2}{\sqrt{\|\boldsymbol{u}_\tau + t\boldsymbol{v}_\tau\|^2 + \rho^2} + \sqrt{\|\boldsymbol{u}_\tau\|^2 + \rho^2}} \right| \tag{5.140}$$

a.e. on Γ_3 and, therefore,

$$F_{\mathrm{b}} \frac{\sqrt{\|\boldsymbol{u}_\tau + t\boldsymbol{v}_\tau\|^2 + \rho^2} - \sqrt{\|\boldsymbol{u}_\tau\|^2 + \rho^2}}{t} \to F_{\mathrm{b}} \frac{\boldsymbol{u}_\tau \cdot \boldsymbol{v}_\tau}{\sqrt{\|\boldsymbol{u}_\tau\|^2 + \rho^2}} \tag{5.141}$$

$$\text{as } t \to 0, \text{ a.e. on } \Gamma_3.$$

Then, using (5.140) and the inequalities

$$\boldsymbol{u}_\tau \cdot \boldsymbol{v}_\tau \le \|\boldsymbol{u}_\tau\|\,\|\boldsymbol{v}_\tau\|, \quad \boldsymbol{u}_\tau \cdot \boldsymbol{v}_\tau + t\,\|\boldsymbol{v}_\tau\|^2 \le \|\boldsymbol{u}_\tau + t\boldsymbol{v}_\tau\|\,\|\boldsymbol{v}_\tau\|,$$

an elementary calculus shows that

$$\left| F_{\mathrm{b}}\, \frac{\sqrt{\|\boldsymbol{u}_\tau + t\boldsymbol{v}_\tau\|^2 + \rho^2} - \sqrt{\|\boldsymbol{u}_\tau\|^2 + \rho^2}}{t} \right|$$

$$\le F_{\mathrm{b}}\, \frac{\|\boldsymbol{u}_\tau + t\boldsymbol{v}_\tau\|\,\|\boldsymbol{v}_\tau\| + \|\boldsymbol{u}_\tau\|\,\|\boldsymbol{v}_\tau\|}{\sqrt{\|\boldsymbol{u}_\tau + t\boldsymbol{v}_\tau\|^2 + \rho^2} + \sqrt{\|\boldsymbol{u}_\tau\|^2 + \rho^2}}$$

$$\le F_{\mathrm{b}}\, \frac{\|\boldsymbol{u}_\tau + t\boldsymbol{v}_\tau\|\,\|\boldsymbol{v}_\tau\|}{\sqrt{\|\boldsymbol{u}_\tau + t\boldsymbol{v}_\tau\|^2 + \rho^2}} + F_{\mathrm{b}}\, \frac{\|\boldsymbol{u}_\tau\|\,\|\boldsymbol{v}_\tau\|}{\sqrt{\|\boldsymbol{u}_\tau\|^2 + \rho^2}} \qquad \text{a.e. on } \Gamma_3.$$

Using the inequalities

$$\frac{\|\boldsymbol{u}_\tau + t\boldsymbol{v}_\tau\|}{\sqrt{\|\boldsymbol{u}_\tau + t\boldsymbol{v}_\tau\|^2 + \rho^2}} \le 1, \qquad \frac{\|\boldsymbol{u}_\tau\|}{\sqrt{\|\boldsymbol{u}_\tau\|^2 + \rho^2}} \le 1$$

yields

$$\left| F_{\mathrm{b}}\, \frac{\sqrt{\|\boldsymbol{u}_\tau + t\boldsymbol{v}_\tau\|^2 + \rho^2} - \sqrt{\|\boldsymbol{u}_\tau\|^2 + \rho^2}}{t} \right| \le 2F_{\mathrm{b}}\, \|\boldsymbol{v}_\tau\| \quad \text{a.e. on } \Gamma_3. \tag{5.142}$$

Next, we use (5.139), (5.141), (5.142) and Lebesgue's convergence theorem to deduce that

$$\lim_{t \to 0} \frac{j_{1\rho}(\boldsymbol{u} + t\boldsymbol{v}) - j_{1\rho}(\boldsymbol{u})}{t} = \int_{\Gamma_3} F_{\mathrm{b}}\, \frac{\boldsymbol{u}_\tau \cdot \boldsymbol{v}_\tau}{\sqrt{\|\boldsymbol{u}_\tau\|^2 + \rho^2}}\, da \tag{5.143}$$

for all $\boldsymbol{u}, \boldsymbol{v} \in V_1$. Since

$$\boldsymbol{v} \mapsto \int_{\Gamma_3} F_{\mathrm{b}}\, \frac{\boldsymbol{u}_\tau \cdot \boldsymbol{v}_\tau}{\sqrt{\|\boldsymbol{u}_\tau\|^2 + \rho^2}}\, da$$

is a linear continuous functional on V_1, from (5.143) and Definition 1.31 we deduce that $j_{1\rho}$ is Gâteaux differentiable and, moreover,

$$(\nabla j_{1\rho}(\boldsymbol{u}), \boldsymbol{v})_V = \int_{\Gamma_3} F_{\mathrm{b}}\, \frac{\boldsymbol{u}_\tau \cdot \boldsymbol{v}_\tau}{\sqrt{\|\boldsymbol{u}_\tau\|^2 + \rho^2}}\, da \quad \forall\, \boldsymbol{u}, \boldsymbol{v} \in V_1. \tag{5.144}$$

We now combine equalities (5.144), (5.135) and (5.134) to see that $\nabla j_{1\rho} = P_\rho$, which concludes the proof. ■

Lemma 5.29 *Assume (5.133). Then for each $\rho > 0$ the operator $P_\rho : V_1 \to V_1$ is monotone and Lipschitz continuous.*

Proof Let $\rho > 0$. We start with some properties of the function ϕ_ρ defined by (5.135). Let $\boldsymbol{u}, \boldsymbol{v} \in \mathbb{R}^d$. It is easy to see that

$$\boldsymbol{u}\cdot\boldsymbol{v} + \rho^2 \le \|\boldsymbol{u}\|\cdot\|\boldsymbol{v}\| + \rho^2 \le \sqrt{\|\boldsymbol{u}\|^2+\rho^2}\cdot\sqrt{\|\boldsymbol{v}\|^2+\rho^2},$$

which implies that

$$\boldsymbol{u}\cdot\boldsymbol{v} - \|\boldsymbol{u}\|^2 \le \sqrt{\|\boldsymbol{u}\|^2+\rho^2}\cdot\sqrt{\|\boldsymbol{v}\|^2+\rho^2} - (\|\boldsymbol{u}\|^2+\rho^2).$$

We divide this inequality by $\sqrt{\|\boldsymbol{u}\|^2+\rho^2} > 0$ to find that

$$\frac{\boldsymbol{u}\cdot(\boldsymbol{v}-\boldsymbol{u})}{\sqrt{\|\boldsymbol{u}\|^2+\rho^2}} \le \sqrt{\|\boldsymbol{v}\|^2+\rho^2} - \sqrt{\|\boldsymbol{u}\|^2+\rho^2}$$

and, therefore,

$$\phi_\rho(\boldsymbol{u})\cdot(\boldsymbol{v}-\boldsymbol{u}) \le \sqrt{\|\boldsymbol{v}\|^2+\rho^2} - \sqrt{\|\boldsymbol{u}\|^2+\rho^2}. \tag{5.145}$$

A similar argument shows that

$$\phi_\rho(\boldsymbol{v})\cdot(\boldsymbol{u}-\boldsymbol{v}) \le \sqrt{\|\boldsymbol{u}\|^2+\rho^2} - \sqrt{\|\boldsymbol{v}\|^2+\rho^2} \tag{5.146}$$

and, adding (5.145) and (5.146), we find that

$$(\phi_\rho(\boldsymbol{u}) - \phi_\rho(\boldsymbol{v}))\cdot(\boldsymbol{u}-\boldsymbol{v}) \ge 0. \tag{5.147}$$

Also, by the definition (5.135) of the function ϕ_ρ we deduce that

$$\phi_\rho(\boldsymbol{u}) - \phi_\rho(\boldsymbol{v}) = \frac{\boldsymbol{u}-\boldsymbol{v}}{\sqrt{\|\boldsymbol{u}\|^2+\rho^2}} + \frac{\sqrt{\|\boldsymbol{v}\|^2+\rho^2} - \sqrt{\|\boldsymbol{u}\|^2+\rho^2}}{\sqrt{\|\boldsymbol{u}\|^2+\rho^2}\cdot\sqrt{\|\boldsymbol{v}\|^2+\rho^2}}\,\boldsymbol{v}$$

$$= \frac{\boldsymbol{u}-\boldsymbol{v}}{\sqrt{\|\boldsymbol{u}\|^2+\rho^2}} + \frac{(\|\boldsymbol{v}\|+\|\boldsymbol{u}\|)(\|\boldsymbol{v}\|-\|\boldsymbol{u}\|)}{\sqrt{\|\boldsymbol{u}\|^2+\rho^2}+\sqrt{\|\boldsymbol{v}\|^2+\rho^2}}\cdot\frac{\boldsymbol{v}}{\sqrt{\|\boldsymbol{u}\|^2+\rho^2}\cdot\sqrt{\|\boldsymbol{v}\|^2+\rho^2}}.$$

Therefore,

$$\|\phi_\rho(\boldsymbol{u}) - \phi_\rho(\boldsymbol{v})\| \le \frac{\|\boldsymbol{u}-\boldsymbol{v}\|}{\sqrt{\|\boldsymbol{u}\|^2+\rho^2}}$$

$$+\frac{\|\boldsymbol{v}\|+\|\boldsymbol{u}\|}{\sqrt{\|\boldsymbol{u}\|^2+\rho^2}+\sqrt{\|\boldsymbol{v}\|^2+\rho^2}}\cdot\frac{\|\boldsymbol{v}\|}{\sqrt{\|\boldsymbol{u}\|^2+\rho^2}\cdot\sqrt{\|\boldsymbol{v}\|^2+\rho^2}}\,\|\boldsymbol{u}-\boldsymbol{v}\|,$$

and, using the inequalities

$$\frac{1}{\sqrt{\|\boldsymbol{u}\|^2+\rho^2}} \le \frac{1}{\rho}, \qquad \frac{\|\boldsymbol{v}\|+\|\boldsymbol{u}\|}{\sqrt{\|\boldsymbol{u}\|^2+\rho^2}+\sqrt{\|\boldsymbol{v}\|^2+\rho^2}} \le 1, \qquad \frac{\|\boldsymbol{v}\|}{\sqrt{\|\boldsymbol{v}\|^2+\rho^2}} \le 1,$$

we deduce that

$$\|\phi_\rho(\boldsymbol{u}) - \phi_\rho(\boldsymbol{v})\| \leq \frac{2}{\rho}\,\|\boldsymbol{u} - \boldsymbol{v}\|. \tag{5.148}$$

Consider now two arbitrary elements $\boldsymbol{u}$, $\boldsymbol{v} \in V_1$. It follows from the definition (5.134) that

$$(P_\rho \boldsymbol{u} - P_\rho \boldsymbol{v}, \boldsymbol{u} - \boldsymbol{v})_V = \int_{\Gamma_3} F_{\rm b}\,(\phi_\rho(\boldsymbol{u}_\tau) - \phi_\rho(\boldsymbol{v}_\tau)) \cdot (\boldsymbol{u}_\tau - \boldsymbol{v}_\tau)\,da$$

and, using (5.147) and (5.133), we obtain

$$(P_\rho \boldsymbol{u} - P_\rho \boldsymbol{v}, \boldsymbol{u} - \boldsymbol{v})_V \geq 0,$$

which shows that P_ρ is a monotone operator.

Finally, from (5.134), (5.133) and (5.148) it follows that

$$\begin{aligned}
|(P_\rho \boldsymbol{u} - P_\rho \boldsymbol{v}, \boldsymbol{w})_V| &\leq \int_{\Gamma_3} F_{\rm b}\,\|\phi_\rho(\boldsymbol{u}_\tau) - \phi_\rho(\boldsymbol{v}_\tau)\| \cdot \|\boldsymbol{w}_\tau\|\,da \\
&\leq \frac{2}{\rho}\int_{\Gamma_3} F_{\rm b}\,\|\boldsymbol{u}_\tau - \boldsymbol{v}_\tau\|\,\|\boldsymbol{w}_\tau\|\,da \\
&\leq \frac{2}{\rho}\,\|F_{\rm b}\|_{L^\infty(\Gamma_3)}\int_{\Gamma_3} \|\boldsymbol{u} - \boldsymbol{v}\|\,\|\boldsymbol{w}\|\,da \\
&\leq \frac{2}{\rho}\,\|F_{\rm b}\|_{L^\infty(\Gamma_3)}\|\boldsymbol{u} - \boldsymbol{v}\|_{L^2(\Gamma_3)^d}\,\|\boldsymbol{w}\|_{L^2(\Gamma_3)^d}
\end{aligned}$$

for all $\boldsymbol{w} \in V_1$. Next, we use (4.13) to deduce that

$$\|P_\rho \boldsymbol{u} - P_\rho \boldsymbol{v}\|_V \leq \frac{2}{\rho}\,c_0^2\,\|F_{\rm b}\|_{L^\infty(\Gamma_3)}\|\boldsymbol{u} - \boldsymbol{v}\|_V. \tag{5.149}$$

Therefore, P_ρ is a Lipschitz continuous operator, which concludes the proof of the lemma. ■

Lemma 5.30 *Assume (5.133). Then the family of functionals $(j_{1\rho})$ and j_1 satisfy condition (2.39) on the space V_1.*

Proof Let $\rho > 0$. We note that for all $\boldsymbol{v} \in V_1$ we have

$$\left|\sqrt{\|\boldsymbol{v}_\tau\|^2 + \rho^2} - \rho - \|\boldsymbol{v}_\tau\|\right| = \rho + \|\boldsymbol{v}_\tau\| - \sqrt{\|\boldsymbol{v}_\tau\|^2 + \rho^2} \leq \rho \quad \text{a.e. on } \Gamma_3,$$

which implies that

$$\int_{\Gamma_3} F_{\rm b}\left|\sqrt{\|\boldsymbol{v}_\tau\|^2 + \rho^2} - \rho - \|\boldsymbol{v}_\tau\|\right| da \leq \rho\int_{\Gamma_3} F_{\rm b}\,da. \tag{5.150}$$

On the other hand, by the definitions (5.136) and (5.109) of the functionals $j_{1\rho}$ and j_1 we obtain

$$|j_{1\rho}(\boldsymbol{v}) - j_1(\boldsymbol{v})| \leq \int_{\Gamma_3} F_{\rm b}\left|\sqrt{\|\boldsymbol{v}_\tau\|^2 + \rho^2} - \rho - \|\boldsymbol{v}_\tau\|\right| da. \tag{5.151}$$

We combine (5.151) and (5.150) to find that

$$|j_{1\rho}(\boldsymbol{v}) - j_1(\boldsymbol{v})| \leq \rho \int_{\Gamma_3} F_b \, da \qquad \forall \, \boldsymbol{v} \in V_1,$$

which concludes the proof. ■

We are now in position to present the proof of Theorem 5.27.

Proof Lemmas 5.28 and 5.29 combined with Proposition 1.32 imply that the family of functionals $(j_{1\rho})$ satisfies condition (2.38) on the space $X = V_1$. Moreover, Lemma 5.30 shows that condition (2.39) holds, too. Let A_1 be the operator given by (5.117) and recall that, as shown in the proof of Theorem 5.22, $A_1 : V_1 \to V_1$ is strongly monotone and Lipschitz continuous. Let $\rho > 0$. Since $P_\rho = \nabla j_{1\rho}$ it is easy to see that equation (5.137) is equivalent to the nonlinear equation

$$A_1 \boldsymbol{u}_\rho + \nabla j_{1\rho}(\boldsymbol{u}_\rho) = \boldsymbol{f}_1. \tag{5.152}$$

Theorem 5.27 is now a direct consequence of Theorem 2.12. ■

A careful examination of the proof of Theorem 5.27 shows that we need assumption (5.133) only in Lemma 5.29, in order to prove the estimate (5.149). Note that Lemmas 5.28 and 5.30 still remain valid if we replace assumption (5.133) by its weaker version (5.105).

The interest in Theorem 5.27 is twofold; first, it provides the existence and the uniqueness of the solution to the variational inequality (5.115) and, therefore, in the case when $F_b \in L^\infty(\Gamma_3)$, it recovers the existence and uniqueness result obtained in Theorem 5.22; second, it shows that the solution of (5.115) represents the strong limit of the sequence of solutions $\boldsymbol{u}_\rho$ of the nonlinear equation (5.137), as $\rho \to 0$.

The convergence result (5.138) is important from the mechanical point of view, since it shows that the weak solution of the elastic contact problem with Coulomb's law may be approached as closely as one wishes by the solution of an elastic contact problem with regularized friction, with a sufficiently small regularization parameter. It is also important from a numerical point of view since it shows that solving the variational inequality (5.115) leads to solving the nonlinear equation (5.137) for a "small" parameter ρ and, to this end, various numerical methods can be used, see e.g. [48].

5.2.5 Dual variational formulation

In this section we study the dual formulation of the elastic problem with given normal stress and Coulomb's law of dry friction, i.e. Problem 5.17. A similar study can be carried out for Problem 5.18 but, in order to avoid repetitions, we skip it. Below we assume that (4.33), (5.7), (5.103) and (5.104) hold and we use the notation (4.9), (5.106) and (5.107).

Recall that the primal variational formulation of this problem in terms of displacements is Problem 5.19 and its unique solvability was given by Theorem 5.21. Our aim in this section is to study a variational formulation of Problem 5.17 in which the unknown is the stress tensor. To this end we consider the set of admissible stress fields Σ defined by

$$\Sigma = \{\, \boldsymbol{\tau} \in Q \;:\; (\boldsymbol{\tau}, \boldsymbol{\varepsilon}(\boldsymbol{v}))_Q + j(\boldsymbol{v}) \geq (\boldsymbol{f}, \boldsymbol{v})_V \quad \forall\, \boldsymbol{v} \in V \,\} \tag{5.153}$$

and, as usual, we assume that $(\boldsymbol{u}, \boldsymbol{\sigma})$ represents a regular solution to (5.91)–(5.96). Then, using the arguments presented on page 146 it follows that (5.113) holds, that is

$$(\boldsymbol{\sigma}, \boldsymbol{\varepsilon}(\boldsymbol{v}) - \boldsymbol{\varepsilon}(\boldsymbol{u}))_Q + j(\boldsymbol{v}) - j(\boldsymbol{u}) \geq (\boldsymbol{f}, \boldsymbol{v} - \boldsymbol{u})_V \quad \forall\, \boldsymbol{v} \in V. \tag{5.154}$$

We take $\boldsymbol{v} = 2\boldsymbol{u}$ and $\boldsymbol{v} = \boldsymbol{0}_V$ in (5.154) and use the equalities $j(2\boldsymbol{u}) = 2j(\boldsymbol{u})$, $j(\boldsymbol{0}_V) = 0$ to obtain

$$(\boldsymbol{\sigma}, \boldsymbol{\varepsilon}(\boldsymbol{u}))_Q + j(\boldsymbol{u}) = (\boldsymbol{f}, \boldsymbol{u})_V. \tag{5.155}$$

Then we combine (5.154) and (5.155) to see that

$$(\boldsymbol{\sigma}, \boldsymbol{\varepsilon}(\boldsymbol{v}))_Q + j(\boldsymbol{v}) \geq (\boldsymbol{f}, \boldsymbol{v})_V \qquad \forall\, \boldsymbol{v} \in V,$$

which implies that

$$\boldsymbol{\sigma} \in \Sigma. \tag{5.156}$$

Also, equality (5.155) and definition (5.153) imply that

$$(\boldsymbol{\tau} - \boldsymbol{\sigma}, \boldsymbol{\varepsilon}(\boldsymbol{u}))_Q \geq 0 \qquad \forall\, \boldsymbol{\tau} \in \Sigma, \tag{5.157}$$

and assumption (4.33) on the operator $\mathcal{F}$ combined with the constitutive law (5.91) implies that

$$\boldsymbol{\varepsilon}(\boldsymbol{u}) = \mathcal{F}^{-1}\boldsymbol{\sigma}, \tag{5.158}$$

where $\mathcal{F}^{-1}$ denotes the inverse of $\mathcal{F}$.

We now use (5.156)–(5.158) to derive the following variational formulation of Problem 5.17 in terms of the stress field.

Problem 5.31 *Find a stress field $\boldsymbol{\sigma}$ such that*

$$\boldsymbol{\sigma} \in \Sigma, \qquad (\mathcal{F}^{-1}\boldsymbol{\sigma}, \boldsymbol{\tau} - \boldsymbol{\sigma})_Q \geq 0 \quad \forall\, \boldsymbol{\tau} \in \Sigma. \tag{5.159}$$

Problem 5.31 is called the *dual variational formulation* of the frictional contact problem 5.17. Note that this formulation is similar to the dual formulation of the Signorini frictionless contact problem, i.e. Problem 5.11 on page 133. The only difference arises in the choice of the convex Σ which is different from the convex Σ_0 used in the study of Problem 5.11.

We have the following existence and uniqueness result.

Theorem 5.32 *Assume (4.33), (5.7), (5.103) and (5.104). Then there exists a unique solution to Problem 5.31.*

Proof We take $\boldsymbol{\tau} = \boldsymbol{\varepsilon}(\boldsymbol{f})$ and use (4.11) to obtain

$$(\boldsymbol{\tau}, \boldsymbol{\varepsilon}(\boldsymbol{v}))_Q = (\boldsymbol{f}, \boldsymbol{v})_V \qquad \forall\, \boldsymbol{v} \in V.$$

Therefore, since (5.103) and (5.104) imply that $j(\boldsymbol{v}) \geq 0$ for all $\boldsymbol{v} \in V$, the previous equality yields

$$(\boldsymbol{\tau}, \boldsymbol{\varepsilon}(\boldsymbol{v}))_Q + j(\boldsymbol{v}) \geq (\boldsymbol{f}, \boldsymbol{v})_V \qquad \forall\, \boldsymbol{v} \in V$$

which shows that $\boldsymbol{\tau} \in \Sigma$, i.e. Σ is not empty. On the other hand, it is easy to see that Σ is a closed convex subset of Q and recall that $\mathcal{F}^{-1} : Q \to Q$ is a strongly monotone Lipschitz continuous operator. Theorem 5.32 is now a direct consequence of Theorem 2.1. ■

A brief comparison between the primal and dual variational formulation of Problem 5.17 shows that, despite the fact that the primal variational formulation is in the form of an elliptic variational inequality of the second kind for the displacement field, (5.114), the dual variational formulation is in the form of an elliptic variational inequality of the first kind for the stress field, (5.159). This allows us to obtain the unique solvability of Problem 5.31 by using the abstract result provided by Theorem 2.1, as shown above. In contrast, the proof of the solvability of the primal variational formulation was based on the abstract result provided by Theorem 2.8, as shown on page 147.

We present now a technical result which will be used later in this section.

Lemma 5.33 *Assume (5.103), (5.104) and let j be the functional given by (5.107). Then, for each $\boldsymbol{u} \in V$ there exists an element $\boldsymbol{d}(\boldsymbol{u}) \in V$ such that*

$$j(\boldsymbol{v}) - j(\boldsymbol{u}) \geq (\boldsymbol{d}(\boldsymbol{u}), \boldsymbol{v} - \boldsymbol{u})_V \qquad \forall\, \boldsymbol{v} \in V. \tag{5.160}$$

Proof Let $\boldsymbol{u} \in V$ and denote by $\widetilde{\boldsymbol{d}}(\boldsymbol{u})$ the function defined by

$$\widetilde{\boldsymbol{d}}(\boldsymbol{u}) = \begin{cases} \dfrac{\boldsymbol{u}_\tau}{\|\boldsymbol{u}_\tau\|} & \text{if} \quad \boldsymbol{u}_\tau \neq \boldsymbol{0}, \\ \boldsymbol{0} & \text{if} \quad \boldsymbol{u}_\tau = \boldsymbol{0}. \end{cases} \qquad \text{a.e. on } \Gamma_3. \tag{5.161}$$

It is easy to see that $\widetilde{\boldsymbol{d}}(\boldsymbol{u}) \in L^\infty(\Gamma_3)^d$ and, moreover, for all $\boldsymbol{v} \in V$ we have

$$\widetilde{\boldsymbol{d}}(\boldsymbol{u}) \cdot (\boldsymbol{v}_\tau - \boldsymbol{u}_\tau) \leq \|\boldsymbol{v}_\tau\| - \|\boldsymbol{u}_\tau\| \quad \text{a.e. on } \Gamma_3.$$

Therefore, since $\mu\, F \geq 0$ a.e. on Γ_3, we deduce that

$$\int_{\Gamma_3} \mu\, F\, \|\boldsymbol{v}_\tau\|\, \mathrm{da} - \int_{\Gamma_3} \mu\, F\, \|\boldsymbol{u}_\tau\|\, \mathrm{da} \geq \int_{\Gamma_3} \mu\, F\, \widetilde{\boldsymbol{d}}(\boldsymbol{u}) \cdot (\boldsymbol{v}_\tau - \boldsymbol{u}_\tau)\, \mathrm{da}. \tag{5.162}$$

Next, we apply the Riesz representation theorem to obtain that there exists a unique element $\boldsymbol{d}(\boldsymbol{u}) \in V$ such that

$$\int_{\Gamma_3} \mu F \widetilde{\boldsymbol{d}}(\boldsymbol{u}) \cdot \boldsymbol{v}_\tau \, da = (\boldsymbol{d}(\boldsymbol{u}), \boldsymbol{v})_V \quad \forall \boldsymbol{v} \in V. \tag{5.163}$$

Then, combining (5.107), (5.162) and (5.163) we obtain the inequality (5.160), which concludes the proof. ■

Lemma 5.33 shows that the functional j is subdifferentiable on V. This result can be obtained directly from a general theorem on the subdifferentiability of convex lower semicontinuous functions. Nevertheless, for the convenience of the reader, we chose to provide here a direct proof of this result.

The link between the solutions $\boldsymbol{\sigma}$ and $\boldsymbol{u}$ of the variational problems 5.19 and 5.31 is provided by the following result.

Theorem 5.34 *Assume (4.33), (5.7), (5.103) and (5.104).*

(1) *If $\boldsymbol{u}$ is the solution to Problem 5.19 obtained in Theorem 5.21 and $\boldsymbol{\sigma}$ denotes the function given by*

$$\boldsymbol{\sigma} = \mathcal{F}\boldsymbol{\varepsilon}(\boldsymbol{u}), \tag{5.164}$$

then $\boldsymbol{\sigma}$ is a solution to Problem 5.31.

(2) *Conversely, if $\boldsymbol{\sigma}$ is the solution to Problem 5.31 obtained in Theorem 5.32 then there exists a unique element $\boldsymbol{u} \in V$ such that (5.164) holds and, moreover, $\boldsymbol{u}$ is a solution to Problem 5.19.*

Proof (1) Assume that $\boldsymbol{u}$ satisfies (5.114) and let $\boldsymbol{\sigma}$ be given by (5.164). Then, using the properties (4.33) of the operator $\mathcal{F}$, it follows that $\boldsymbol{\sigma} \in Q$. Moreover, by (5.114) we have

$$(\boldsymbol{\sigma}, \boldsymbol{\varepsilon}(\boldsymbol{v}) - \boldsymbol{\varepsilon}(\boldsymbol{u}))_Q + j(\boldsymbol{v}) - j(\boldsymbol{u}) \geq (\boldsymbol{f}, \boldsymbol{v} - \boldsymbol{u})_V \qquad \forall \boldsymbol{v} \in V. \tag{5.165}$$

We test this inequality with $\boldsymbol{v} = 2\boldsymbol{u}$ and $\boldsymbol{v} = \boldsymbol{0}_V$. Since $j(2\boldsymbol{u}) = 2j(\boldsymbol{u})$ and $j(\boldsymbol{0}_V) = 0$ we obtain

$$(\boldsymbol{\sigma}, \boldsymbol{\varepsilon}(\boldsymbol{u}))_Q + j(\boldsymbol{u}) = (\boldsymbol{f}, \boldsymbol{u})_V. \tag{5.166}$$

We combine now (5.165) and (5.166) to see that

$$(\boldsymbol{\sigma}, \boldsymbol{\varepsilon}(\boldsymbol{v}))_Q + j(\boldsymbol{v}) \geq (\boldsymbol{f}, \boldsymbol{v})_V \qquad \forall \boldsymbol{v} \in V,$$

which shows that

$$\boldsymbol{\sigma} \in \Sigma. \tag{5.167}$$

Next, using (5.164) and (5.166) we have

$$(\mathcal{F}^{-1}\boldsymbol{\sigma},\boldsymbol{\tau}-\boldsymbol{\sigma})_Q=(\boldsymbol{\tau}-\boldsymbol{\sigma},\boldsymbol{\varepsilon}(\boldsymbol{u}))_Q=(\boldsymbol{\tau},\boldsymbol{\varepsilon}(\boldsymbol{u}))_Q+j(\boldsymbol{u})-(\boldsymbol{f},\boldsymbol{u})_V\quad\forall\,\boldsymbol{\tau}\in Q$$

and, by the definition (5.153) of the set Σ we deduce that

$$(\mathcal{F}^{-1}\boldsymbol{\sigma},\boldsymbol{\tau}-\boldsymbol{\sigma})_Q\geq 0\qquad\forall\,\boldsymbol{\tau}\in\Sigma. \tag{5.168}$$

We combine now (5.167) and (5.168) to conclude that $\boldsymbol{\sigma}$ is a solution to Problem 5.31.

(2) Assume that $\boldsymbol{\sigma}$ satisfies (5.159) and consider an element $\boldsymbol{z}\in Q$ such that (5.55) holds. Then, using (5.153) we deduce that $\boldsymbol{\sigma}\pm\boldsymbol{z}\in\Sigma$ and, testing in (5.159) with $\boldsymbol{\tau}=\boldsymbol{\sigma}\pm\boldsymbol{z}$, we obtain that (5.56) holds. The existence of a unique element $\boldsymbol{u}\in V$ which satisfies (5.164) follows now by the arguments used on page 135, in the proof of Theorem 5.13.

Next, we denote by $\boldsymbol{\tau}(\boldsymbol{u})\in Q$ the element defined by

$$\boldsymbol{\tau}(\boldsymbol{u})=\boldsymbol{\varepsilon}(\boldsymbol{f}-\boldsymbol{d}(\boldsymbol{u})), \tag{5.169}$$

where $\boldsymbol{d}(\boldsymbol{u})\in V$ is the element defined in Lemma 5.33. We use (5.169), (5.160) and the definition (4.11) of the inner product on V to see that

$$(\boldsymbol{\tau}(\boldsymbol{u}),\boldsymbol{\varepsilon}(\boldsymbol{v})-\boldsymbol{\varepsilon}(\boldsymbol{u}))_Q+j(\boldsymbol{v})-j(\boldsymbol{u})\geq(\boldsymbol{f},\boldsymbol{v}-\boldsymbol{u})_V\quad\forall\,\boldsymbol{v}\in V. \tag{5.170}$$

We replace $\boldsymbol{v}=2\boldsymbol{u}$ and $\boldsymbol{v}=\boldsymbol{0}_V$ in this inequality to find

$$(\boldsymbol{\tau}(\boldsymbol{u}),\boldsymbol{\varepsilon}(\boldsymbol{u}))_Q+j(\boldsymbol{u})=(\boldsymbol{f},\boldsymbol{u})_V. \tag{5.171}$$

Using now (5.170) and (5.171) it follows that

$$(\boldsymbol{\tau}(\boldsymbol{u}),\boldsymbol{\varepsilon}(\boldsymbol{v}))_Q+j(\boldsymbol{v})\geq(\boldsymbol{f},\boldsymbol{v})_V\quad\forall\,\boldsymbol{v}\in V$$

and by (5.153) we deduce that $\boldsymbol{\tau}(\boldsymbol{u})\in\Sigma$. So, we can test with $\boldsymbol{\tau}=\boldsymbol{\tau}(\boldsymbol{u})$ in (5.159) and, using (5.164), we obtain

$$(\boldsymbol{\tau}(\boldsymbol{u}),\boldsymbol{\varepsilon}(\boldsymbol{u}))_Q\geq(\boldsymbol{\sigma},\boldsymbol{\varepsilon}(\boldsymbol{u}))_Q. \tag{5.172}$$

We now combine (5.171) and (5.172) to see that

$$(\boldsymbol{f},\boldsymbol{u})_V\geq(\boldsymbol{\sigma},\boldsymbol{\varepsilon}(\boldsymbol{u}))_Q+j(\boldsymbol{u})$$

and we note that the converse inequality follows from (5.153), since $\boldsymbol{\sigma}\in\Sigma$ and $\boldsymbol{u}\in V$. Therefore, we conclude that

$$(\boldsymbol{\sigma},\boldsymbol{\varepsilon}(\boldsymbol{u}))_Q+j(\boldsymbol{u})=(\boldsymbol{f},\boldsymbol{u})_V. \tag{5.173}$$

Using now (5.167), the definition (5.153) of the set Σ and (5.173), we deduce that

$$(\boldsymbol{\sigma},\boldsymbol{\varepsilon}(\boldsymbol{v})-\boldsymbol{\varepsilon}(\boldsymbol{u}))_Q+j(\boldsymbol{v})-j(\boldsymbol{u})\geq(\boldsymbol{f},\boldsymbol{v}-\boldsymbol{u})_V\quad\forall\,\boldsymbol{v}\in V. \tag{5.174}$$

Therefore, by (5.164) and (5.174) we find that $\boldsymbol{u}$ is a solution to Problem 5.19, which concludes the proof. ∎

The mechanical interpretation of the results in Theorem 5.34 combined with the uniqueness part of Theorems 5.21 and 5.32 is the following.

(1) If the displacement field $\boldsymbol{u}$ is the solution of the primal variational formulation of the frictional contact problem (5.91)–(5.96) (i.e. Problem 5.19), then the stress field $\boldsymbol{\sigma}$ associated with $\boldsymbol{u}$ by the elastic constitutive law $\boldsymbol{\sigma} = \mathcal{F}\boldsymbol{\varepsilon}(\boldsymbol{u})$ is the solution of the dual variational formulation of the frictional contact problem (5.91)–(5.96) (i.e. Problem 5.31).

(2) If the stress field $\boldsymbol{\sigma}$ is the solution of the dual variational formulation of the frictional contact problem (5.91)–(5.96) (i.e. Problem 5.31), then there exists a unique displacement field $\boldsymbol{u} \in V$ associated with $\boldsymbol{\sigma}$ by the elastic constitutive law $\boldsymbol{\sigma} = \mathcal{F}\boldsymbol{\varepsilon}(\boldsymbol{u})$ and, moreover, $\boldsymbol{u}$ is the solution of the primal variational formulation of the frictional contact problem (5.91)–(5.96) (i.e. Problem 5.19).

Under the assumptions of Theorem 5.34 it also results that, if the displacement field $\boldsymbol{u}$ is the solution of the primal variational formulation and the stress field $\boldsymbol{\sigma}$ is the solution of the dual variational formulation, then $\boldsymbol{u}$ and $\boldsymbol{\sigma}$ are connected by the elastic constitutive law $\boldsymbol{\sigma} = \mathcal{F}\boldsymbol{\varepsilon}(\boldsymbol{u})$. For this reason, we refer to the pair $(\boldsymbol{u}, \boldsymbol{\sigma})$ given by Theorems 5.21 and 5.32 as a *weak solution* for the frictional contact problem (5.91)–(5.96) and we conclude that this problem has a unique weak solution.

We end this subsection with the remark that the functional j given by (5.107) is a continuous seminorm on the space V. This property was used several times in the proof of Theorem 5.34 and represents one of the main ingredients in the study of the dual variational formulation of the contact problem (5.91)–(5.96).

5.2.6 Minimization

We consider again the contact problem with given normal stress and Coulomb friction, i.e. problem (5.91)–(5.96) and we assume that (5.73), (5.7), (5.103) and (5.104) hold. Also, we consider the *energy function* $J : V \to \mathbb{R}$ and the *complementary energy function* $J^* : \Sigma \to \mathbb{R}$ defined by

$$J(\boldsymbol{v}) = \frac{1}{2}\, a(\boldsymbol{v}, \boldsymbol{v}) + j(\boldsymbol{v}) - (\boldsymbol{f}, \boldsymbol{v})_V \qquad \forall\, \boldsymbol{v} \in V,$$

$$J^*(\boldsymbol{\tau}) = \frac{1}{2}\, a^*(\boldsymbol{\tau}, \boldsymbol{\tau}) \qquad \forall\, \boldsymbol{\tau} \in \Sigma.$$

Recall that, here and below, we use the notation (5.74), (5.75) for the bilinear forms a and a^*, (5.153) for the set of admissible stress fields and, again, j and $\boldsymbol{f}$ are given by (5.107) and (5.106), respectively.

We have the following results.

Theorem 5.35 *Assume (5.73), (5.7), (5.103) and (5.104). Then:*

(1) *the element $\boldsymbol{u}$ is a solution to Problem 5.19 iff*

$$\boldsymbol{u} \in V, \qquad J(\boldsymbol{v}) \geq J(\boldsymbol{u}) \quad \forall \boldsymbol{v} \in V; \tag{5.175}$$

(2) *there exists a unique solution $\boldsymbol{u}$ of the minimization problem (5.175) (and, therefore, there exists a unique solution $\boldsymbol{u}$ to Problem 5.19);*

(3) *the element $\boldsymbol{\sigma}$ is a solution to Problem 5.31 iff*

$$\boldsymbol{\sigma} \in \Sigma, \qquad J^*(\boldsymbol{\tau}) \geq J^*(\boldsymbol{\sigma}) \quad \forall \boldsymbol{\tau} \in \Sigma; \tag{5.176}$$

(4) *there exists a unique solution $\boldsymbol{\sigma}$ of the minimization problem (5.176) (and, therefore, there exists a unique solution $\boldsymbol{\sigma}$ to Problem 5.31).*

Proof The proof of Theorem 5.35 is obtained by using arguments similar to those used in the proof of Theorem 5.14. The main ingredients are the equivalence, existence and uniqueness results provided by Theorem 1.37. ■

The mechanical interpretation of Theorem 5.35 is the following. Assume that the elastic operator $\mathcal{F}$ satisfies (5.73). Then:

(1) a displacement field $\boldsymbol{u}$ is a solution to the primal variational formulation of the frictional contact problem (5.91)–(5.96) if and only if $\boldsymbol{u}$ is a minimizer of the energy function J on the space V;

(2) the energy function J has a unique minimizer on V and, therefore, there exists a unique solution to the primal variational formulation of the frictional contact problem (5.91)–(5.96);

(3) a stress field $\boldsymbol{\sigma}$ is a solution to the dual variational formulation of the frictional contact problem (5.91)–(5.96) if and only if $\boldsymbol{\sigma}$ is a minimizer of the complementary energy function J^* on the set of admissible stress fields Σ;

(4) the energy function J^* has a unique minimizer on the set of admissible stress fields and, therefore, there exists a unique solution to the dual variational formulation of the frictional contact problem (5.91)–(5.96).

5.3 A frictional contact problem with normal compliance

For the problem studied in this section we assume that the behavior of the material is elastic and the contact is modelled with the normal compliance condition, associated with Coulomb's law of dry friction. This problem leads to an elliptic quasivariational inequality for the displacement field.

5.3.1 Problem statement

The classical model of the process is the following.

Problem 5.36 *Find a displacement field* $\boldsymbol{u} : \Omega \to \mathbb{R}^d$ *and a stress field* $\boldsymbol{\sigma} : \Omega \to \mathbb{S}^d$ *such that*

$$\boldsymbol{\sigma} = \mathcal{F}\boldsymbol{\varepsilon}(\boldsymbol{u}) \qquad \text{in } \Omega, \tag{5.177}$$

$$\operatorname{Div} \boldsymbol{\sigma} + \boldsymbol{f}_0 = \boldsymbol{0} \qquad \text{in } \Omega, \tag{5.178}$$

$$\boldsymbol{u} = \boldsymbol{0} \qquad \text{on } \Gamma_1, \tag{5.179}$$

$$\boldsymbol{\sigma}\boldsymbol{\nu} = \boldsymbol{f}_2 \qquad \text{on } \Gamma_2, \tag{5.180}$$

$$-\sigma_\nu = p_\nu(u_\nu - g_a) \qquad \text{on } \Gamma_3, \tag{5.181}$$

$$\left.\begin{aligned} &\|\boldsymbol{\sigma}_\tau\| \le p_\tau(u_\nu - g_a), \\ &\boldsymbol{\sigma}_\tau = -\,p_\tau(u_\nu - g_a)\,\frac{\boldsymbol{u}_\tau}{\|\boldsymbol{u}_\tau\|} \quad \text{if} \quad \boldsymbol{u}_\tau \ne \boldsymbol{0} \end{aligned}\right\} \qquad \text{on } \Gamma_3. \tag{5.182}$$

The equations and boundary conditions involved in Problem 5.36 have the same meaning as those involved in the elastic contact problems studied in the previous sections of this chapter. The difference arises from the fact that here we use the normal compliance condition (5.181) associated with the friction law (5.182), see (4.83), (4.86) and (4.94). Recall that here and everywhere in this section g_a represents the gap function.

In the study of Problem 5.36 we assume that the elasticity operator satisfies (4.33), the densities of body forces and surface tractions have the regularity (5.7) and the gap function satisfies condition (5.8). Also, the normal compliance functions p_e $(e = \nu, \tau)$ satisfy

$$\left.\begin{aligned} &\text{(a) } p_e : \Gamma_3 \times \mathbb{R} \to \mathbb{R}_+; \\ &\text{(b) there exists } L_e > 0 \text{ such that} \\ &\qquad |p_e(\boldsymbol{x}, r_1) - p_e(\boldsymbol{x}, r_2)| \le L_e\,|r_1 - r_2| \\ &\qquad\qquad \forall\, r_1,\, r_2 \in \mathbb{R},\ \text{a.e. } \boldsymbol{x} \in \Gamma_3; \\ &\text{(c) the mapping } \boldsymbol{x} \mapsto p_e(\boldsymbol{x}, r) \text{ is measurable on } \Gamma_3, \\ &\qquad \text{for any } r \in \mathbb{R}; \\ &\text{(d) } p_e(\boldsymbol{x}, r) = 0 \text{ for all } r \le 0,\ \text{a.e. } \boldsymbol{x} \in \Gamma_3. \end{aligned}\right\} \tag{5.183}$$

We note that assumptions (5.183) on the functions p_ν and p_τ are less restrictive than assumption (5.27) used for the normal compliance function p_ν in Section 5.1. Indeed, in contrast with (5.27), no monotonicity assumption on p_e is required in (5.183) and, moreover, $p_e(\boldsymbol{x},\cdot)$ could vanish for $r > 0$, for $e = \nu, \tau$. Assumption (5.183)(b) essentially requires the functions to grow at most linearly at infinity. It is easily seen that the functions (4.79) and (4.80) satisfy this condition. We also observe that if the functions p_ν and p_τ are related by equalities (4.95) or (4.96) with $\mu > 0$ and p_ν satisfies condition (5.183)(b), then p_τ also satisfies condition (5.183)(b) with $L_\tau = \mu L_\nu$. As in (5.27), condition (5.183)(c) is introduced here for mathematical reasons since, combined with condition (5.183)(b) and (d), it guarantees that the mapping $\boldsymbol{x} \mapsto p_e(\boldsymbol{x}, \zeta(\boldsymbol{x}))$ belongs to $L^2(\Gamma_3)$ whenever $\zeta \in L^2(\Gamma_3)$. Moreover, note that $\zeta \mapsto p_e(\boldsymbol{x}, \zeta(\boldsymbol{x}))$ is a Lipschitz continuous operator from $L^2(\Gamma_3)$ to $L^2(\Gamma_3)$, for $e = \nu, \tau$. And, finally, condition (5.183)(d) shows that when there is separation between the body and the foundation then the normal and tangential parts of the stress tensor vanish on the boundary.

We turn to the variational formulation of Problem 5.36. To this end we use the space V, (4.9), and the element $\boldsymbol{f} \in V$ given by (5.9). Also, everywhere in this section we consider the functional $j : V \times V \to \mathbb{R}$ defined by

$$\begin{aligned} j(\boldsymbol{u}, \boldsymbol{v}) = & \int_{\Gamma_3} p_\nu(u_\nu - g_{\rm a})|v_\nu| \, da \\ & + \int_{\Gamma_3} p_\tau(u_\nu - g_{\rm a})\|\boldsymbol{v}_\tau\| \, da \quad \forall\, \boldsymbol{u}, \boldsymbol{v} \in V. \end{aligned} \tag{5.184}$$

Note that, by assumptions (5.8) and (5.183), the integrals in (5.184) are well defined.

Assume in what follows that $(\boldsymbol{u}, \boldsymbol{\sigma})$ are regular functions which satisfy (5.177)–(5.182) and let $\boldsymbol{v} \in V$. Then, using (5.181) it is easy to see that

$$\sigma_\nu(v_\nu - u_\nu) = -p_\nu(u_\nu - g_{\rm a})v_\nu + p_\nu(u_\nu - g_{\rm a})u_\nu \quad \text{a.e. on } \Gamma_3$$

and, using (5.183)(a), (d) it follows that

$$\begin{aligned} -p_\nu(u_\nu - g_{\rm a})v_\nu &\geq -p_\nu(u_\nu - g_{\rm a})\,|v_\nu|, \\ p_\nu(u_\nu - g_{\rm a})u_\nu &= p_\nu(u_\nu - g_{\rm a})\,|u_\nu| \end{aligned}$$

a.e. on Γ_3. We conclude from the above that

$$\sigma_\nu(v_\nu - u_\nu) \geq -p_\nu(u_\nu - g_{\rm a})\,|v_\nu| + p_\nu(u_\nu - g_{\rm a})\,|u_\nu| \quad \text{a.e. on } \Gamma_3,$$

which implies that

$$\int_{\Gamma_3} \sigma_\nu(v_\nu - u_\nu)\, da$$
$$\geq -\int_{\Gamma_3} p_\nu(u_\nu - g_a)\,|v_\nu|\, da + \int_{\Gamma_3} p_\nu(u_\nu - g_a)\,|u_\nu|\, da. \quad (5.185)$$

On the other hand, (5.182) and arguments similar to those used to obtain (5.112) yield

$$\int_{\Gamma_3} \boldsymbol{\sigma}_\tau \cdot (\boldsymbol{v}_\tau - \boldsymbol{u}_\tau)\, da$$
$$\geq -\int_{\Gamma_3} p_\tau(u_\nu - g_a)\,\|\boldsymbol{v}_\tau\|\, da + \int_{\Gamma_3} p_\tau(u_\nu - g_a)\,\|\boldsymbol{u}_\tau\|\, da. \quad (5.186)$$

We now use standard arguments based on Green's formula (4.22), inequalities (5.185), (5.186), definitions (5.184), (5.9) and the constitutive law (5.177) to obtain the following variational formulation of the frictional contact problem 5.36.

Problem 5.37 *Find a displacement field $\boldsymbol{u} \in V$ such that*

$$(\mathcal{F}\boldsymbol{\varepsilon}(\boldsymbol{u}), \boldsymbol{\varepsilon}(\boldsymbol{v}) - \boldsymbol{\varepsilon}(\boldsymbol{u}))_Q + j(\boldsymbol{u}, \boldsymbol{v}) - j(\boldsymbol{u}, \boldsymbol{u}) \geq (\boldsymbol{f}, \boldsymbol{v} - \boldsymbol{u})_V \quad \forall \boldsymbol{v} \in V. \quad (5.187)$$

We remark that, unlike the variational inequalities (5.114) and (5.115) studied in Section 5.2, the functional j depends now on the solution and, therefore, (5.187) represents an elliptic quasivariational inequality for the displacement field. Therefore, its solvability is based on the abstract results in Section 2.3, obtained by using fixed point arguments.

5.3.2 The Banach fixed point argument

We state and prove the following existence and uniqueness result.

Theorem 5.38 *Assume (4.33), (5.7), (5.8) and (5.183). Then there exists a constant L_0 which depends only on Ω, Γ_1, Γ_3 and $\mathcal{F}$ such that Problem 5.37 has a unique solution, if $L_\nu + L_\tau < L_0$.*

Proof We use Theorem 2.19 on page 49 with the choice $K = X = V$. To this end we consider the operator $A : V \to V$ defined by (5.16) and we recall that, as shown in the proof of Theorem 5.3, A is a strongly monotone Lipschitz continuous operator on the space V. Also, we note that for all $\boldsymbol{\eta} \in V$ the functional $j(\boldsymbol{\eta}, \cdot) : V \to \mathbb{R}$ is a continuous seminorm on V and, therefore, it satisfies (2.59)(a). Let $\boldsymbol{\eta}_1$, $\boldsymbol{\eta}_2$, $\boldsymbol{v}_1$, $\boldsymbol{v}_2 \in V$; using (5.184) and

(5.183)(b) we find that

$$
\begin{aligned}
&j(\boldsymbol{\eta}_1, \boldsymbol{v}_2) - j(\boldsymbol{\eta}_1, \boldsymbol{v}_1) + j(\boldsymbol{\eta}_2, \boldsymbol{v}_1) - j(\boldsymbol{\eta}_2, \boldsymbol{v}_2) \\
&\quad = \int_{\Gamma_3} \big(p_\nu(\eta_{1\nu} - g_a) - p_\nu(\eta_{2\nu} - g_a)\big)\big(|v_{2\nu}| - |v_{1\nu}|\big)\, da \\
&\quad + \int_{\Gamma_3} \big(p_\tau(\eta_{1\nu} - g_a) - p_\tau(\eta_{2\nu} - g_a)\big)\big(\|\boldsymbol{v}_{2\tau}\| - \|\boldsymbol{v}_{1\tau}\|\big)\, da \\
&\quad \le \int_{\Gamma_3} L_\nu |\eta_{1\nu} - \eta_{2\nu}|\, \big||v_{1\nu}| - |v_{2\nu}|\big|\, da \\
&\quad + \int_{\Gamma_3} L_\tau |\eta_{1\nu} - \eta_{2\nu}|\, \big|\|\boldsymbol{v}_{1\tau}\| - \|\boldsymbol{v}_{2\tau}\|\big|\, da.
\end{aligned}
$$

Since

$$
\begin{aligned}
&|\eta_{1\nu} - \eta_{2\nu}| \le \|\boldsymbol{\eta}_1 - \boldsymbol{\eta}_2\|, \quad \big|\,|v_{1\nu}| - |v_{2\nu}|\,\big| \le |v_{1\nu} - v_{2\nu}| \le \|\boldsymbol{v}_1 - \boldsymbol{v}_2\|, \\
&\big|\,\|\boldsymbol{v}_{1\tau}\| - \|\boldsymbol{v}_{2\tau}\|\,\big| \le \|\boldsymbol{v}_{1\tau} - \boldsymbol{v}_{2\tau}\| \le \|\boldsymbol{v}_1 - \boldsymbol{v}_2\| \quad \text{a.e. on } \Gamma_3,
\end{aligned}
$$

it follows from the above that

$$
\begin{aligned}
&j(\boldsymbol{\eta}_1, \boldsymbol{v}_2) - j(\boldsymbol{\eta}_1, \boldsymbol{v}_1) + j(\boldsymbol{\eta}_2, \boldsymbol{v}_1) - j(\boldsymbol{\eta}_2, \boldsymbol{v}_2) \\
&\qquad \le \int_{\Gamma_3} (L_\nu + L_\tau)\|\boldsymbol{\eta}_1 - \boldsymbol{\eta}_2\|\, \|\boldsymbol{v}_1 - \boldsymbol{v}_2\|\, da
\end{aligned}
$$

and, therefore,

$$
\begin{aligned}
&j(\boldsymbol{\eta}_1, \boldsymbol{v}_2) - j(\boldsymbol{\eta}_1, \boldsymbol{v}_1) + j(\boldsymbol{\eta}_2, \boldsymbol{v}_1) - j(\boldsymbol{\eta}_2, \boldsymbol{v}_2) \\
&\qquad \le (L_\nu + L_\tau)\|\boldsymbol{\eta}_1 - \boldsymbol{\eta}_2\|_{L^2(\Gamma_3)}\, \|\boldsymbol{v}_1 - \boldsymbol{v}_2\|_{L^2(\Gamma_3)}. \qquad (5.188)
\end{aligned}
$$

We now combine (5.188) and (4.13) to see that

$$
\begin{aligned}
&j(\boldsymbol{\eta}_1, \boldsymbol{v}_2) - j(\boldsymbol{\eta}_1, \boldsymbol{v}_1) + j(\boldsymbol{\eta}_2, \boldsymbol{v}_1) - j(\boldsymbol{\eta}_2, \boldsymbol{v}_2) \\
&\qquad \le c_0^2\,(L_\nu + L_\tau)\,\|\boldsymbol{\eta}_1 - \boldsymbol{\eta}_2\|_V\, \|\boldsymbol{v}_1 - \boldsymbol{v}_2\|_V. \qquad (5.189)
\end{aligned}
$$

Let

$$
L_0 = \frac{m_{\mathcal{F}}}{c_0^2}\,. \qquad (5.190)
$$

Clearly L_0 depends only on Ω, Γ_1, Γ_3 and $\mathcal{F}$. It follows from (5.189) that j satisfies condition (2.59)(b) with $\alpha = c_0^2\,(L_\nu + L_\tau)$ and, moreover, it follows from (5.18) that A satisfies condition (1.48) with $m = m_{\mathcal{F}}$. Assume now that $L_\nu + L_\tau < L_0$; then we obtain $c_0^2\,(L_\nu + L_\tau) < m_{\mathcal{F}}$ which implies that $m > \alpha$. Theorem 5.38 follows now from Theorem 2.19. ■

5.3.3 *The Schauder fixed point argument*

We now prove the existence of a solution to the quasivariational inequality (5.187) without any smallness assumption on the normal compliance functions.

Theorem 5.39 *Assume (4.33), (5.7), (5.8) and (5.183). Then there exists at least one solution to Problem 5.37.*

Proof We use Theorem 2.22 on page 51 with the choice $K = X = V$. To this end, we note that (2.63) holds. Moreover, for all $\boldsymbol{\eta} \in V$ the functional $j(\boldsymbol{\eta}, \cdot) : V \to \mathbb{R}$ is a continuous seminorm on V and, therefore, it satisfies condition (2.64). Consider now two sequences $\{\boldsymbol{\eta}_n\} \subset V$, $\{\boldsymbol{u}_n\} \subset V$ such that

$$\boldsymbol{\eta}_n \rightharpoonup \boldsymbol{\eta} \quad \text{in } V, \quad \boldsymbol{u}_n \rightharpoonup \boldsymbol{u} \quad \text{in } V \tag{5.191}$$

and let $\boldsymbol{v}$ be an arbitrary element in V. Similar computations to those used in the proof of (5.188) lead to the inequalities

$$|j(\boldsymbol{\eta}_n, \boldsymbol{v}) - j(\boldsymbol{\eta}, \boldsymbol{v})| \le (L_\nu + L_\tau)\|\boldsymbol{\eta}_n - \boldsymbol{\eta}\|_{L^2(\Gamma_3)^d}\|\boldsymbol{v}\|_{L^2(\Gamma_3)^d}, \tag{5.192}$$

$$\begin{aligned} &|j(\boldsymbol{\eta}_n, \boldsymbol{u}_n) - j(\boldsymbol{\eta}, \boldsymbol{u}_n)| \\ &\qquad \le (L_\nu + L_\tau)\|\boldsymbol{\eta}_n - \boldsymbol{\eta}\|_{L^2(\Gamma_3)^d}\|\boldsymbol{u}_n\|_{L^2(\Gamma_3)^d}, \end{aligned} \tag{5.193}$$

$$|j(\boldsymbol{\eta}, \boldsymbol{u}_n) - j(\boldsymbol{\eta}, \boldsymbol{u})| \le (L_\nu + L_\tau)\|\boldsymbol{\eta}\|_{L^2(\Gamma_3)^d}\|\boldsymbol{u}_n - \boldsymbol{u}\|_{L^2(\Gamma_3)^d}, \tag{5.194}$$

for all $n \in \mathbb{N}$. Using (5.191) and the compactness property (4.7) we see that

$$\boldsymbol{\eta}_n \to \boldsymbol{\eta} \quad \text{in } L^2(\Gamma_3)^d, \quad \boldsymbol{u}_n \to \boldsymbol{u} \quad \text{in } L^2(\Gamma_3)^d$$

and, therefore, (5.192)–(5.194) yield

$$j(\boldsymbol{\eta}_n, \boldsymbol{v}) \to j(\boldsymbol{\eta}, \boldsymbol{v}), \tag{5.195}$$

$$j(\boldsymbol{\eta}_n, \boldsymbol{u}_n) - j(\boldsymbol{\eta}, \boldsymbol{u}_n) \to 0, \tag{5.196}$$

$$j(\boldsymbol{\eta}, \boldsymbol{u}_n) \to j(\boldsymbol{\eta}, \boldsymbol{u}), \tag{5.197}$$

as $n \to \infty$. We now write

$$j(\boldsymbol{\eta}_n, \boldsymbol{v}) - j(\boldsymbol{\eta}_n, \boldsymbol{u}_n) = j(\boldsymbol{\eta}_n, \boldsymbol{v}) - j(\boldsymbol{\eta}, \boldsymbol{u}_n) + j(\boldsymbol{\eta}, \boldsymbol{u}_n) - j(\boldsymbol{\eta}_n, \boldsymbol{u}_n)$$

for all $n \in \mathbb{N}$, then we use the convergences (5.195)–(5.197) to see that

$$\lim_{n\to\infty} \left[j(\boldsymbol{\eta}_n, \boldsymbol{v}) - j(\boldsymbol{\eta}_n, \boldsymbol{u}_n)\right] = j(\boldsymbol{\eta}, \boldsymbol{v}) - j(\boldsymbol{\eta}, \boldsymbol{u}),$$

which shows that the functional j satisfies conditions (2.65). Theorem 5.39 follows now from Theorem 2.22. ■

A pair of functions $(\boldsymbol{u}, \boldsymbol{\sigma})$ which satisfies (5.177) and (5.187) is called a *weak solution* of the frictionless contact problem (5.177)–(5.182). Note that Theorem 5.38 provides the existence of a unique weak solution to the problem (5.177)–(5.182) under a smallness assumption on the normal compliance functions. Also, Theorem 5.39 provides the existence of a weak solution to the problem without any smallness assumption but, in turn, it does not guarantee its unique weak solvability. Finally, as in the study of the contact problems presented on the previous sections of this chapter, it can be proved that the regularity of the weak solution is $\boldsymbol{u} \in V$ and $\boldsymbol{\sigma} \in Q_1$.

5.3.4 Convergence results

We now study the dependence of the solution with respect to perturbations of the normal compliance functions p_ν and p_τ. To this end, we assume in what follows that (4.33), (5.7), (5.8) and (5.183) hold. Moreover, we assume that

$$L_\nu + L_\tau < L_0,$$

where L_0 is defined by (5.190) and we denote by $\boldsymbol{u}$ the solution of Problem 5.37 obtained in Theorem 5.38.

For each $\rho > 0$ let p_e^ρ be a perturbation of p_e which satisfies (5.183) with the Lipschitz constant L_e^ρ $(e = \nu, \tau)$. We introduce the functional $j_\rho : V \times V \to \mathbb{R}$ defined by

$$\begin{aligned} j_\rho(\boldsymbol{u}, \boldsymbol{v}) = & \int_{\Gamma_3} p_\nu^\rho(u_\nu - g_{\mathrm{a}})|v_\nu|\, da \\ & + \int_{\Gamma_3} p_\tau^\rho(u_\nu - g_{\mathrm{a}})\|\boldsymbol{v}_\tau\|\, da \quad \forall\, \boldsymbol{u}, \boldsymbol{v} \in V \end{aligned} \tag{5.198}$$

and we consider the following variational problem.

Problem 5.40 *Find a displacement field $\boldsymbol{u}_\rho \in V$ such that*

$$\begin{aligned} (\mathcal{F}\boldsymbol{\varepsilon}(\boldsymbol{u}_\rho), \boldsymbol{\varepsilon}(\boldsymbol{v}) - \boldsymbol{\varepsilon}(\boldsymbol{u}_\rho))_Q + j_\rho(\boldsymbol{u}_\rho, \boldsymbol{v}) - j_\rho(\boldsymbol{u}_\rho, \boldsymbol{u}_\rho) & \\ \geq (\boldsymbol{f}, \boldsymbol{v} - \boldsymbol{u}_\rho)_V \quad \forall\, \boldsymbol{v} \in V. & \end{aligned} \tag{5.199}$$

Assume that

$$L_\nu^\rho + L_\tau^\rho < L_0 \quad \forall\, \rho > 0.$$

Then it follows from Theorem 5.38 that, for each $\rho > 0$, Problem 5.40 has a unique solution $\boldsymbol{u}_\rho \in V$.

Consider now the following assumption on the normal compliance functions p_e^ρ and p_e, for $e = \nu, \tau$.

$$\left.\begin{array}{l} \text{There exists } G_e : \mathbb{R}_+ \to \mathbb{R}_+ \text{ such that} \\ \text{(a) } |p_e^\rho(\boldsymbol{x}, r) - p_e(\boldsymbol{x}, r)| \le G_e(\rho) \\ \quad \forall\, r \in \mathbb{R}, \text{ a.e. } \boldsymbol{x} \in \Gamma_3, \text{ for each } \rho > 0. \\ \text{(b) } \lim\limits_{\rho \to 0} G_e(\rho) = 0. \end{array}\right\} \tag{5.200}$$

We have the following convergence result.

Theorem 5.41 *Under assumption (5.200) the solution $\boldsymbol{u}_\rho$ of Problem 5.40 converges to the solution $\boldsymbol{u}$ of Problem 5.37, i.e.,*

$$\boldsymbol{u}_\rho \to \boldsymbol{u} \quad \text{in } V \quad \text{as} \quad \rho \to 0. \tag{5.201}$$

Proof The convergence result (5.201) is a direct consequence of Theorem 2.27. Indeed, assume that $\rho > 0$ is given and let $\boldsymbol{u}_1, \boldsymbol{u}_2 \in V$. We use (5.198) and (5.184) to see that

$$\begin{aligned} & j_\rho(\boldsymbol{u}_1, \boldsymbol{u}_2) - j_\rho(\boldsymbol{u}_1, \boldsymbol{u}_1) + j(\boldsymbol{u}_2, \boldsymbol{u}_1) - j(\boldsymbol{u}_2, \boldsymbol{u}_2) \\ & = \int_{\Gamma_3} (p_\nu^\rho(u_{1\nu} - g_{\rm a}) - p_\nu(u_{2\nu} - g_{\rm a}))(|u_{2\nu}| - |u_{1\nu}|)\, da \\ & \quad + \int_{\Gamma_3} (p_\tau^\rho(u_{1\nu} - g_{\rm a}) - p_\tau(u_{2\nu} - g_{\rm a}))(\|\boldsymbol{u}_{2\tau}\| - \|\boldsymbol{u}_{1\tau}\|)\, da \\ & \le \int_{\Gamma_3} |p_\nu^\rho(u_{1\nu} - g_{\rm a}) - p_\nu(u_{2\nu} - g_{\rm a})|\, \big|\, |u_{2\nu}| - |u_{1\nu}|\, \big|\, da \\ & \quad + \int_{\Gamma_3} |p_\tau^\rho(u_{1\nu} - g_{\rm a}) - p_\tau(u_{2\nu} - g_{\rm a})|\, \big|\, \|\boldsymbol{u}_{2\tau}\| - \|\boldsymbol{u}_{1\tau}\|\, \big|\, da \end{aligned}$$

and, therefore,

$$\begin{aligned} & j_\rho(\boldsymbol{u}_1, \boldsymbol{u}_2) - j_\rho(\boldsymbol{u}_1, \boldsymbol{u}_1) + j(\boldsymbol{u}_2, \boldsymbol{u}_1) - j(\boldsymbol{u}_2, \boldsymbol{u}_2) \\ & \le \int_{\Gamma_3} |p_\nu^\rho(u_{1\nu} - g_{\rm a}) - p_\nu(u_{2\nu} - g_{\rm a})|\, \|\boldsymbol{u}_1 - \boldsymbol{u}_2\|\, da \\ & \quad + \int_{\Gamma_3} |p_\tau^\rho(u_{1\nu} - g_{\rm a}) - p_\tau(u_{2\nu} - g_{\rm a})|\, \|\boldsymbol{u}_1 - \boldsymbol{u}_2\|\, da. \end{aligned} \tag{5.202}$$

On the other hand, for $e = \nu, \tau$, using (5.200)(a) and (5.183)(b) we have

$$\begin{aligned} & |p_e^\rho(u_{1\nu} - g_{\rm a}) - p_e(u_{2\nu} - g_{\rm a})| \\ & \le |p_e^\rho(u_{1\nu} - g_{\rm a}) - p_e(u_{1\nu} - g_{\rm a})| + |p_e(u_{1\nu} - g_{\rm a}) - p_e(u_{2\nu} - g_{\rm a})| \\ & \le G_e(\rho) + L_e |u_{1\nu} - u_{2\nu}| \le G_e(\rho) + L_e \|\boldsymbol{u}_1 - \boldsymbol{u}_2\| \end{aligned}$$

a.e. on Γ_3 and, therefore, (4.13) yields

$$\int_{\Gamma_3} |p_e^\rho(u_{1\nu} - g_a) - p_e(u_{2\nu} - g_a)| \, \|\boldsymbol{u}_1 - \boldsymbol{u}_2\| \, da$$
$$\leq \operatorname{meas}(\Gamma_3)^{1/2} c_0 \, G_e(\rho) \, \|\boldsymbol{u}_1 - \boldsymbol{u}_2\|_V + c_0^2 \, L_e \, \|\boldsymbol{u}_1 - \boldsymbol{u}_2\|_V^2. \quad (5.203)$$

We write (5.203) for $e = \nu, \tau$, add the resulting inequalities and use (5.202) to see that

$$\begin{aligned} j_\rho(\boldsymbol{u}_1, \boldsymbol{u}_2) - j_\rho(\boldsymbol{u}_1, \boldsymbol{u}_1) + j(\boldsymbol{u}_2, \boldsymbol{u}_1) - j(\boldsymbol{u}_2, \boldsymbol{u}_2) \\ \leq \operatorname{meas}(\Gamma_3)^{1/2} c_0 \, (G_\nu(\rho) + G_\tau(\rho)) \|\boldsymbol{u}_1 - \boldsymbol{u}_2\|_V \\ + c_0^2 (L_\nu + L_\tau) \|\boldsymbol{u}_1 - \boldsymbol{u}_2\|_V^2. \end{aligned} \quad (5.204)$$

Denote by $G : \mathbb{R}_+ \to \mathbb{R}_+$ the function given by

$$G(\rho) = \operatorname{meas}(\Gamma_3)^{1/2} c_0 \, (G_\nu(\rho) + G_\tau(\rho)) \quad \forall \, \rho \in \mathbb{R}_+ \quad (5.205)$$

and recall that, as shown on page 165, the functional j satisfies condition (2.59)(b) with

$$\alpha = c_0^2 (L_\nu + L_\tau). \quad (5.206)$$

Therefore, it follows from (5.204)–(5.206) that condition (2.74)(a) is satisfied and, moreover, assumption (5.200)(b) implies that (2.74)(b) holds, too. The convergence (5.201) is now a consequence of Theorem 2.27. ■

Let $\alpha > 0$, $\mu > 0$, $\delta > 0$ and let $c_\nu^\rho > 0$, $c_\nu > 0$ be such that $c_\nu^\rho \to c_\nu$ as $\rho \to 0$. Assume that the function p_ν is given by (4.80) and the function p_τ is given by (4.95). Also, assume that p_ν^ρ and p_τ^ρ are given by

$$p_\nu^\rho(r) = \begin{cases} c_\nu^\rho r_+ & \text{if } r \leq \alpha, \\ c_\nu^\rho \alpha & \text{if } r > \alpha, \end{cases}$$

$$p_\tau^\rho = \mu p_\nu^\rho. \quad (5.207)$$

Then it is easy to check that assumption (5.200) is valid. A similar conclusion arises when p_τ is given by (4.96) and p_τ^ρ is the function

$$p_\tau^\rho = \mu p_\nu^\rho (1 - \delta p_\nu^\rho)_+. \quad (5.208)$$

We conclude that the convergence result in Theorem 5.41 is valid in the two examples above.

Nevertheless, we note that the uniform estimate (5.200)(a) is not valid in the case when the function p_ν is given by (4.79) and p_ν^ρ is given by

$$p_\nu^\rho(r) = c_\nu^\rho r_+. \quad (5.209)$$

Therefore, there is a need to relax condition (5.200) in order to obtain a more general convergence result, applicable to the normal compliance functions (4.79) and (5.209). One option is to consider the following assumption, for $e = \nu, \tau$.

$$\left.\begin{array}{l}\text{There exist } G_e : \mathbb{R}_+ \to \mathbb{R}_+ \text{ and } \beta_e \in \mathbb{R}_+ \text{ such that}\\ \text{(a) } |p_e^\rho(\boldsymbol{x}, r) - p_e(\boldsymbol{x}, r)| \le G_e(\rho)(|r| + \beta_e)\\ \qquad \forall\, r \in \mathbb{R},\ \text{a.e. } \boldsymbol{x} \in \Gamma_3,\ \text{for each } \rho > 0;\\ \text{(b) } \lim_{\rho \to 0} G_e(\rho) = 0.\end{array}\right\} \tag{5.210}$$

It is easy to see that condition (5.210) is more general than condition (5.200). Moreover, it is satisfied by the functions (4.79) and (5.209) and the corresponding tangential functions given by (4.95), (5.207) or (4.96), (5.208). In addition, we have the following convergence result.

Theorem 5.42 *Under assumption (5.210) the solution $\boldsymbol{u}_\rho$ of Problem 5.40 converges to the solution $\boldsymbol{u}$ of Problem 5.37, i.e. the convergence (5.201) holds.*

Proof Assume that (5.210) holds and let $\rho > 0$. We take $\boldsymbol{v} = \boldsymbol{u}$ in (5.199) and $\boldsymbol{v} = \boldsymbol{u}_\rho$ in (5.187), then we add the resulting inequalities to obtain

$$\begin{aligned}&(\mathcal{F}\boldsymbol{\varepsilon}(\boldsymbol{u}_\rho) - \mathcal{F}\boldsymbol{\varepsilon}(\boldsymbol{u}), \boldsymbol{\varepsilon}(\boldsymbol{u}_\rho) - \boldsymbol{\varepsilon}(\boldsymbol{u}))_Q\\ &\qquad \le j_\rho(\boldsymbol{u}_\rho, \boldsymbol{u}) - j_\rho(\boldsymbol{u}_\rho, \boldsymbol{u}_\rho) + j(\boldsymbol{u}, \boldsymbol{u}_\rho) - j(\boldsymbol{u}, \boldsymbol{u}).\end{aligned}$$

Therefore, using (4.33)(c) and arguments similar to those used in the proof of (5.202) yields

$$\begin{aligned}m_\mathcal{F}\, \|\boldsymbol{u}_\rho - \boldsymbol{u}\|_V^2 &\le \int_{\Gamma_3} |p_\nu^\rho(u_{\rho\nu} - g_\mathrm{a}) - p_\nu(u_\nu - g_\mathrm{a})|\|\boldsymbol{u}_\rho - \boldsymbol{u}\|\, da\\ &+ \int_{\Gamma_3} |p_\tau^\rho(u_{\rho\nu} - g_\mathrm{a}) - p_\tau(u_\nu - g_\mathrm{a})|\|\boldsymbol{u}_\rho - \boldsymbol{u}\|\, da.\end{aligned} \tag{5.211}$$

Next, for $e = \nu, \tau$ we use (5.210)(a) and (5.183)(b) to see that

$$\begin{aligned}|p_e^\rho(u_{\rho\nu} - g_\mathrm{a}) - p_e(u_\nu - g_\mathrm{a})| &\le |p_e^\rho(u_{\rho\nu} - g_\mathrm{a}) - p_e(u_{\rho\nu} - g_\mathrm{a})|\\ &+ |p_e(u_{\rho\nu} - g_\mathrm{a}) - p_e(u_\nu - g_\mathrm{a})|\\ &\le G_e(\rho)(|u_{\rho\nu} - g_\mathrm{a}| + \beta_e) + L_e|u_{\rho\nu} - u_\nu|\\ &\le G_e(\rho)(\|\boldsymbol{u}_\rho\| + g_\mathrm{a} + \beta_e) + L_e\|\boldsymbol{u}_\rho - \boldsymbol{u}\|\end{aligned}$$

a.e. on Γ_3 and, therefore, using (4.13) we obtain

$$\int_{\Gamma_3} |p_e^\rho(u_{\rho\nu} - g_{\rm a}) - p_e(u_\nu - g_{\rm a})| \|\boldsymbol{u}_\rho - \boldsymbol{u}\| \, da$$

$$\leq c_0 G_e(\rho)(\|g_{\rm a}\|_{L^2(\Gamma_3)} + \text{meas}\,(\Gamma_3)^{1/2}\beta_e)\|\boldsymbol{u}_\rho - \boldsymbol{u}\|_V$$

$$+ c_0^2 G_e(\rho)\|\boldsymbol{u}_\rho\|_V \|\boldsymbol{u}_\rho - \boldsymbol{u}\|_V + c_0^2 L_e \|\boldsymbol{u}_\rho - \boldsymbol{u}\|_V^2. \qquad (5.212)$$

We write (5.212) for $e = \nu, \tau$, add the resulting inequalities and use (5.211) to see that

$$m_{\mathcal{F}} \|\boldsymbol{u}_\rho - \boldsymbol{u}\|_V$$

$$\leq c_0(G_\nu(\rho) + G_\tau(\rho))\big(\|g_{\rm a}\|_{L^2(\Gamma_3)} + \text{meas}\,(\Gamma_3)^{1/2}(\beta_\nu + \beta_\tau)\big)$$

$$+ c_0^2(G_\nu(\rho) + G_\tau(\rho))\|\boldsymbol{u}_\rho\|_V + c_0^2(L_\nu + L_\tau)\|\boldsymbol{u}_\rho - \boldsymbol{u}\|_V.$$

Since $c_0^2(L_\nu + L_\tau) < m_{\mathcal{F}}$ if follows from the above that

$$\|\boldsymbol{u}_\rho - \boldsymbol{u}\|_V \leq c_1(G_\nu(\rho) + G_\tau(\rho))(1 + \|\boldsymbol{u}_\rho\|_V), \qquad (5.213)$$

where c_1 is a positive constant which does not depend on ρ.

On the other hand, we take $\boldsymbol{v} = \boldsymbol{0}_V$ in (5.199) and use (4.33)(c) to see that

$$m_{\mathcal{F}} \|\boldsymbol{u}_\rho\|_V^2 \leq (\boldsymbol{f}, \boldsymbol{u}_\rho)_V - (\mathcal{F}\boldsymbol{0}_Q, \boldsymbol{\varepsilon}(\boldsymbol{u}_\rho))_Q,$$

which implies that

$$\|\boldsymbol{u}_\rho\|_V \leq \frac{1}{m_{\mathcal{F}}}(\|\boldsymbol{f}\|_V + \|\mathcal{F}\boldsymbol{0}_Q\|_Q). \qquad (5.214)$$

We combine now (5.213) and (5.214) and use assumption (5.210)(b) to obtain (5.201), which concludes the proof. ∎

We note that, unlike the proof of Theorem 5.41 which was based on the abstract convergence result provided by Theorem 2.27, the proof of Theorem 5.42 is a direct one, i.e. it does not follow as a consequence of any abstract convergence result.

Finally, note that the convergence result (5.201) is of importance from a mechanical point of view, since it states that small errors in the estimation of the normal compliance functions lead to small errors in the solution. We conclude that the weak solution of Problem 5.36 depends continuously on the frictional contact conditions.

6
Analysis of elastic-viscoplastic contact problems

In this chapter we illustrate the use of the abstract results obtained in Chapter 3 in the study of quasistatic frictionless and frictional contact problems. We model the material's behavior either with a nonlinear viscoelastic constitutive law with short or long memory, or with a viscoplastic constitutive law. The contact is either bilateral or modelled with the normal compliance condition with or without unilateral constraint. The friction is modelled with Coulomb's law and its versions. For each problem we provide a variational formulation which is in the form of a nonlinear equation or variational inequality for the displacement or the velocity field. Then we use the abstract existence and uniqueness results presented in Chapter 3 to prove the unique weak solvability of the corresponding contact problems. For some of the problems we also provide convergence results. Everywhere in this chapter we use the function spaces introduced in Section 4.1 and the equations and boundary conditions described in Section 4.4.

6.1 Bilateral frictionless contact problems

In this section we study two frictionless contact problems for viscoelastic materials in which the contact is bilateral. The main feature of these problems consists of the fact that their variational formulation is in the form of a nonlinear equation for the displacement field, which is either evolutionary or involves a history-dependent operator. Therefore, the unique solvability of the models is proved by using the abstract results presented in Section 3.1.

6.1.1 Contact of materials with short memory

In the first problem we use a viscoelastic constitutive law with short memory and the contact condition (4.102). Therefore, the classical model of the process is the following.

Problem 6.1 *Find a displacement field* $\boldsymbol{u} : \Omega \times [0,T] \to \mathbb{R}^d$ *and a stress field* $\boldsymbol{\sigma} : \Omega \times [0,T] \to \mathbb{S}^d$ *such that*

$$\boldsymbol{\sigma} = \mathcal{A}\boldsymbol{\varepsilon}(\dot{\boldsymbol{u}}) + \mathcal{B}\boldsymbol{\varepsilon}(\boldsymbol{u}) \quad \text{in} \quad \Omega \times (0,T), \tag{6.1}$$

$$\operatorname{Div} \boldsymbol{\sigma} + \boldsymbol{f}_0 = \boldsymbol{0} \quad \text{in} \quad \Omega \times (0,T), \tag{6.2}$$

$$\boldsymbol{u} = \boldsymbol{0} \quad \text{on} \quad \Gamma_1 \times (0,T), \tag{6.3}$$

$$\boldsymbol{\sigma\nu} = \boldsymbol{f}_2 \quad \text{on} \quad \Gamma_2 \times (0,T), \tag{6.4}$$

$$-\sigma_\nu = F \quad \text{on} \quad \Gamma_3 \times (0,T), \tag{6.5}$$

$$\boldsymbol{\sigma}_\tau = \boldsymbol{0} \quad \text{on} \quad \Gamma_3 \times (0,T), \tag{6.6}$$

$$\boldsymbol{u}(0) = \boldsymbol{u}_0 \quad \text{in} \quad \Omega. \tag{6.7}$$

Here (6.1) represents the viscoelastic constitutive law (4.39), introduced on page 95. Also, (6.2) represents the equation of equilibrium (4.98), (6.3) is the displacement boundary condition (4.99) and (6.4) is the traction boundary condition (4.100). Next, (6.5) and (6.6) represent the contact and the frictionless condition (4.102) and (4.109), respectively. Finally, (6.7) represents the initial condition (4.43), in which $\boldsymbol{u}_0$ is a given displacement field.

We turn to the variational formulation of Problem 6.1. To this end we assume in what follows that the viscosity and elasticity operators satisfy conditions (4.40) and (4.41), respectively. The densities of body forces and surface tractions have the regularity

$$\boldsymbol{f}_0 \in C([0,T]; L^2(\Omega)^d), \quad \boldsymbol{f}_2 \in C([0,T]; L^2(\Gamma_2)^d), \tag{6.8}$$

and the prescribed normal stress is such that

$$F \in C([0,T]; L^2(\Gamma_3)) \ \text{and} \ F(\boldsymbol{x},t) \geq 0 \ \ \forall\, t \in [0,T], \ \text{a.e.}\ \boldsymbol{x} \in \Gamma_3. \tag{6.9}$$

Finally, we assume that the initial displacement satisfies

$$\boldsymbol{u}_0 \in V, \tag{6.10}$$

where V denotes the space (4.9).

Next, we consider the function $\boldsymbol{f} : [0,T] \to V$ defined by

$$(\boldsymbol{f}(t), \boldsymbol{v})_V = \int_\Omega \boldsymbol{f}_0(t) \cdot \boldsymbol{v}\, \mathrm{d}x + \int_{\Gamma_2} \boldsymbol{f}_2(t) \cdot \boldsymbol{v}\, \mathrm{d}a$$
$$- \int_{\Gamma_3} F(t)\, v_\nu\, \mathrm{d}a \qquad \forall\, \boldsymbol{v} \in V,\ t \in [0,T]. \tag{6.11}$$

We recall that the definition of $\boldsymbol{f}$ follows by using Riesz's representation theorem. Note also that assumptions (6.8) and (6.9) imply that the integrals in (6.11) are well defined and, moreover,

$$\boldsymbol{f} \in C([0,T];V). \tag{6.12}$$

Assume now that Problem 6.1 has a regular solution $(\boldsymbol{u},\boldsymbol{\sigma})$ and let $t \in [0,T]$. Using Green's formula (4.22), the equilibrium equation (6.2) and the boundary conditions (6.3)–(6.6) we obtain that

$$(\boldsymbol{\sigma}(t),\boldsymbol{\varepsilon}(\boldsymbol{v}))_Q = (\boldsymbol{f}(t),\boldsymbol{v})_V \qquad \forall\,\boldsymbol{v} \in V. \tag{6.13}$$

Then we use the constitutive law (6.1), the initial condition (6.7) and equality (6.13) to derive the following variational formulation of the frictionless contact problem 6.1.

Problem 6.2 *Find a displacement field $\boldsymbol{u} : [0,T] \to V$ such that*

$$(\mathcal{A}\boldsymbol{\varepsilon}(\dot{\boldsymbol{u}}(t)),\boldsymbol{\varepsilon}(\boldsymbol{v}))_Q + (\mathcal{B}\boldsymbol{\varepsilon}(\boldsymbol{u}(t)),\boldsymbol{\varepsilon}(\boldsymbol{v}))_Q = (\boldsymbol{f}(t),\boldsymbol{v})_V \quad \forall\,\boldsymbol{v} \in V,\ t \in [0,T], \tag{6.14}$$

$$\boldsymbol{u}(0) = \boldsymbol{u}_0. \tag{6.15}$$

The unique solvability of Problem 6.2 is given by the following existence and uniqueness result.

Theorem 6.3 *Assume (4.40), (4.41), (6.8), (6.9) and (6.10). Then there exists a unique solution $\boldsymbol{u} \in C^1([0,T];V)$ to Problem 6.2.*

Proof We apply the Riesz representation theorem to define the operators $A : V \to V$ and $B : V \to V$ by

$$(A\boldsymbol{u},\boldsymbol{v})_V = (\mathcal{A}\boldsymbol{\varepsilon}(\boldsymbol{u}),\boldsymbol{\varepsilon}(\boldsymbol{v}))_Q \quad \forall\,\boldsymbol{u},\,\boldsymbol{v} \in V, \tag{6.16}$$

$$(B\boldsymbol{u},\boldsymbol{v})_V = (\mathcal{B}\boldsymbol{\varepsilon}(\boldsymbol{u}),\boldsymbol{\varepsilon}(\boldsymbol{v}))_Q \quad \forall\,\boldsymbol{u},\,\boldsymbol{v} \in V. \tag{6.17}$$

Then we use assumption (4.40) to see that A is a strongly monotone Lipschitz continuous operator. Also, it follows from (4.41) that B is a Lipschitz continuous operator and recall the regularity (6.12) and (6.10). We apply Theorem 3.3 to obtain the existence of a unique function $\boldsymbol{u} \in C^1([0,T];V)$ which satisfies

$$A\dot{\boldsymbol{u}}(t) + B\boldsymbol{u}(t) = \boldsymbol{f}(t) \quad \forall\,t \in [0,T], \tag{6.18}$$

$$\boldsymbol{u}(0) = \boldsymbol{u}_0. \tag{6.19}$$

Then we use (6.16)–(6.19) to see that $\boldsymbol{u}$ is the unique solution to the Cauchy problem (6.14)–(6.15), which completes the proof. ■

Let $\boldsymbol{u}$ be the solution of Problem 6.2 obtained in Theorem 6.3 and let $\boldsymbol{\sigma} = \mathcal{A}\boldsymbol{\varepsilon}(\dot{\boldsymbol{u}}) + \mathcal{B}\boldsymbol{\varepsilon}(\boldsymbol{u})$. The properties (4.40) and (4.41) of the operators $\mathcal{A}$ and $\mathcal{B}$ combined with the regularity of $\boldsymbol{u}$ yield $\boldsymbol{\sigma} \in C([0,T];Q)$. Moreover, (6.14) implies that

$$(\boldsymbol{\sigma}(t), \boldsymbol{\varepsilon}(\boldsymbol{v}))_Q = (\boldsymbol{f}(t), \boldsymbol{v})_V \quad \forall \boldsymbol{v} \in V,\ t \in [0,T]$$

and, taking $\boldsymbol{v} = \boldsymbol{\varphi} \in C_0^\infty(\Omega)^d \subset V$ in this equality, we obtain

$$(\boldsymbol{\sigma}(t), \boldsymbol{\varepsilon}(\boldsymbol{\varphi}))_Q = (\boldsymbol{f}(t), \boldsymbol{\varphi})_V \quad \forall \boldsymbol{\varphi} \in C_0^\infty(\Omega)^d,\ t \in [0,T].$$

We now use (6.11) and Definition 4.2 of the divergence operator to see that

$$(\mathrm{Div}\, \boldsymbol{\sigma}(t), \boldsymbol{\varphi})_{L^2(\Omega)^d} + (\boldsymbol{f}_0(t), \boldsymbol{\varphi})_{L^2(\Omega)^d} = 0 \quad \forall \boldsymbol{\varphi} \in C_0^\infty(\Omega)^d,\ t \in [0,T]$$

and, by the density of the space $C_0^\infty(\Omega)^d$ in $L^2(\Omega)^d$ it follows that

$$\mathrm{Div}\, \boldsymbol{\sigma}(t) + \boldsymbol{f}_0(t) = \boldsymbol{0} \quad \text{in } \Omega,\ \forall t \in [0,T]. \tag{6.20}$$

Using the regularity $\boldsymbol{f}_0 \in C([0,T];L^2(\Omega)^d)$ in (6.8), it follows from (6.20) that $\mathrm{Div}\, \boldsymbol{\sigma} \in C([0,T];L^2(\Omega)^d)$ and, therefore, $\boldsymbol{\sigma} \in C([0,T];Q_1)$.

A pair of functions $(\boldsymbol{u}, \boldsymbol{\sigma})$ which satisfies (6.1), (6.14) and (6.15) is called a *weak solution* of the frictionless contact problem (6.1)–(6.7). We conclude by Theorem 6.3 that problem (6.1)–(6.7) has a unique weak solution. Moreover, the solution satisfies $\boldsymbol{u} \in C^1([0,T];V)$ and $\boldsymbol{\sigma} \in C([0,T];Q_1)$.

6.1.2 Contact of materials with long memory

In the second problem we use a viscoelastic constitutive law with long memory and the contact condition (4.101). The classical model of the process is the following.

Problem 6.4 *Find a displacement field* $\boldsymbol{u} : \Omega \times [0,T] \to \mathbb{R}^d$ *and a stress field* $\boldsymbol{\sigma} : \Omega \times [0,T] \to \mathbb{S}^d$ *such that, for all* $t \in [0,T]$,

$$\boldsymbol{\sigma}(t) = \mathcal{F}\boldsymbol{\varepsilon}(\boldsymbol{u}(t)) + \int_0^t \mathcal{R}(t-s)\boldsymbol{\varepsilon}(\boldsymbol{u}(s))\,\mathrm{d}s \quad \text{in} \quad \Omega, \tag{6.21}$$

$$\mathrm{Div}\, \boldsymbol{\sigma}(t) + \boldsymbol{f}_0(t) = \boldsymbol{0} \quad \text{in} \quad \Omega, \tag{6.22}$$

$$\boldsymbol{u}(t) = \boldsymbol{0} \quad \text{on} \quad \Gamma_1, \tag{6.23}$$

$$\boldsymbol{\sigma}(t)\boldsymbol{\nu} = \boldsymbol{f}_2(t) \quad \text{on} \quad \Gamma_2, \tag{6.24}$$

$$u_\nu(t) = 0 \quad \text{on} \quad \Gamma_3, \tag{6.25}$$

$$\boldsymbol{\sigma}_\tau(t) = \boldsymbol{0} \quad \text{on} \quad \Gamma_3. \tag{6.26}$$

Here (6.21) represents the viscoelastic constitutive law with long memory (4.48), introduced on page 97. Also, (6.22) represents the equation of equilibrium (4.98), (6.23) is the displacement boundary condition (4.99) and (6.24) is the traction boundary condition (4.100). Next, (6.25) and (6.26) represent the bilateral contact and the frictionless condition (4.101) and (4.109), respectively.

We turn to the variational formulation of Problem 6.2. To this end we assume in what follows that the elasticity operator satisfies condition (4.33) and the relaxation operator satisfies condition (4.47). Moreover, the densities of body forces and surface tractions have the regularity (6.8). We use the space V_1 defined by (4.19). Also, we introduce the function $\boldsymbol{f}_1 : [0,T] \to V_1$ defined by

$$(\boldsymbol{f}_1(t), \boldsymbol{v})_V = \int_\Omega \boldsymbol{f}_0(t) \cdot \boldsymbol{v}\, \mathrm{d}x + \int_{\Gamma_2} \boldsymbol{f}_2(t) \cdot \boldsymbol{v}\, \mathrm{d}a \quad \forall\, \boldsymbol{v} \in V_1,\ t \in [0,T] \quad (6.27)$$

and we remark that (6.8) yields

$$\boldsymbol{f}_1 \in C([0,T]; V_1). \quad (6.28)$$

The variational formulation of Problem 6.4, obtained by standard arguments, is the following.

Problem 6.5 *Find a displacement field $\boldsymbol{u} : [0,T] \to V_1$ such that, for all $t \in [0,T]$,*

$$(\mathcal{F}\boldsymbol{\varepsilon}(\boldsymbol{u}(t)), \boldsymbol{\varepsilon}(\boldsymbol{v}))_Q + \Big(\int_0^t \mathcal{R}(t-s)\boldsymbol{\varepsilon}(\boldsymbol{u}(s))\, \mathrm{d}s, \boldsymbol{\varepsilon}(\boldsymbol{v}) \Big)_Q$$
$$= (\boldsymbol{f}_1(t), \boldsymbol{v})_V \quad \forall\, \boldsymbol{v} \in V_1. \quad (6.29)$$

We proceed with the following existence and uniqueness result.

Theorem 6.6 *Assume (4.33), (4.47) and (6.8). Then there exists a unique solution $\boldsymbol{u} \in C([0,T]; V_1)$ to Problem 6.5.*

Proof We apply the Riesz representation theorem to define the operators $A_1 : V_1 \to V_1$ and $\mathcal{S}_1 : C([0,T]; V_1) \to C([0,T]; V_1)$ by

$$(A_1 \boldsymbol{u}, \boldsymbol{v})_V = (\mathcal{F}\boldsymbol{\varepsilon}(\boldsymbol{u}), \boldsymbol{\varepsilon}(\boldsymbol{v}))_Q \quad \forall\, \boldsymbol{u},\ \boldsymbol{v} \in V_1,$$

$$(\mathcal{S}_1 \boldsymbol{u}(t), \boldsymbol{v})_V = \Big(\int_0^t \mathcal{R}(t-s)\boldsymbol{\varepsilon}(\boldsymbol{u}(s))\, \mathrm{d}s, \boldsymbol{\varepsilon}(\boldsymbol{v}) \Big)_Q$$
$$\forall\, \boldsymbol{u} \in C([0,T]; V_1),\ \boldsymbol{v} \in V_1.$$

Then it is easy to see that equality (6.29) is equivalent to the equation

$$A_1\boldsymbol{u}(t) + \mathcal{S}_1\boldsymbol{u}(t) = \boldsymbol{f}_1(t) \quad \forall\, t \in [0,T].$$

We use (4.33) to see that the operator A_1 is strongly monotone and Lipschitz continuous. In addition, a simple calculation based on (4.46) and (4.47) shows that

$$\|\mathcal{S}_1\boldsymbol{u}(t) - \mathcal{S}_1\boldsymbol{v}(t)\|_V \le d \max_{s\in[0,T]} \|\mathcal{R}(s)\|_{\mathbf{Q}_\infty} \int_0^t \|\boldsymbol{u}(s) - \boldsymbol{v}(s)\|_V \,\mathrm{d}s$$

$$\forall\, \boldsymbol{u},\, \boldsymbol{v} \in C([0,T];V_1),\ \forall\, t \in [0,T].$$

Since the hypotheses of Theorem 3.4 are fulfilled with $X = V_1$, we apply this theorem to conclude the proof. ■

A pair of functions $(\boldsymbol{u}, \boldsymbol{\sigma})$ which satisfies (6.21), (6.29) is called a *weak solution* of the frictionless contact problem (6.21)–(6.26). We conclude by Theorem 6.6 that problem (6.21)–(6.26) has a unique weak solution. Moreover, as in the case of problem (6.1)–(6.7), it follows that the solution satisfies $\boldsymbol{u} \in C([0,T];V_1)$ and $\boldsymbol{\sigma} \in C([0,T];Q_1)$.

6.2 Viscoelastic contact problems with long memory

In this section we present two mathematical models of contact for viscoelastic materials with long memory. The main feature of these models consists of the fact that their variational formulation is given by a history-dependent quasivariational inequality for the displacement field.

6.2.1 Frictionless contact with unilateral constraint

In the first problem the contact is frictionless and it is modelled with normal compliance and unilateral constraint. The classical formulation of the problem is the following.

Problem 6.7 *Find a displacement field* $\boldsymbol{u} : \Omega \times [0,T] \to \mathbb{R}^d$ *and a stress field* $\boldsymbol{\sigma} : \Omega \times [0,T] \to \mathbb{S}^d$ *such that, for all* $t \in [0,T]$,

$$\boldsymbol{\sigma}(t) = \mathcal{F}\boldsymbol{\varepsilon}(\boldsymbol{u}(t)) + \int_0^t \mathcal{R}(t-s)\boldsymbol{\varepsilon}(\boldsymbol{u}(s))\,\mathrm{d}s \qquad in \quad \Omega, \tag{6.30}$$

$$\mathrm{Div}\,\boldsymbol{\sigma}(t) + \boldsymbol{f}_0(t) = \boldsymbol{0} \qquad in \quad \Omega, \tag{6.31}$$

$$\boldsymbol{u}(t) = \boldsymbol{0} \qquad on \quad \Gamma_1, \tag{6.32}$$

$$\boldsymbol{\sigma}(t)\boldsymbol{\nu} = \boldsymbol{f}_2(t) \qquad on \quad \Gamma_2, \tag{6.33}$$

$$\left.\begin{array}{l} u_\nu \le g, \quad \sigma_\nu(t) + p(u_\nu(t)) \le 0, \\ (u_\nu(t) - g)(\sigma_\nu(t) + p(u_\nu(t))) = 0 \end{array}\right\} \quad on \quad \Gamma_3, \tag{6.34}$$

$$\boldsymbol{\sigma}_\tau(t) = \mathbf{0} \quad on \quad \Gamma_3. \tag{6.35}$$

The equations and boundary conditions in Problem 6.7 have a similar meaning to those in Problem 6.2. The difference arises from the fact that here we replace the bilateral contact condition (6.25) with the contact condition with normal compliance and unilateral constraint (6.34), introduced on page 113. Recall that in this condition $g \ge 0$ is a given bound for the normal displacement and p is a prescribed function.

We assume in what follows that the elasticity operator satisfies condition (4.33) and the relaxation operator satisfies condition (4.47). Moreover, the densities of body forces and surface tractions have the regularity (6.8). Also, we assume that the normal compliance function p is such that

$$\left.\begin{array}{l} \text{(a) } p : \Gamma_3 \times \mathbb{R} \to \mathbb{R}_+; \\ \text{(b) there exists } L_p > 0 \text{ such that} \\ \qquad |p(\boldsymbol{x}, r_1) - p(\boldsymbol{x}, r_2)| \le L_p |r_1 - r_2| \\ \qquad \forall\, r_1,\, r_2 \in \mathbb{R}, \text{ a.e. } \boldsymbol{x} \in \Gamma_3; \\ \text{(c) } (p(\boldsymbol{x}, r_1) - p(\boldsymbol{x}, r_2))(r_1 - r_2) \ge 0 \\ \qquad \forall\, r_1,\, r_2 \in \mathbb{R}, \text{ a.e. } \boldsymbol{x} \in \Gamma_3; \\ \text{(d) the mapping } \boldsymbol{x} \mapsto p(\boldsymbol{x}, r) \text{ is measurable on } \Gamma_3, \\ \qquad \text{for any } r \in \mathbb{R}; \\ \text{(e) } p(\boldsymbol{x}, r) = 0 \text{ for all } r \le 0, \text{ a.e. } \boldsymbol{x} \in \Gamma_3. \end{array}\right\} \tag{6.36}$$

Next, we introduce the set of admissible displacements U, the contact operator $P : V \to V$ and the function $\boldsymbol{f} : [0, T] \to V$ defined by

$$U = \{\, \boldsymbol{v} \in V : v_\nu \le g \text{ a.e. on } \Gamma_3 \,\}, \tag{6.37}$$

$$(P\boldsymbol{u}, \boldsymbol{v})_V = \int_{\Gamma_3} p(u_\nu) v_\nu \,\mathrm{d}a \quad \forall\, \boldsymbol{u},\, \boldsymbol{v} \in V, \tag{6.38}$$

$$(\boldsymbol{f}(t), \boldsymbol{v})_V = \int_\Omega \boldsymbol{f}_0(t) \cdot \boldsymbol{v} \,\mathrm{d}x + \int_{\Gamma_2} \boldsymbol{f}_2(t) \cdot \boldsymbol{v} \,\mathrm{d}a$$
$$\forall\, \boldsymbol{u},\, \boldsymbol{v} \in V,\ t \in [0, T]. \tag{6.39}$$

Assume in what follows that $(\boldsymbol{u}, \boldsymbol{\sigma})$ are sufficiently regular functions which satisfy (6.30)–(6.35) and, again, let $t \in [0, T]$ be given. Then, using standard arguments based on Green's formula, it follows that

$$(\boldsymbol{\sigma}(t), \boldsymbol{\varepsilon}(\boldsymbol{v}) - \boldsymbol{\varepsilon}(\boldsymbol{u}(t)))_Q + (P\boldsymbol{u}(t), \boldsymbol{v} - \boldsymbol{u}(t))_V$$
$$\ge (\boldsymbol{f}(t), \boldsymbol{v} - \boldsymbol{u}(t))_V \quad \forall\, \boldsymbol{v} \in U. \tag{6.40}$$

In addition, we note that the first inequality in (6.34) and definition (6.37) imply that $\boldsymbol{u}(t) \in U$. We combine the results above with the constitutive law (6.30) to deduce the following variational formulation of Problem 6.7.

Problem 6.8 *Find a displacement field* $\boldsymbol{u} : [0,T] \to V$ *such that, for all* $t \in [0,T]$, *the inequality below holds:*

$$\begin{aligned} &\boldsymbol{u}(t) \in U, \quad (\mathcal{F}\boldsymbol{\varepsilon}(\boldsymbol{u}(t)), \boldsymbol{\varepsilon}(\boldsymbol{v}) - \boldsymbol{\varepsilon}(\boldsymbol{u}(t)))_Q \\ &\qquad + \Big(\int_0^t \mathcal{R}(t-s)\boldsymbol{\varepsilon}(\boldsymbol{u}(s))\,\mathrm{d}s, \boldsymbol{\varepsilon}(\boldsymbol{v}) - \boldsymbol{\varepsilon}(\boldsymbol{u}(t)) \Big)_Q \\ &\qquad + (P\boldsymbol{u}(t), \boldsymbol{v} - \boldsymbol{u}(t))_V \geq (\boldsymbol{f}(t), \boldsymbol{v} - \boldsymbol{u}(t))_V \quad \forall\, \boldsymbol{v} \in U. \end{aligned} \tag{6.41}$$

In the study of Problem 6.8 we have the following existence and uniqueness result.

Theorem 6.9 *Assume (4.33), (4.47), (6.8) and (6.36). Then Problem 6.8 has a unique solution* $\boldsymbol{u} \in C([0,T];U)$.

Proof We apply Theorem 3.6 with $X = Y = V$, $K = U$, $\varphi(\boldsymbol{u},\boldsymbol{v}) = (\boldsymbol{u},\boldsymbol{v})_V$ and $j \equiv 0$. First, we note that assumptions (3.36) and (3.38) are clearly satisfied. Also, (3.39) holds with $\alpha = 0$. Next, we use the Riesz representation theorem to define the operators $A : V \to V$ and $\mathcal{S} : C([0,T];V) \to C([0,T];V)$ by the equalities

$$(A\boldsymbol{u}, \boldsymbol{v})_V = (\mathcal{F}\boldsymbol{\varepsilon}(\boldsymbol{u}), \boldsymbol{\varepsilon}(\boldsymbol{v}))_Q + (P\boldsymbol{u}, \boldsymbol{v})_V \quad \forall\, \boldsymbol{u},\, \boldsymbol{v} \in V, \tag{6.42}$$

$$\begin{aligned} (\mathcal{S}\boldsymbol{u}(t), \boldsymbol{v})_V &= \Big(\int_0^t \mathcal{R}(t-s)\boldsymbol{\varepsilon}(\boldsymbol{u}(s))\,\mathrm{d}s, \boldsymbol{\varepsilon}(\boldsymbol{v}) \Big)_Q \\ &\qquad \forall\, \boldsymbol{u} \in C([0,T];V),\ \boldsymbol{v} \in V. \end{aligned} \tag{6.43}$$

We use (4.33), (6.36) and (4.13) to see that the operator A verifies condition (3.37). Moreover, since $\alpha = 0$ it follows that (3.40) holds, too. In addition, using (4.46) and (4.47) we have

$$\begin{aligned} \|\mathcal{S}\boldsymbol{u}_1(t) - \mathcal{S}\boldsymbol{u}_2(t)\|_V &\leq d \max_{s\in[0,T]} \|\mathcal{R}(s)\|_{\mathbf{Q}_\infty} \int_0^t \|\boldsymbol{u}_1(s) - \boldsymbol{u}_2(s)\|_V\,\mathrm{d}s \\ &\forall\, \boldsymbol{u}_1,\, \boldsymbol{u}_2 \in C([0,T];V),\ \forall\, t \in [0,T]. \end{aligned}$$

This inequality shows that the operator (6.43) satisfies condition (3.41) with

$$L_{\mathcal{S}} = d\,\|\mathcal{R}\|_{C([0,T];\mathbf{Q}_\infty)} = d \max_{s\in[0,T]} \|\mathcal{R}(s)\|_{\mathbf{Q}_\infty}. \tag{6.44}$$

Finally, using (6.8) and (6.39) we deduce that $\boldsymbol{f} \in C([0,T];V)$ and, therefore, (3.42) holds, as well. It follows now from Theorem 3.6 that there exists a unique function $\boldsymbol{u} \in C([0,T],V)$ which satisfies the inequality

$$\begin{aligned} \boldsymbol{u}(t) \in U, \quad & (A\boldsymbol{u}(t), \boldsymbol{v} - \boldsymbol{u}(t))_V + (\mathcal{S}\boldsymbol{u}(t), \boldsymbol{v} - \boldsymbol{u}(t))_V \\ & \geq (\boldsymbol{f}(t), \boldsymbol{v} - \boldsymbol{u}(t))_V \quad \forall \boldsymbol{v} \in U, \end{aligned} \tag{6.45}$$

for all $t \in [0,T]$. Theorem 6.9 is now a consequence of (6.42), (6.43) and (6.45). ■

A pair of functions $(\boldsymbol{u}, \boldsymbol{\sigma})$ which satisfies (6.30), (6.41) for all $t \in [0,T]$ is called a *weak solution* of Problem 6.7. Theorem 6.9 provides the unique weak solvability of Problem 6.7. Moreover, following the arguments used on page 176 it results that $\boldsymbol{\sigma} \in C([0,T]; Q_1)$.

6.2.2 Frictional contact with normal compliance

In the second problem the contact is modelled with normal compliance and a time-dependent version of Coulomb's law of dry friction. The classical formulation of the problem is the following.

Problem 6.10 *Find a displacement field* $\boldsymbol{u} : \Omega \times [0,T] \to \mathbb{R}^d$ *and a stress field* $\boldsymbol{\sigma} : \Omega \times [0,T] \to \mathbb{S}^d$ *such that, for all* $t \in [0,T]$,

$$\boldsymbol{\sigma}(t) = \mathcal{F}\boldsymbol{\varepsilon}(\boldsymbol{u}(t)) + \int_0^t \mathcal{R}(t-s)\boldsymbol{\varepsilon}(\boldsymbol{u}(s))\,ds \quad in \quad \Omega, \tag{6.46}$$

$$\operatorname{Div} \boldsymbol{\sigma}(t) + \boldsymbol{f}_0(t) = \boldsymbol{0} \quad in \quad \Omega, \tag{6.47}$$

$$\boldsymbol{u}(t) = \boldsymbol{0} \quad on \quad \Gamma_1, \tag{6.48}$$

$$\boldsymbol{\sigma}(t)\boldsymbol{\nu} = \boldsymbol{f}_2(t) \quad on \quad \Gamma_2, \tag{6.49}$$

$$-\sigma_\nu(t) = p_\nu(u_\nu(t) - g_a) \quad on \quad \Gamma_3, \tag{6.50}$$

$$\left.\begin{aligned} & \|\boldsymbol{\sigma}_\tau(t)\| \leq p_\tau(u_\nu(t) - g_a), \\ & \boldsymbol{\sigma}_\tau(t) = -\,p_\tau(u_\nu(t) - g_a)\,\frac{\boldsymbol{u}_\tau(t)}{\|\boldsymbol{u}_\tau(t)\|} \quad \text{if} \quad \boldsymbol{u}_\tau(t) \neq \boldsymbol{0} \end{aligned}\right\} \quad on \quad \Gamma_3. \tag{6.51}$$

The equations and boundary conditions in Problem 6.10 have a similar meaning to those in Problem 6.7. The difference arises from the fact that here we replace the frictionless contact conditions (6.34)–(6.35) with the frictional contact condition (6.50)–(6.51), which represents the time-dependent version of the frictional contact condition (5.181)–(5.182).

In the study of Problem 6.10 we assume that the elasticity operator $\mathcal{F}$ and the relaxation tensor $\mathcal{R}$ satisfy the conditions (4.33) and (4.47), respectively. We also assume that the densities of body forces and surface tractions have the regularity (6.8). Finally, the normal compliance functions satisfy condition (5.183) and the gap function verifies (5.8).

We use the notation (6.39) and, in addition, we introduce the functional $j : V \times V \to \mathbb{R}$ defined by

$$j(\boldsymbol{u}, \boldsymbol{v}) = \int_{\Gamma_3} p_\nu(u_\nu - g_{\rm a})|v_\nu|\,{\rm d}a + \int_{\Gamma_3} p_\tau(u_\nu - g_{\rm a})\|\boldsymbol{v}_\tau\|\,{\rm d}a \quad \forall\,\boldsymbol{u}, \boldsymbol{v} \in V. \tag{6.52}$$

Assume in what follows that $(\boldsymbol{u}, \boldsymbol{\sigma})$ are sufficiently regular functions which satisfy (6.46)–(6.51) and let $t \in [0, T]$ be given. We use arguments similar to those used in Section 5.3 to see that

$$(\boldsymbol{\sigma}(t), \boldsymbol{\varepsilon}(\boldsymbol{v}) - \boldsymbol{\varepsilon}(\boldsymbol{u}(t)))_Q + j(\boldsymbol{u}(t), \boldsymbol{v}) - j(\boldsymbol{u}(t), \boldsymbol{u}(t)) \geq (\boldsymbol{f}(t), \boldsymbol{v} - \boldsymbol{u}(t))_V \quad \forall\,\boldsymbol{v} \in V. \tag{6.53}$$

Then we substitute the constitutive law (6.46) in (6.53) to derive the following variational formulation of Problem 6.10.

Problem 6.11 *Find a displacement field $\boldsymbol{u} : [0, T] \to V$ such that, for all $t \in [0, T]$, the inequality below holds:*

$$(\mathcal{F}\boldsymbol{\varepsilon}(\boldsymbol{u}(t)), \boldsymbol{\varepsilon}(\boldsymbol{v}) - \boldsymbol{\varepsilon}(\boldsymbol{u}(t)))_Q + \Big(\int_0^t \mathcal{R}(t-s)\boldsymbol{\varepsilon}(\boldsymbol{u}(s))\,{\rm d}s, \boldsymbol{\varepsilon}(\boldsymbol{v}) - \boldsymbol{\varepsilon}(\boldsymbol{u}(t))\Big)_Q + j(\boldsymbol{u}(t), \boldsymbol{v}) - j(\boldsymbol{u}(t), \boldsymbol{u}(t)) \geq (\boldsymbol{f}(t), \boldsymbol{v} - \boldsymbol{u}(t))_V \quad \forall\,\boldsymbol{v} \in V. \tag{6.54}$$

In the study of Problem 6.11 we have the following existence and uniqueness result.

Theorem 6.12 *Assume (4.33), (4.47), (5.8), (5.183) and (6.8). Then there exists a constant L_0 which depends only on $\Omega, \Gamma_1, \Gamma_3$ and $\mathcal{F}$ such that Problem 6.11 has a unique solution $\boldsymbol{u} \in C([0, T]; V)$, if $L_\nu + L_\tau < L_0$.*

Proof We use similar arguments to those used in the proof of Theorem 6.9. The difference arises from the fact that here we choose $K = V$ and the functional j is defined by (6.52). Also, besides the operator $\mathcal{S} : C([0, T]; V) \to C([0, T]; V)$ defined by (6.43) we now use the operator $A : V \to V$ given by

$$(A\boldsymbol{v}, \boldsymbol{w})_V = (\mathcal{F}\boldsymbol{\varepsilon}(\boldsymbol{v}), \boldsymbol{\varepsilon}(\boldsymbol{w}))_Q \quad \forall\,\boldsymbol{v}, \boldsymbol{w} \in V. \tag{6.55}$$

We check in what follows the assumptions of Theorem 3.6. First, we note that (3.36) and (3.38) hold, since $K = V$ and $\varphi(\boldsymbol{u}, \boldsymbol{v}) = (\boldsymbol{u}, \boldsymbol{v})_V$. Next, we use (4.33) to see that the operator A verifies condition (3.37) with $m = m_{\mathcal{F}}$. In addition, as shown in the proof of Theorem 5.38, it follows that the functional j satisfies assumption (3.39) with $\alpha = c_0^2\,(L_\nu + L_\tau)$.

Let L_0 be given by (5.190) which, clearly, depends only on $\Omega, \Gamma_1, \Gamma_3$ and $\mathcal{F}$. Assume now that $L_\nu + L_\tau < L_0$; then we obtain $c_0^2\,(L_\nu + L_\tau) < m_{\mathcal{F}}$ which implies that $m > \alpha$ and, therefore, (3.40) holds. We recall that the operator (6.43) satisfies condition (3.41) with L_S given by (6.44). Finally, using (6.8) and (6.39) we deduce that $\boldsymbol{f} \in C([0,T];V)$ and, therefore, (3.42) holds.

It follows from the above that we are in position to apply Theorem 3.6. Thus, we deduce that there exists a unique function $\boldsymbol{u} \in C([0,T],V)$ which satisfies the inequality

$$(A\boldsymbol{u}(t), \boldsymbol{v} - \boldsymbol{u}(t))_V + (\mathcal{S}\boldsymbol{u}(t), \boldsymbol{v} - \boldsymbol{u}(t))_V \\ + j(\boldsymbol{u}(t), \boldsymbol{v}) - j(\boldsymbol{u}(t), \boldsymbol{u}(t)) \geq (\boldsymbol{f}(t), \boldsymbol{v} - \boldsymbol{u}(t))_V \quad \forall\, \boldsymbol{v} \in V, \tag{6.56}$$

for all $t \in [0,T]$. Theorem 6.12 is now a consequence of (6.55), (6.43) and (6.56). ■

A pair of functions $(\boldsymbol{u}, \boldsymbol{\sigma})$ which satisfies (6.46), (6.54) for all $t \in [0,T]$ is called a *weak solution* to Problem 6.10. We conclude that Theorem 6.12 provides the unique weak solvability of Problem 6.10. Moreover, using arguments on page 176 it follows that $\boldsymbol{\sigma} \in C([0,T];Q_1)$.

6.2.3 A convergence result

We now study the dependence of the solution to Problem 6.11 with respect to perturbations of both the relaxation operator $\mathcal{R}$ and the normal compliance functions p_ν and p_τ. To this end, we assume in what follows that (4.33), (4.47), (5.8), (5.183) and (6.8) hold. Moreover, we assume that

$$L_\nu + L_\tau < L_0,$$

where L_0 is defined by (5.190) and we denote by $\boldsymbol{u}$ the solution of Problem 6.11 obtained in Theorem 6.12.

For each $\rho > 0$ let $\mathcal{R}_\rho$ be a perturbation of the relaxation operator $\mathcal{R}$ which satisfies (4.47) and let p_e^ρ be a perturbation of p_e which satisfies (5.183) with the Lipschitz constant L_e^ρ $(e = \nu, \tau)$. We introduce the functional $j_\rho : V \times V \to \mathbb{R}$ defined by

$$j_\rho(\boldsymbol{u}, \boldsymbol{v}) = \int_{\Gamma_3} p_\nu^\rho(u_\nu - g_{\rm a})|v_\nu|\,{\rm d}a \\ + \int_{\Gamma_3} p_\tau^\rho(u_\nu - g_{\rm a})\|\boldsymbol{v}_\tau\|\,{\rm d}a \quad \forall\, \boldsymbol{u}, \boldsymbol{v} \in V$$

and we consider the following variational problem.

Problem 6.13 *Find a displacement field $\boldsymbol{u}_\rho : [0,T] \to V$ such that, for all $t \in [0,T]$, the inequality below holds:*

$$(\mathcal{F}\boldsymbol{\varepsilon}(\boldsymbol{u}_\rho(t)), \boldsymbol{\varepsilon}(\boldsymbol{v}) - \boldsymbol{\varepsilon}(\boldsymbol{u}_\rho(t)))_Q$$
$$+\Big(\int_0^t \mathcal{R}_\rho(t-s)(\boldsymbol{\varepsilon}(\boldsymbol{u}_\rho(s)))\,\mathrm{d}s, \boldsymbol{\varepsilon}(\boldsymbol{v}) - \boldsymbol{\varepsilon}(\boldsymbol{u}_\rho(t))\Big)_Q$$
$$+j_\rho(\boldsymbol{u}_\rho(t), \boldsymbol{v}) - j_\rho(\boldsymbol{u}_\rho(t), \boldsymbol{u}_\rho(t)) \geq (\boldsymbol{f}(t), \boldsymbol{v} - \boldsymbol{u}_\rho(t))_V \quad \forall \boldsymbol{v} \in V.$$

Assume that

$$L_\nu^\rho + L_\tau^\rho < L_0 \quad \forall \rho > 0.$$

Then it follows from Theorem 6.12 that, for each $\rho > 0$, Problem 6.13 has a unique solution $\boldsymbol{u}_\rho \in C([0,T];V)$.

Consider now the following assumptions on the relaxation operators $\mathcal{R}_\rho$ and $\mathcal{R}$ as well as the normal compliance functions p_e^ρ and p_e, for $e = \nu, \tau$.

$$\lim_{\rho \to 0} \|\mathcal{R}_\rho - \mathcal{R}\|_{C([0,T];\mathbf{Q}_\infty)} = 0. \tag{6.57}$$

$$\left.\begin{array}{l} \text{There exists } G_e : \mathbb{R}_+ \to \mathbb{R}_+ \text{ such that} \\ \text{(a) } |p_e^\rho(\boldsymbol{x}, r) - p_e(\boldsymbol{x}, r)| \leq G_e(\rho) \\ \quad\ \forall r \in \mathbb{R}, \text{ a.e. } \boldsymbol{x} \in \Gamma_3, \text{ for each } \rho > 0. \\ \text{(b) } \lim_{\rho \to 0} G_e(\rho) = 0. \end{array}\right\} \tag{6.58}$$

We have the following convergence result.

Theorem 6.14 *Under assumptions (6.57) and (6.58), the solution $\boldsymbol{u}_\rho$ of Problem 6.13 converges to the solution $\boldsymbol{u}$ of Problem 6.11, i.e.,*

$$\boldsymbol{u}_\rho \to \boldsymbol{u} \quad \textit{in } C([0,T];V) \quad \textit{as} \quad \rho \to 0. \tag{6.59}$$

Proof It follows from the proof of Theorem 6.12 that $\boldsymbol{u}$ is a solution of Problem 6.11 iff $\boldsymbol{u}$ solves inequality (6.56), for all $t \in [0,T]$. A similar comment can be applied to the solution of Problem 6.13. Therefore, the convergence result (6.59) will be obtained as a direct consequence of Theorem 3.9 with $K = X = Y = V$. To prove it, we only have to check the assumptions (3.59), (3.60) and (3.61) for the functions φ, j and the operator $\mathcal{S}$, respectively.

First, we note that the history-dependent quasivariational inequalities associated with Problems 6.11 and 6.13 involve the same function φ, i.e. $\varphi_\rho(\boldsymbol{u}, \boldsymbol{v}) = \varphi(\boldsymbol{u}, \boldsymbol{v}) = (\boldsymbol{u}, \boldsymbol{v})_V$, see (6.56). Therefore, (3.59) holds. Also, using (6.58) it follows that condition (3.60) is satisfied, too, as results from the proof of Theorem 5.41 on page 168. Next, let $t \in [0,T]$. It follows from (6.43) that

$$(\mathcal{S}_\rho \boldsymbol{u}(t), \boldsymbol{v})_V = \Big(\int_0^t \mathcal{R}_\rho(t-s)\boldsymbol{\varepsilon}(\boldsymbol{u}(s))\,\mathrm{d}s, \boldsymbol{\varepsilon}(\boldsymbol{v})\Big)_Q$$

$$(\mathcal{S}\boldsymbol{u}(t), \boldsymbol{v})_V = \Big(\int_0^t \mathcal{R}(t-s)\boldsymbol{\varepsilon}(\boldsymbol{u}(s))\,\mathrm{d}s, \boldsymbol{\varepsilon}(\boldsymbol{v})\Big)_Q$$

for all $\boldsymbol{u} \in C([0,T];V)$ and $\boldsymbol{v} \in V$. Therefore, using (4.46) yields

$$\begin{aligned}
&|(\mathcal{S}_\rho \boldsymbol{u}(t) - \mathcal{S}\boldsymbol{u}(t), \boldsymbol{v})_V| \\
&\quad \le d\Big(\int_0^t \|\mathcal{R}_\rho(t-s) - \mathcal{R}(t-s)\|_{\mathbf{Q}_\infty} \|\boldsymbol{\varepsilon}(\boldsymbol{u}(s))\|_Q\,\mathrm{d}s\Big)\|\boldsymbol{\varepsilon}(\boldsymbol{v})\|_Q \\
&\quad \le d \max_{s\in[0,T]} \|\mathcal{R}_\rho(s) - \mathcal{R}(s)\|_{\mathbf{Q}_\infty}\Big(\int_0^t \|\boldsymbol{u}(s)\|_V\,\mathrm{d}s\Big)\|\boldsymbol{v}\|_V
\end{aligned}$$

for all $\boldsymbol{u} \in C([0,T];V)$ and $\boldsymbol{v} \in V$. This inequality shows that

$$\begin{aligned}
&\|\mathcal{S}_\rho \boldsymbol{u}(t) - \mathcal{S}\boldsymbol{u}(t)\|_V \\
&\quad \le d\,\|\mathcal{R}_\rho - \mathcal{R}\|_{C([0,T];\mathbf{Q}_\infty)} \int_0^T \|\boldsymbol{u}(s)\|_V\,\mathrm{d}s
\end{aligned}$$

for all $\boldsymbol{u} \in C([0,T];V)$ and, therefore, (3.61)(a) holds with

$$H(\rho) = d\,\|\mathcal{R}_\rho - \mathcal{R}\|_{C([0,T];\mathbf{Q}_\infty)} \quad \text{and} \quad J(\boldsymbol{u}) = \int_0^T \|\boldsymbol{u}(s)\|_V\,\mathrm{d}s.$$

We use (6.44) to see that

$$L_{\mathcal{S}_\rho} = d\,\|\mathcal{R}_\rho\|_{C([0,T];\mathbf{Q}_\infty)}$$

and, keeping in mind (6.57), it follows that (3.61)(b) and (3.61)(c) hold, too. We conclude from the above that (3.59), (3.60) and (3.61) hold, which completes the proof. ■

In addition to the mathematical interest in the convergence result (6.59), it is of importance from a mechanical point of view, since it states that the weak solution of problem (6.46)–(6.51) depends continuously on the relaxation operator and the normal compliance functions, as well.

6.3 Viscoelastic contact problems with short memory

In this section we present several mathematical models of frictional contact for viscoelastic materials with short memory. The main feature of these models consists of the fact that their variational formulation is in the form of an evolutionary quasivariational inequality for the displacement field or, equivalently, a history-dependent quasivariational inequality for the velocity field.

6.3.1 *Contact with normal compliance*

For the first problem studied in this section we assume that the contact is modelled with the normal compliance condition, associated with Coulomb's law of dry friction. The classical model of the process is the following.

Problem 6.15 *Find a displacement field* $\boldsymbol{u} : \Omega \times [0,T] \to \mathbb{R}^d$ *and a stress field* $\boldsymbol{\sigma} : \Omega \times [0,T] \to \mathbb{S}^d$ *such that*

$$\boldsymbol{\sigma} = \mathcal{A}\boldsymbol{\varepsilon}(\dot{\boldsymbol{u}}) + \mathcal{B}\boldsymbol{\varepsilon}(\boldsymbol{u}) \quad \textit{in} \quad \Omega \times (0,T), \tag{6.60}$$

$$\mathrm{Div}\,\boldsymbol{\sigma} + \boldsymbol{f}_0 = \boldsymbol{0} \quad \textit{in} \quad \Omega \times (0,T), \tag{6.61}$$

$$\boldsymbol{u} = \boldsymbol{0} \quad \textit{on} \quad \Gamma_1 \times (0,T), \tag{6.62}$$

$$\boldsymbol{\sigma\nu} = \boldsymbol{f}_2 \quad \textit{on} \quad \Gamma_2 \times (0,T), \tag{6.63}$$

$$-\sigma_\nu = p_\nu(u_\nu - g_{\rm a}) \quad \textit{on} \quad \Gamma_3 \times (0,T), \tag{6.64}$$

$$\left.\begin{array}{l} \|\boldsymbol{\sigma}_\tau\| \le p_\tau(u_\nu - g_{\rm a}), \\ \boldsymbol{\sigma}_\tau = -\,p_\tau(u_\nu - g_{\rm a})\,\dfrac{\dot{\boldsymbol{u}}_\tau}{\|\dot{\boldsymbol{u}}_\tau\|} \quad \text{if} \quad \dot{\boldsymbol{u}}_\tau \ne \boldsymbol{0} \end{array}\right\} \quad \textit{on} \quad \Gamma_3 \times (0,T), \tag{6.65}$$

$$\boldsymbol{u}(0) = \boldsymbol{u}_0 \quad \textit{in} \quad \Omega. \tag{6.66}$$

Note that Problem 6.15 represents the viscoelastic version of Problem 5.36 studied in Section 5.3. Here, (6.65) is the quasistatic version of the friction law (5.182).

We turn to the variational formulation of Problem 6.15. To this end, as usual, we assume in what follows that the viscosity and elasticity operators satisfy (4.40) and (4.41), respectively. The densities of body forces and surface tractions have the regularity (6.8), the normal compliance functions p_e $(e = \nu, \tau)$ satisfy (5.183) and the initial displacement satisfies (6.10).

Assume now that $(\boldsymbol{u}, \boldsymbol{\sigma})$ are sufficiently regular functions which satisfy (6.60)–(6.66) and let $t \in [0,T]$. Then, it follows that

$$\begin{aligned} (\boldsymbol{\sigma}(t), \boldsymbol{\varepsilon}(\boldsymbol{v}) - \boldsymbol{\varepsilon}(\dot{\boldsymbol{u}}(t)))_Q + j(\boldsymbol{u}(t), \boldsymbol{v}) - j(\boldsymbol{u}(t), \dot{\boldsymbol{u}}(t)) \\ \ge (\boldsymbol{f}(t), \boldsymbol{v} - \dot{\boldsymbol{u}}(t))_V \quad \forall\, \boldsymbol{v} \in V, \end{aligned}$$

where, recall, $j : V \times V \to \mathbb{R}$ and $\boldsymbol{f} : [0,T] \to V$ are defined by (6.52) and (6.39), respectively. We combine this inequality with the constitutive law (6.60) and the initial condition (6.66) to obtain the following variational formulation of Problem 6.15.

Problem 6.16 *Find a displacement field* $\boldsymbol{u} : [0,T] \to V$ *such that*

$$(\mathcal{A}\boldsymbol{\varepsilon}(\dot{\boldsymbol{u}}(t)), \boldsymbol{\varepsilon}(\boldsymbol{v}) - \boldsymbol{\varepsilon}(\dot{\boldsymbol{u}}(t)))_Q + (\mathcal{B}\boldsymbol{\varepsilon}(\boldsymbol{u}(t)), \boldsymbol{\varepsilon}(\boldsymbol{v}) - \boldsymbol{\varepsilon}(\dot{\boldsymbol{u}}(t)))_Q$$
$$+ j(\boldsymbol{u}(t), \boldsymbol{v}) - j(\boldsymbol{u}(t), \dot{\boldsymbol{u}}(t)) \geq (\boldsymbol{f}(t), \boldsymbol{v} - \dot{\boldsymbol{u}}(t))_V$$
$$\forall \boldsymbol{v} \in V, \ t \in [0,T], \tag{6.67}$$

$$\boldsymbol{u}(0) = \boldsymbol{u}_0. \tag{6.68}$$

We remark that the functional j in (6.67) depends on the solution and, therefore, (6.67) represents an evolutionary quasivariational inequality for the displacement field. Moreover, we remark on the presence of the term $(\mathcal{A}\boldsymbol{\varepsilon}(\dot{\boldsymbol{u}}(t)), \boldsymbol{\varepsilon}(\boldsymbol{v}) - \boldsymbol{\varepsilon}(\boldsymbol{u}(t)))_Q$, governed by the viscosity operator $\mathcal{A}$. This justifies the terminology of *viscosity term* introduced on page 75.

The unique solvability of Problem 6.16 is given by the following existence and uniqueness result.

Theorem 6.17 *Assume (4.40), (4.41), (5.183), (6.8) and (6.10). Then there exists a unique solution* $\boldsymbol{u} \in C^1([0,T]; V)$ *to Problem 6.16.*

Proof We consider the operators $A : V \to V$ and $B : V \to V$ defined by equalities (6.16) and (6.17), respectively. We use assumption (4.40) to see that A is a strongly monotone Lipschitz continuous operator. Also, it follows from (4.41) that B is a Lipschitz continuous operator and recall the inequality (5.189) which shows that the functional j satisfies assumption (3.72) on the space $X = V$. Finally, recall the regularity $\boldsymbol{f} \in C([0,T]; V)$, $\boldsymbol{u}_0 \in V$. Therefore, it follows from Theorem 3.10 that there exists a unique function $\boldsymbol{u} \in C^1([0,T]; V)$ such that

$$(A\dot{\boldsymbol{u}}(t), \boldsymbol{v} - \dot{\boldsymbol{u}}(t))_V + (B\boldsymbol{u}(t), \boldsymbol{v} - \dot{\boldsymbol{u}}(t))_V + j(\boldsymbol{u}(t), \boldsymbol{v})$$
$$- j(\boldsymbol{u}(t), \dot{\boldsymbol{u}}(t)) \geq (\boldsymbol{f}(t), \boldsymbol{v} - \dot{\boldsymbol{u}}(t))_V \quad \forall \boldsymbol{v} \in V, \ t \in [0,T], \tag{6.69}$$

$$\boldsymbol{u}(0) = \boldsymbol{u}_0. \tag{6.70}$$

We now use (6.16), (6.17), (6.69) and (6.70) to see that $\boldsymbol{u}$ is the unique solution to the variational problem 6.16. ■

We proceed with a different variational formulation to Problem 6.15 in which the unknown is the velocity field. To this end we consider the operator $\mathcal{S} : C([0,T]; V) \to C([0,T]; V)$ defined by

$$\mathcal{S}\boldsymbol{v}(t) = \int_0^t \boldsymbol{v}(s)\, ds + \boldsymbol{u}_0 \qquad \forall \boldsymbol{v} \in C([0,T]; V), \ t \in [0,T]. \tag{6.71}$$

Let $\boldsymbol{w} = \dot{\boldsymbol{u}}$ denote the velocity field. Then, the initial condition $\boldsymbol{u}(0) = \boldsymbol{u}_0$ and (6.71) imply that $\boldsymbol{u} = \mathcal{S}\boldsymbol{w}$. Therefore, (6.67) leads to the following variational problem.

Problem 6.18 *Find a velocity field* $\boldsymbol{w} : [0,T] \to V$ *such that*

$$\begin{aligned}&(\mathcal{A}\boldsymbol{\varepsilon}(\boldsymbol{w}(t)), \boldsymbol{\varepsilon}(\boldsymbol{v}) - \boldsymbol{\varepsilon}(\boldsymbol{w}(t)))_Q + (\mathcal{B}\boldsymbol{\varepsilon}(\mathcal{S}\boldsymbol{w}(t)), \boldsymbol{\varepsilon}(\boldsymbol{v}) - \boldsymbol{\varepsilon}(\boldsymbol{w}(t)))_Q \\ &+ j(\mathcal{S}\boldsymbol{w}(t), \boldsymbol{v}) - j(\mathcal{S}\boldsymbol{w}(t)), \boldsymbol{w}(t)) \geq (\boldsymbol{f}(t), \boldsymbol{v} - \boldsymbol{w}(t))_V \\ &\forall \boldsymbol{v} \in V,\ t \in [0,T]. \qquad (6.72)\end{aligned}$$

In the study of Problem 6.18 we have the following result.

Theorem 6.19 *Assume (4.40), (4.41), (5.183), (6.8) and (6.10). Then Problem 6.18 has a unique solution* $\boldsymbol{w} \in C([0,T];V)$.

Proof Besides the operators A and $\mathcal{S}$ defined by (6.16) and (6.71), respectively, we consider the function $\varphi : V \times V \to \mathbb{R}$ defined by

$$\varphi(\boldsymbol{u}, \boldsymbol{v}) = (\mathcal{B}\boldsymbol{\varepsilon}(\boldsymbol{u}), \boldsymbol{\varepsilon}(\boldsymbol{v}))_Q + j(\boldsymbol{u}, \boldsymbol{v}) \qquad \forall\, \boldsymbol{u}, \boldsymbol{v} \in V. \qquad (6.73)$$

Then, Problem 6.18 is equivalent to the problem of finding a velocity field $\boldsymbol{w} : [0,T] \to V$ such that, for all $t \in [0,T]$,

$$\begin{aligned}&(A\boldsymbol{w}(t), \boldsymbol{v} - \boldsymbol{w}(t))_V + \varphi(\mathcal{S}\boldsymbol{w}(t), \boldsymbol{v}) - \varphi(\mathcal{S}\boldsymbol{w}(t), \boldsymbol{w}(t)) \\ &\geq (\boldsymbol{f}(t), \boldsymbol{v} - \boldsymbol{w}(t))_V \quad \forall\, \boldsymbol{v} \in V. \qquad (6.74)\end{aligned}$$

It is straightforward to see that inequality (6.74) represents a quasivariational inequality of the form (3.35) in which $X = Y = K = V$ and the corresponding function j vanishes. Therefore, in order to use Theorem 3.6 we note that by assumption (4.40) it follows that A is a strongly monotone Lipschitz continuous operator and assumption (6.8) implies that $\boldsymbol{f} \in C([0,T];V)$. Moreover, from (4.41), (5.183) and (4.13) it follows that the function φ defined by (6.73) satisfies (3.38)(a) and, in addition,

$$\begin{aligned}&\varphi(\boldsymbol{u}_1, \boldsymbol{v}_2) - \varphi(\boldsymbol{u}_1, \boldsymbol{v}_1) + \varphi(\boldsymbol{u}_2, \boldsymbol{v}_1) - \varphi(\boldsymbol{u}_2, \boldsymbol{v}_2) \\ &\leq L_{\mathcal{B}} \|\boldsymbol{u}_1 - \boldsymbol{u}_2\|_V \, \|\boldsymbol{v}_1 - \boldsymbol{v}_2\|_V \\ &\quad + c_0^2 (L_\nu + L_\tau)\, \|\boldsymbol{u}_1 - \boldsymbol{u}_2\|_V \, \|\boldsymbol{v}_1 - \boldsymbol{v}_2\|_V \\ &\forall\, \boldsymbol{u}_1, \boldsymbol{u}_2, \boldsymbol{v}_1, \boldsymbol{v}_2 \in V.\end{aligned}$$

The rest of the assumptions in Theorem 3.6 are clearly satisfied. Therefore, we conclude by this theorem that inequality (6.74) has a unique solution $\boldsymbol{w} \in C([0,T];V)$, which completes the proof. ■

Note that the link between the solutions $\boldsymbol{u}$ and $\boldsymbol{w}$ of Problems 6.16 and 6.18 is given by equalities $\boldsymbol{w} = \dot{\boldsymbol{u}}$ and $\boldsymbol{u} = \mathcal{S}\boldsymbol{w}$. This remark allows us to prove Theorem 6.17 by using Theorem 6.19. Moreover, when the velocity field and the displacement field are known, then the stress field $\boldsymbol{\sigma}$ can be easily obtained by using the constitutive law (6.60).

A brief comparison between Theorems 5.38 and 6.12 on the one hand, and Theorem 6.17 on the other hand, shows that:

(a) to provide the unique weak solvability of the elastic problem with normal compliance (5.177)–(5.182) as well as that of the viscoelastic problem with normal compliance (6.46)–(6.51), we need the smallness assumption $L_\nu + L_\tau < L_0$ on the normal compliance functions where, recall, L_0 is defined by (5.190);

(b) in contrast, to solve the viscoelastic problem with normal compliance (6.60)–(6.66), such an assumption is not needed. This feature arises from the presence of the viscosity term in the variational inequality (6.67).

We conclude from the above that the viscosity term has a smoothing effect in solving frictional contact problems with normal compliance.

6.3.2 Contact with normal damped response

The classical formulation of the second problem of frictional contact we consider in this section is the following.

Problem 6.20 *Find the displacement* $\boldsymbol{u} : \Omega \times [0,T] \to \mathbb{R}^d$ *and the stress field* $\boldsymbol{\sigma} : \Omega \times [0,T] \to \mathbb{S}^d$ *such that*

$$\boldsymbol{\sigma}(t) = \mathcal{A}\boldsymbol{\varepsilon}(\dot{\boldsymbol{u}}(t)) + \mathcal{B}\boldsymbol{\varepsilon}(\boldsymbol{u}(t)) \quad \text{in} \quad \Omega \times (0,T), \qquad (6.75)$$

$$\operatorname{Div} \boldsymbol{\sigma}(t) + \boldsymbol{f}_0(t) = \boldsymbol{0} \quad \text{in} \quad \Omega \times (0,T), \qquad (6.76)$$

$$\boldsymbol{u}(t) = \boldsymbol{0} \quad \text{on} \quad \Gamma_1 \times (0,T), \qquad (6.77)$$

$$\boldsymbol{\sigma}(t)\boldsymbol{\nu} = \boldsymbol{f}_2(t) \quad \text{on} \quad \Gamma_2 \times (0,T), \qquad (6.78)$$

$$-\sigma_\nu(t) = p_\nu(\dot{u}_\nu(t)) \quad \text{on} \quad \Gamma_3 \times (0,T), \qquad (6.79)$$

$$\left.\begin{array}{l} \|\boldsymbol{\sigma}_\tau(t)\| \le p_\tau(\dot{u}_\nu(t)), \\ \boldsymbol{\sigma}_\tau(t) = -\,p_\tau(\dot{u}_\nu(t))\,\dfrac{\dot{\boldsymbol{u}}_\tau(t)}{\|\dot{\boldsymbol{u}}_\tau(t)\|} \quad \text{if} \ \ \dot{\boldsymbol{u}}_\tau(t) \ne \boldsymbol{0} \end{array}\right\} \quad \text{on} \quad \Gamma_3 \times (0,T), \qquad (6.80)$$

$$\boldsymbol{u}(0) = \boldsymbol{u}_0 \quad \text{in} \quad \Omega \times (0,T). \qquad (6.81)$$

Note that the equations and conditions in Problem 6.20 have the same signification as those in Problem 6.15 except for the fact that here we replace the frictional contact conditions with normal compliance (6.64), (6.65) with the frictional contact conditions with normal damped response, (6.79), (6.80). Here, again, p_ν and p_τ represent given contact functions.

In the study of Problem 6.20 we assume that the viscosity and elasticity operators satisfy (4.40) and (4.41), respectively, the densities of body forces and surface tractions have the regularity (6.8) and the initial displacement verifies (6.10). In addition, we assume that the contact functions p_e ($e = \nu, \tau$) satisfy

$$\left.\begin{array}{l}
\text{(a) } p_e : \Gamma_3 \times \mathbb{R} \to \mathbb{R}_+;\\
\text{(b) there exists } L_e > 0 \text{ such that}\\
\qquad |p_e(\boldsymbol{x}, r_1) - p_e(\boldsymbol{x}, r_2)| \le L_e\, |r_1 - r_2|\\
\qquad\quad \forall\, r_1, r_2 \in \mathbb{R}, \text{ a.e. } \boldsymbol{x} \in \Gamma_3;\\
\text{(c) the mapping } \boldsymbol{x} \mapsto p_e(\boldsymbol{x}, r) \text{ is measurable on } \Gamma_3,\\
\qquad \text{for any } r \in \mathbb{R};\\
\text{(d) the mapping } \boldsymbol{x} \mapsto p_e(\boldsymbol{x}, 0) \text{ belongs to } L^2(\Gamma_3).
\end{array}\right\} \tag{6.82}$$

Note that the difference between assumptions (5.183) and (6.82) arises in condition (6.82)(d) which is more general that (5.183)(d). We also note that the results involving condition (5.183) presented in the previous sections still hold if we replace it with condition (6.82). Nevertheless, in the study of problems with normal compliance we made the choice to use (5.183) for physical reasons. Indeed, condition (5.183)(d) associated with the normal compliance condition shows that when there is separation between the body and the foundation, both the normal and the tangential reactions vanish, which represents an accurate description of the contact process. In contrast, this situation could not arise for contact models with normal damped response. Indeed, as mentioned on page 113, the normal damped response condition models situations when the surfaces are lubricated. And, in this case, even when there is separation between the body and the foundation, the fluid could fill the gap and, therefore, the normal and tangential reaction could not vanish. To conclude, condition (5.183)(d) is not appropriate in this case and, for this reason, we replace it by condition (6.82)(d).

We use notation (6.39) for the function $\boldsymbol{f}$. Moreover, we consider the functional $j : V \times V \to \mathbb{R}$ defined by

$$j(\boldsymbol{u}, \boldsymbol{v}) = \int_{\Gamma_3} p_\nu(u_\nu) v_\nu \, da + \int_{\Gamma_3} p_\tau(u_\nu) \|\boldsymbol{v}_\tau\| \, da$$
$$\forall\, \boldsymbol{u}, \boldsymbol{v} \in V. \tag{6.83}$$

Then it is easy to derive the following variational formulation of the contact Problem 6.20.

Problem 6.21 *Find a displacement field* $\boldsymbol{u} : [0, T] \to V$ *such that*

$$(\mathcal{A}\boldsymbol{\varepsilon}(\dot{\boldsymbol{u}}(t)), \boldsymbol{\varepsilon}(\boldsymbol{v}) - \boldsymbol{\varepsilon}(\dot{\boldsymbol{u}}(t)))_Q + (\mathcal{B}\boldsymbol{\varepsilon}(\boldsymbol{u}(t)), \boldsymbol{\varepsilon}(\boldsymbol{v}) - \boldsymbol{\varepsilon}(\dot{\boldsymbol{u}}(t)))_Q$$
$$+ j(\dot{\boldsymbol{u}}(t), \boldsymbol{v}) - j(\dot{\boldsymbol{u}}(t), \dot{\boldsymbol{u}}(t)) \ge (\boldsymbol{f}(t), \boldsymbol{v} - \dot{\boldsymbol{u}}(t))_V$$
$$\forall\, \boldsymbol{v} \in V, \ t \in [0, T], \tag{6.84}$$

$$\boldsymbol{u}(0) = \boldsymbol{u}_0. \tag{6.85}$$

The well-posedness of Problem 6.21 is given by the following result.

Theorem 6.22 *Assume (4.40), (4.41), (6.8), (6.10) and (6.82). Then, there exists* $L_0 > 0$ *which depends only on* Ω, Γ_1, Γ_3 *and* $\mathcal{A}$ *such that Problem 6.21 has a unique solution* $\boldsymbol{u} \in C^1([0,T];V)$, *if* $L_\nu + L_\tau < L_0$.

Proof We consider the operators $A : V \to V$ and $B : V \to V$ defined by equalities (6.16) and (6.17), respectively. We use assumption (4.40) to see that A is a strongly monotone Lipschitz continuous operator. Moreover, it satisfies condition (3.11) on the space $X = V$ with $m_A = m_{\mathcal{A}}$. Also, it follows from (4.41) that B is a Lipschitz continuous operator and recall the regularity $\boldsymbol{f} \in C([0,T];V)$ and $\boldsymbol{u}_0 \in V$. Finally, we use (6.82) and (4.13) to see that the function j defined by (6.83) satisfies condition (3.72)(a) and, in addition,

$$\begin{aligned} &j(\boldsymbol{u}_1, \boldsymbol{v}_2) - j(\boldsymbol{u}_1, \boldsymbol{v}_1) + j(\boldsymbol{u}_2, \boldsymbol{v}_1) - j(\boldsymbol{u}_2, \boldsymbol{v}_2) \\ &\qquad \le c_0^2 (L_\nu + L_\tau) \, \|\boldsymbol{u}_1 - \boldsymbol{u}_2\|_V \, \|\boldsymbol{v}_1 - \boldsymbol{v}_2\|_V \quad \forall\, \boldsymbol{u}_1, \boldsymbol{u}_2, \boldsymbol{v}_1, \boldsymbol{v}_2 \in V. \end{aligned}$$

This inequality shows that j satisfies condition (3.72)(b) with $\alpha = c_0^2(L_\nu + L_\tau)$. Let $L_0 = m_{\mathcal{A}}/c_0^2$ and note that L_0 depends only on Ω, Γ_1, Γ_3 and $\mathcal{A}$. Then, if $L_\nu + L_\tau < L_0$ we have $m_{\mathcal{A}} > \alpha$ and it follows from Theorem 3.11 that there exists a unique function $\boldsymbol{u} \in C^1([0,T];V)$ such that

$$\begin{aligned} &(A\dot{\boldsymbol{u}}(t), \boldsymbol{v} - \dot{\boldsymbol{u}}(t))_V + (B\boldsymbol{u}(t), \boldsymbol{v} - \dot{\boldsymbol{u}}(t))_V + j(\dot{\boldsymbol{u}}(t), \boldsymbol{v}) \\ &\quad - j(\dot{\boldsymbol{u}}(t), \dot{\boldsymbol{u}}(t)) \ge (\boldsymbol{f}(t), \boldsymbol{v} - \dot{\boldsymbol{u}}(t))_V \quad \forall\, \boldsymbol{v} \in V,\ t \in [0,T], \end{aligned} \tag{6.86}$$

$$\boldsymbol{u}(0) = \boldsymbol{u}_0. \tag{6.87}$$

We now use (6.16), (6.17), (6.86) and (6.87) to see that $\boldsymbol{u}$ is the unique solution to the variational problem 6.21, which concludes the proof. ∎

We proceed with a different variational formulation to Problem 6.20 in which the unknown is the velocity field. To this end we consider the operator $\mathcal{S} : C([0,T];V) \to C([0,T];V)$ defined by (6.71). Let $\boldsymbol{w} = \dot{\boldsymbol{u}}$ denote the velocity field. Then the initial condition $\boldsymbol{u}(0) = \boldsymbol{u}_0$ and (6.71) imply that $\boldsymbol{u} = \mathcal{S}\boldsymbol{w}$. Therefore, (6.84) leads to the following variational problem.

Problem 6.23 *Find a velocity field* $\boldsymbol{w} : [0,T] \to V$ *such that*

$$\begin{aligned} &(\mathcal{A}\boldsymbol{\varepsilon}(\boldsymbol{w}(t)), \boldsymbol{\varepsilon}(\boldsymbol{v}) - \boldsymbol{\varepsilon}(\boldsymbol{w}(t)))_Q + (\mathcal{B}\boldsymbol{\varepsilon}(\mathcal{S}\boldsymbol{w}(t)), \boldsymbol{\varepsilon}(\boldsymbol{v}) - \boldsymbol{\varepsilon}(\boldsymbol{w}(t)))_Q \\ &\quad + j(\boldsymbol{w}(t), \boldsymbol{v}) - j(\boldsymbol{w}(t), \boldsymbol{w}(t)) \ge (\boldsymbol{f}(t), \boldsymbol{v} - \boldsymbol{w}(t))_V \\ &\qquad \forall\, \boldsymbol{v} \in V,\ t \in [0,T]. \end{aligned} \tag{6.88}$$

In the study of Problem 6.23 we have the following result.

Theorem 6.24 *Assume (4.40), (4.41), (6.8), (6.10) and (6.82). Then there exists* $L_0 > 0$ *which depends only on* Ω, Γ_1, Γ_3 *and* $\mathcal{A}$ *such that Problem 6.23 has a unique solution* $\boldsymbol{w} \in C([0,T];V)$, *if* $L_\nu + L_\tau < L_0$.

Proof Besides the operators A and $\mathcal{S}$ defined by (6.16) and (6.71), respectively, we consider the function $\varphi : V \times V \to \mathbb{R}$, defined by

$$\varphi(\boldsymbol{u}, \boldsymbol{v}) = (\mathcal{B}\boldsymbol{\varepsilon}(\boldsymbol{u}), \boldsymbol{\varepsilon}(\boldsymbol{v}))_Q \qquad \forall\, \boldsymbol{u}, \boldsymbol{v} \in V. \tag{6.89}$$

Then Problem 6.23 is equivalent to the problem of finding a velocity field $\boldsymbol{w} : [0, T] \to V$ such that, for all $t \in [0, T]$,

$$\begin{aligned}(A\boldsymbol{w}(t), \boldsymbol{v} - \boldsymbol{w}(t))_V + \varphi(\mathcal{S}\boldsymbol{w}(t), \boldsymbol{v}) - \varphi(\mathcal{S}\boldsymbol{w}(t), \boldsymbol{w}(t))\\ + j(\boldsymbol{w}(t), \boldsymbol{v}) - j(\boldsymbol{w}(t), \boldsymbol{w}(t)) \geq (\boldsymbol{f}(t), \boldsymbol{v} - \boldsymbol{w}(t))_V \quad \forall\, \boldsymbol{v} \in V.\end{aligned} \tag{6.90}$$

It is easy to see that inequality (6.90) represents a quasivariational inequality of the form (3.35) in which $X = Y = K = V$. Therefore, in order to use Theorem 3.6 we note that by assumption (4.40) it follows that A is a strongly monotone Lipschitz continuous operator, i.e. it satisfies conditions (3.37) with $m = m_{\mathcal{A}}$. Note also that assumption (6.8) implies $\boldsymbol{f} \in C([0, T]; V)$ and, therefore, $\boldsymbol{f}$ satisfies (3.42). Moreover, from (4.41) it follows that the function φ defined by (6.89) satisfies (3.38) and, clearly the operator $\mathcal{S}$ verifies condition (3.41). Finally, the function j defined by (6.83) satisfies (3.39) with $\alpha = c_0^2(L_\nu + L_\tau)$, as shown in the proof of Theorem 6.22. Applying Theorem 3.6 we conclude that if $c_0^2(L_\nu + L_\tau) < m_{\mathcal{A}}$ then inequality (6.90) has a unique solution $\boldsymbol{w} \in C([0, T]; V)$ and we may take $L_0 = m_{\mathcal{A}}/c_0^2$. ■

Note that the link between the solutions $\boldsymbol{u}$ and $\boldsymbol{w}$ of Problems 6.21 and 6.23 are given by the equalities $\boldsymbol{w} = \dot{\boldsymbol{u}}$ and $\boldsymbol{u} = \mathcal{S}\boldsymbol{w}$. This remark allows us to prove Theorem 6.22 by using Theorem 6.24. Moreover, when the velocity field and the displacement field are known, then the stress field $\boldsymbol{\sigma}$ can be easily obtained by using the constitutive law (6.75).

A brief comparison between Theorem 6.22 and Theorem 6.17 shows that:

(a) to provide the unique weak solvability of the viscoelastic problem with normal damped response (6.75)–(6.81) we need the smallness assumption $L_\nu + L_\tau < L_0$ on the contact functions, already used in Theorems 5.38 and 6.12;

(b) in contrast, to solve the viscoelastic problem with normal compliance (6.60)–(6.66), such an assumption is not needed.

This feature arises from the fact that in (6.84) the velocity field $\dot{\boldsymbol{u}}$ appears in both the viscosity term and the functional j. In contrast, in (6.69) the velocity field appears only in the viscosity term $(\mathcal{A}\boldsymbol{\varepsilon}(\dot{\boldsymbol{u}}(t)), \boldsymbol{\varepsilon}(\boldsymbol{v}) - \boldsymbol{\varepsilon}(\dot{\boldsymbol{u}}(t)))_Q$.

6.3.3 Other frictional contact problems

In this subsection we study two additional frictional contact problems for viscoelastic materials. In the first problem we assume that the normal stress is prescribed on the contact surface and we use the quasistatic version of Coulomb's law of dry friction. The classical model of the process is the following.

Problem 6.25 *Find a displacement field* $\boldsymbol{u} : \Omega \times [0,T] \to \mathbb{R}^d$ *and a stress field* $\boldsymbol{\sigma} : \Omega \times [0,T] \to \mathbb{S}^d$ *such that*

$$\boldsymbol{\sigma} = \mathcal{A}\boldsymbol{\varepsilon}(\dot{\boldsymbol{u}}) + \mathcal{B}\boldsymbol{\varepsilon}(\boldsymbol{u}) \quad \text{in} \quad \Omega \times (0,T), \tag{6.91}$$

$$\operatorname{Div} \boldsymbol{\sigma} + \boldsymbol{f}_0 = \boldsymbol{0} \quad \text{in} \quad \Omega \times (0,T), \tag{6.92}$$

$$\boldsymbol{u} = \boldsymbol{0} \quad \text{on} \quad \Gamma_1 \times (0,T), \tag{6.93}$$

$$\boldsymbol{\sigma\nu} = \boldsymbol{f}_2 \quad \text{on} \quad \Gamma_2 \times (0,T), \tag{6.94}$$

$$-\sigma_\nu = F \quad \text{on} \quad \Gamma_3 \times (0,T), \tag{6.95}$$

$$\left.\begin{array}{l} \|\boldsymbol{\sigma}_\tau\| \le \mu\,|\sigma_\nu|, \\ \boldsymbol{\sigma}_\tau = -\mu\,|\sigma_\nu|\,\dfrac{\dot{\boldsymbol{u}}_\tau}{\|\dot{\boldsymbol{u}}_\tau\|} \quad \text{if} \quad \dot{\boldsymbol{u}}_\tau \neq \boldsymbol{0} \end{array}\right\} \quad \text{on} \quad \Gamma_3 \times (0,T), \tag{6.96}$$

$$\boldsymbol{u}(0) = \boldsymbol{u}_0 \quad \text{in} \quad \Omega. \tag{6.97}$$

In the second problem we use the contact condition (4.101) associated with Tresca's friction law (4.110), (4.87). Therefore, the classical model of the process is the following.

Problem 6.26 *Find a displacement field* $\boldsymbol{u} : \Omega \times [0,T] \to \mathbb{R}^d$ *and a stress field* $\boldsymbol{\sigma} : \Omega \times [0,T] \to \mathbb{S}^d$ *such that*

$$\boldsymbol{\sigma} = \mathcal{A}\boldsymbol{\varepsilon}(\dot{\boldsymbol{u}}) + \mathcal{B}\boldsymbol{\varepsilon}(\boldsymbol{u}) \quad \text{in} \quad \Omega \times (0,T), \tag{6.98}$$

$$\operatorname{Div} \boldsymbol{\sigma} + \boldsymbol{f}_0 = \boldsymbol{0} \quad \text{in} \quad \Omega \times (0,T), \tag{6.99}$$

$$\boldsymbol{u} = \boldsymbol{0} \quad \text{on} \quad \Gamma_1 \times (0,T), \tag{6.100}$$

$$\boldsymbol{\sigma\nu} = \boldsymbol{f}_2 \quad \text{on} \quad \Gamma_2 \times (0,T), \tag{6.101}$$

$$u_\nu = 0 \quad \text{on} \quad \Gamma_3 \times (0,T), \tag{6.102}$$

$$\left.\begin{array}{l} \|\boldsymbol{\sigma}_\tau\| \le F_b, \\ \boldsymbol{\sigma}_\tau = -F_b\,\dfrac{\dot{\boldsymbol{u}}_\tau}{\|\dot{\boldsymbol{u}}_\tau\|} \quad \text{if} \quad \dot{\boldsymbol{u}}_\tau \neq \boldsymbol{0} \end{array}\right\} \quad \text{on} \quad \Gamma_3 \times (0,T), \tag{6.103}$$

$$\boldsymbol{u}(0) = \boldsymbol{u}_0 \quad \text{in} \quad \Omega. \tag{6.104}$$

Note that Problems 6.25 and 6.26 represent the viscoelastic version of the frictional contact problems 5.17 and 5.18, respectively, studied in Section 5.2. Moreover, Problem 6.25 also represents a frictional version of Problem 6.1 studied in Section 6.1.

We turn to the variational formulation of Problems 6.25 and 6.26. To this end, as usual, we assume in what follows that the viscosity and the elasticity operators satisfy (4.40) and (4.41), respectively. The densities of body forces and surface tractions have the regularity (6.8) and the prescribed normal stress satisfies

$$F \in L^2(\Gamma_3) \quad \text{and} \quad F(\boldsymbol{x}) \geq 0 \text{ a.e. } \boldsymbol{x} \in \Gamma_3. \tag{6.105}$$

The coefficient of friction satisfies

$$\mu \in L^\infty(\Gamma_3) \quad \text{and} \quad \mu(\boldsymbol{x}) \geq 0 \text{ a.e. } \boldsymbol{x} \in \Gamma_3 \tag{6.106}$$

and the friction bound is such that

$$F_{\rm b} \in L^2(\Gamma_3) \quad \text{and} \quad F_{\rm b}(\boldsymbol{x}) \geq 0 \text{ a.e. } \boldsymbol{x} \in \Gamma_3. \tag{6.107}$$

Finally, in the case of Problem 6.25 we assume that the initial displacement satisfies

$$\boldsymbol{u}_0 \in V, \tag{6.108}$$

and in the case of Problem 6.26 we assume that

$$\boldsymbol{u}_0 \in V_1, \tag{6.109}$$

where, recall, V and V_1 denote the spaces given by (4.9) and (4.19), respectively.

In the study of Problem 6.25 we use the function $\boldsymbol{f} : [0,T] \to V$ defined by (6.11) (with the remark that now F does not depend on time) and the seminorm $j : V \to \mathbb{R}$ given by

$$j(\boldsymbol{v}) = \int_{\Gamma_3} \mu F \, \|\boldsymbol{v}_\tau\| \, da \quad \forall \boldsymbol{v} \in V. \tag{6.110}$$

Similarly, in the study of Problem 6.26 we use the function $\boldsymbol{f}_1 : [0,T] \to V_1$ defined by (6.27) and the functional $j_1 : V_1 \to \mathbb{R}$ given by

$$j_1(\boldsymbol{v}) = \int_{\Gamma_3} F_{\rm b} \, \|\boldsymbol{v}_\tau\| \, da \quad \forall \boldsymbol{v} \in V_1. \tag{6.111}$$

Assume now that the mechanical problem 6.25 has a regular solution $(\boldsymbol{u}, \boldsymbol{\sigma})$ and let $t \in [0,T]$. Using arguments similar to those used in the proof of (5.113) we obtain that

$$\begin{aligned}(\boldsymbol{\sigma}(t), \boldsymbol{\varepsilon}(\boldsymbol{v}) - \boldsymbol{\varepsilon}(\dot{\boldsymbol{u}}(t)))_Q + j(\boldsymbol{v}) - j(\dot{\boldsymbol{u}}(t)) \\ \geq (\boldsymbol{f}(t), \boldsymbol{v} - \dot{\boldsymbol{u}}(t))_V \quad \forall \boldsymbol{v} \in V.\end{aligned} \tag{6.112}$$

We now use the constitutive law (6.91), the initial condition (6.97) and inequality (6.112) to derive the following variational formulation of the frictionless contact problem 6.25.

Problem 6.27 *Find a displacement field* $\boldsymbol{u} : [0,T] \to V$ *such that*

$$(\mathcal{A}\boldsymbol{\varepsilon}(\dot{\boldsymbol{u}}(t)), \boldsymbol{\varepsilon}(\boldsymbol{v}) - \boldsymbol{\varepsilon}(\dot{\boldsymbol{u}}(t)))_Q + (\mathcal{B}\boldsymbol{\varepsilon}(\boldsymbol{u}(t)), \boldsymbol{\varepsilon}(\boldsymbol{v}) - \boldsymbol{\varepsilon}(\dot{\boldsymbol{u}}(t)))_Q + j(\boldsymbol{v}) - j(\dot{\boldsymbol{u}}(t)) \geq (\boldsymbol{f}(t), \boldsymbol{v} - \dot{\boldsymbol{u}}(t))_V \quad \forall\, \boldsymbol{v} \in V,\ t \in [0,T], \tag{6.113}$$

$$\boldsymbol{u}(0) = \boldsymbol{u}_0. \tag{6.114}$$

The variational formulation of the frictionless contact problem 6.26 can be obtained by similar arguments and is as follows.

Problem 6.28 *Find a displacement field* $\boldsymbol{u} : [0,T] \to V_1$ *such that*

$$(\mathcal{A}\boldsymbol{\varepsilon}(\dot{\boldsymbol{u}}(t)), \boldsymbol{\varepsilon}(\boldsymbol{v}) - \boldsymbol{\varepsilon}(\dot{\boldsymbol{u}}(t)))_Q + (\mathcal{B}\boldsymbol{\varepsilon}(\boldsymbol{u}(t)), \boldsymbol{\varepsilon}(\boldsymbol{v}) - \boldsymbol{\varepsilon}(\dot{\boldsymbol{u}}(t)))_Q + j_1(\boldsymbol{v}) - j_1(\dot{\boldsymbol{u}}(t)) \geq (\boldsymbol{f}_1(t), \boldsymbol{v} - \dot{\boldsymbol{u}}(t))_V \quad \forall\, \boldsymbol{v} \in V_1,\ t \in [0,T], \tag{6.115}$$

$$\boldsymbol{u}(0) = \boldsymbol{u}_0. \tag{6.116}$$

Using the terminology introduced on page 76, we conclude from the above that the variational formulations of the frictional contact problems 6.25 and 6.26 are given by evolutionary variational inequalities for the displacement field.

The unique solvability of Problem 6.27 is given by the following existence and uniqueness result.

Theorem 6.29 *Assume (4.40), (4.41), (6.8), (6.105), (6.106) and (6.108). Then there exists a unique solution* $\boldsymbol{u} \in C^1([0,T];V)$ *to Problem 6.27.*

Proof We apply the Riesz representation theorem to define the operators $A : V \to V$ and $B : V \to V$ by equalities (6.16) and (6.17), respectively. We use assumption (4.40) to see that A is a strongly monotone Lipschitz continuous operator. Also, it follows from (4.41) that B is a Lipschitz continuous operator and recall the inequality (5.116) which shows that the seminorm j is continuous and, therefore, it is a convex lower semicontinuous function on V. Finally, note that $\boldsymbol{f} \in C([0,T];V)$ and $\boldsymbol{u}_0 \in V$. We now apply Corollary 3.12 to obtain the existence of a unique function $\boldsymbol{u} \in C^1([0,T];V)$, which satisfies

$$(A\dot{\boldsymbol{u}}(t), \boldsymbol{v} - \dot{\boldsymbol{u}}(t))_V + (B\boldsymbol{u}(t), \boldsymbol{v} - \dot{\boldsymbol{u}}(t))_V + j(\boldsymbol{v}) - j(\dot{\boldsymbol{u}}(t)) \geq (\boldsymbol{f}(t), \boldsymbol{v} - \dot{\boldsymbol{u}}(t))_V \quad \forall\, \boldsymbol{v} \in V,\ t \in [0,T], \tag{6.117}$$

$$\boldsymbol{u}(0) = \boldsymbol{u}_0. \tag{6.118}$$

We now use (6.16), (6.17), (6.117) and (6.118) to see that $\boldsymbol{u}$ is the unique solution to the variational problem 6.27. ■

The unique solvability of Problem 6.28 is given by the following existence and uniqueness result.

Theorem 6.30 *Assume (4.40), (4.41), (6.8), (6.107) and (6.109). Then there exists a unique solution* $\boldsymbol{u} \in C^1([0,T]; V_1)$ *to Problem 6.28.*

Proof Define the operators $A_1 : V_1 \to V_1$ and $B_1 : V_1 \to V_1$ by the equalities

$$(A_1\boldsymbol{u}, \boldsymbol{v})_V = (\mathcal{A}\boldsymbol{\varepsilon}(\boldsymbol{u}), \boldsymbol{\varepsilon}(\boldsymbol{v}))_Q \quad \forall\, \boldsymbol{u},\, \boldsymbol{v} \in V_1, \tag{6.119}$$

$$(B_1\boldsymbol{u}, \boldsymbol{v})_V = (\mathcal{B}\boldsymbol{\varepsilon}(\boldsymbol{u}), \boldsymbol{\varepsilon}(\boldsymbol{v}))_Q \quad \forall\, \boldsymbol{u},\, \boldsymbol{v} \in V_1. \tag{6.120}$$

It follows that A_1 is a strongly monotone Lipschitz continuous operator and B_1 is a Lipschitz continuous operator. Also, the functional j_1 defined by (6.111) is a continuous seminorm on the space V_1. Finally, recall the regularity (6.28) and (6.109). Theorem 6.30 is now a consequence of Corollary 3.12. ■

6.3.4 *Convergence results*

We end this section with results concerning the dependence of the solution with respect to the elasticity operator and the frictional conditions. To avoid repetition, we restrict ourselves to the study of Problems 6.21 and 6.27.

We start with Problem 6.21. To this end, we assume in what follows that (4.40), (4.41), (6.8), (6.10) and (6.82) hold. Moreover, we assume that

$$L_\nu + L_\tau < L_0,$$

where $L_0 = m_{\mathcal{A}}/c_0^2$ and we denote by $\boldsymbol{u}$ the solution of Problem 6.21 obtained in Theorem 6.22.

For each $\rho > 0$ let $\mathcal{B}_\rho$ be a perturbation of the elasticity operator $\mathcal{B}$ which satisfies (4.41) and let p_e^ρ be a perturbation of p_e which satisfies (6.82) with the Lipschitz constant L_e^ρ ($e = \nu, \tau$). We introduce the functional $j_\rho : V \times V \to \mathbb{R}$ defined by

$$j_\rho(\boldsymbol{u}, \boldsymbol{v}) = \int_{\Gamma_3} p_\nu^\rho(u_\nu) v_\nu \,\mathrm{d}a + \int_{\Gamma_3} p_\tau^\rho(u_\nu) \|\boldsymbol{v}_\tau\| \,\mathrm{d}a \quad \forall\, \boldsymbol{u}, \boldsymbol{v} \in V$$

and we consider the following variational problem.

Problem 6.31 *Find a displacement field* $\boldsymbol{u}_\rho : [0,T] \to V$ *such that*

$$\begin{aligned} &(\mathcal{A}\boldsymbol{\varepsilon}(\dot{\boldsymbol{u}}_\rho(t)), \boldsymbol{\varepsilon}(\boldsymbol{v}) - \boldsymbol{\varepsilon}(\dot{\boldsymbol{u}}_\rho(t)))_Q + (\mathcal{B}_\rho\boldsymbol{\varepsilon}(\boldsymbol{u}_\rho(t)), \boldsymbol{\varepsilon}(\boldsymbol{v}) - \boldsymbol{\varepsilon}(\dot{\boldsymbol{u}}_\rho(t)))_Q \\ &\quad + j_\rho(\dot{\boldsymbol{u}}_\rho(t), \boldsymbol{v}) - j_\rho(\dot{\boldsymbol{u}}_\rho(t), \dot{\boldsymbol{u}}_\rho(t)) \geq (\boldsymbol{f}(t), \boldsymbol{v} - \dot{\boldsymbol{u}}_\rho(t))_V \\ &\qquad \forall\, \boldsymbol{v} \in V,\ t \in [0,T], \end{aligned}$$

$$\boldsymbol{u}_\rho(0) = \boldsymbol{u}_0.$$

Assume that

$$L^\rho_\nu + L^\rho_\tau < L_0 \quad \forall\, \rho > 0.$$

Then it follows from Theorem 6.22 that, for each $\rho > 0$, Problem 6.31 has a unique solution $\boldsymbol{u}_\rho \in C([0,T];V)$.

Consider now the following assumptions on the elasticity operators $\mathcal{B}_\rho$ and $\mathcal{B}$ as well as on normal compliance functions p^ρ_e and p_e, for $e = \nu, \tau$.

$$\left.\begin{array}{l}\text{There exists } F : \mathbb{R}_+ \to \mathbb{R}_+ \text{ such that}\\[4pt] \text{(a) } \|\mathcal{B}_\rho(\boldsymbol{x},\boldsymbol{\varepsilon}) - \mathcal{B}(\boldsymbol{x},\boldsymbol{\varepsilon})\| \le F(\rho)\\ \qquad \forall\, \boldsymbol{\varepsilon} \in \mathbb{S}^d, \text{ a.e. } \boldsymbol{x} \in \Omega, \text{ for each } \rho > 0;\\[4pt] \text{(b) } \lim\limits_{\rho\to 0} F(\rho) = 0.\end{array}\right\} \qquad (6.121)$$

$$\left.\begin{array}{l}\text{There exists } G_e : \mathbb{R}_+ \to \mathbb{R}_+ \text{ such that}\\[4pt] \text{(a) } |p^\rho_e(\boldsymbol{x},r) - p_e(\boldsymbol{x},r)| \le G_e(\rho)\\ \qquad \forall\, r \in \mathbb{R}, \text{ a.e. } \boldsymbol{x} \in \Gamma_3, \text{ for each } \rho > 0;\\[4pt] \text{(b) } \lim\limits_{\rho\to 0} G_e(\rho) = 0.\end{array}\right\} \qquad (6.122)$$

We have the following convergence result.

Theorem 6.32 *Under assumptions (6.121) and (6.122), the solution* $\boldsymbol{u}_\rho$ *of Problem 6.31 converges to the solution* $\boldsymbol{u}$ *of Problem 6.21, i.e.,*

$$\boldsymbol{u}_\rho \to \boldsymbol{u} \quad \text{in } C^1([0,T];V) \quad \text{as} \quad \rho \to 0. \qquad (6.123)$$

Proof The convergence result (6.123) is a direct consequence of Theorem 3.13. To prove it we only have to check the assumptions (3.82) and (3.83) for the corresponding operator B and the function j, respectively.

Let $\rho > 0$. First, we note that equality (6.17) and assumption (6.121)(a) imply that

$$\begin{aligned}(B_\rho \boldsymbol{v} - B\boldsymbol{v}, \boldsymbol{w})_V &= (\mathcal{B}_\rho \boldsymbol{\varepsilon}(\boldsymbol{v}) - \mathcal{B}\boldsymbol{\varepsilon}(\boldsymbol{v}), \boldsymbol{\varepsilon}(\boldsymbol{w}))_Q\\ &\le \int_\Omega \|\mathcal{B}_\rho \boldsymbol{\varepsilon}(\boldsymbol{v}) - \mathcal{B}\boldsymbol{\varepsilon}(\boldsymbol{v})\|\,\|\boldsymbol{\varepsilon}(\boldsymbol{w})\|\,\mathrm{d}x \le F(\rho)\int_\Omega \|\boldsymbol{\varepsilon}(\boldsymbol{w})\|\,\mathrm{d}x\\ &\le c\,F(\rho)\|\boldsymbol{\varepsilon}(\boldsymbol{w})\|_Q = c\,F(\rho)\|\boldsymbol{w}\|_V\end{aligned}$$

for all $\boldsymbol{v}, \boldsymbol{w} \in V$, where c is a positive constant which depends only on Ω. It follows from this that

$$\|B_\rho \boldsymbol{v} - B\boldsymbol{v}\|_V \le c\,F(\rho) \quad \forall\, \boldsymbol{v} \in V. \qquad (6.124)$$

Inequality (6.124) combined with assumption (6.121)(b) show that condition (3.82) holds.

Next, using (6.122) it follows that condition (3.83) is satisfied, too, as it results from the proof of Theorem 5.41. The convergence (6.123) is now a consequence of Theorem 3.13. ■

We proceed with an example of an elasticity operator $\mathcal{B}$ for which condition (6.121) is satisfied. Let $g > 0$ be given and denote by K the von Mises convex defined by (4.57). We also denote by $\mathcal{P}_K : \mathbb{S}^d \to K$ the projection operator on K and, given a positive function $\beta \in L^\infty(\Omega)$, we consider the elasticity operator $\mathcal{B} : \Omega \times \mathbb{S}^d \to \mathbb{S}^d$ defined by

$$\mathcal{B}(\boldsymbol{x}, \boldsymbol{\varepsilon}) = \beta(\boldsymbol{x})(\boldsymbol{\varepsilon} - \mathcal{P}_K\boldsymbol{\varepsilon}) \quad \forall\, \boldsymbol{\varepsilon} \in \mathbb{S}^d, \text{ a.e. } \boldsymbol{x} \in \Omega. \tag{6.125}$$

We use the regularity $\beta \in L^\infty(\Omega)$ and the nonexpansivity property of the projection operator to see that the elasticity operator (6.125) satisfies condition (4.41).

In the same way, for a given $g_\rho > 0$, we denote by $K(\rho)$ the von Mises convex defined by

$$K_\rho = \{\, \boldsymbol{\varepsilon} \in \mathbb{S}^d \;:\; \|\boldsymbol{\varepsilon}^{\mathrm{D}}\| \le g_\rho\sqrt{2}\,\}.$$

Let $\mathcal{P}_{K_\rho} : \mathbb{S}^d \to K_\rho$ be the projection operator on K_ρ and consider the elasticity operator $\mathcal{B}_\rho : \Omega \times \mathbb{S}^d \to \mathbb{S}^d$ defined by

$$\mathcal{B}_\rho(\boldsymbol{x}, \boldsymbol{\varepsilon}) = \beta(\boldsymbol{x})(\boldsymbol{\varepsilon} - \mathcal{P}_{K_\rho}\boldsymbol{\varepsilon}) \quad \forall\, \boldsymbol{\varepsilon} \in \mathbb{S}^d, \text{ a.e. } \boldsymbol{x} \in \Omega, \tag{6.126}$$

which, again, satisfies condition (4.41).

Let $\boldsymbol{\varepsilon} \in \mathbb{S}^d$. Then, using (6.125), (6.126) and Proposition 4.6 it follows that

$$\begin{aligned} \|\mathcal{B}_\rho(\boldsymbol{x}, \boldsymbol{\varepsilon}) - \mathcal{B}(\boldsymbol{x}, \boldsymbol{\varepsilon})\| &\le |\beta(\boldsymbol{x})|\, \|\mathcal{P}_K\boldsymbol{\varepsilon} - \mathcal{P}_{K_\rho}\boldsymbol{\varepsilon}\| \\ &\le \sqrt{2}\, |g_\rho - g|\, |\beta(\boldsymbol{x})| \quad \text{a.e. } \boldsymbol{x} \in \Omega, \end{aligned}$$

which implies that

$$\|\mathcal{B}_\rho(\boldsymbol{x}, \boldsymbol{\varepsilon}) - \mathcal{B}(\boldsymbol{x}, \boldsymbol{\varepsilon})\| \le \sqrt{2}\, |g_\rho - g|\, \|\beta\|_{L^\infty(\Omega)} \quad \text{a.e. } \boldsymbol{x} \in \Omega. \tag{6.127}$$

It follows now from (6.127) that the operators $\mathcal{B}_\rho$ and $\mathcal{B}$ defined above satisfy condition (6.121)(a) with

$$F(\rho) = \sqrt{2}\, |g_\rho - g|\, \|\beta\|_{L^\infty(\Omega)}.$$

Also, assuming that $g_\rho \to g$ as $\rho \to 0$, we deduce that condition (6.121)(b) holds, too.

We present in what follows examples of functions for which assumption (6.122) is satisfied. Let $d_\nu > 0$ and let $p_0^\rho > 0$, $p_0 > 0$ be such that $p_0^\rho \to p_0$ as $\rho \to 0$. Assume that the function p_ν is given by (4.107) and the function

p_τ is given by (4.95) where $\mu > 0$. Also, assume that p_ν^ρ and p_τ^ρ are given by

$$p_\nu^\rho(r) = d_\nu\, r_+ + p_0^\rho, \tag{6.128}$$

$$p_\tau^\rho = \mu p_\nu^\rho. \tag{6.129}$$

Then it is easy to check that the functions p_e^ρ, p_e satisfy condition (6.82) for $e = \nu, \tau$ and, moreover, assumption (6.122) is valid. A similar conclusion arises when p_ν is given by (4.107), p_τ is given by (4.95),

$$p_\nu^\rho(r) = d_\nu\left(\sqrt{r_+^2 + \rho^2} - \rho\right) + p_0^\rho \tag{6.130}$$

and p_τ^ρ is given by (6.129). We conclude that the convergence result in Theorem 5.41 is valid in the examples above.

We now proceed with Problem 6.27. To this end, we assume in what follows that (4.40), (4.41), (6.8), (6.105), (6.106) and (6.108) hold and we denote by $\boldsymbol{u} \in C^1([0,T];V)$ the solution of Problem 6.27 obtained in Theorem 6.29. Also, for each $\rho > 0$, assume that F_ρ and μ_ρ are given functions which satisfy (6.105) and (6.106), respectively, and $\mathcal{B}_\rho$ is an operator which satisfies (4.41). Denote by $j_\rho : V \to \mathbb{R}$ the functional defined by

$$j_\rho(\boldsymbol{v}) = \int_{\Gamma_3} \mu_\rho\, F_\rho\, \|\boldsymbol{v}_\tau\|\, da \quad \forall\, \boldsymbol{v} \in V, \tag{6.131}$$

and consider the following problem.

Problem 6.33 *Find a displacement field $\boldsymbol{u}_\rho : [0,T] \to V$ such that*

$$\begin{aligned}(\mathcal{A}\boldsymbol{\varepsilon}(\dot{\boldsymbol{u}}_\rho(t)), \boldsymbol{\varepsilon}(\boldsymbol{v}) - \boldsymbol{\varepsilon}(\dot{\boldsymbol{u}}_\rho(t)))_Q + (\mathcal{B}_\rho\boldsymbol{\varepsilon}(\boldsymbol{u}_\rho(t)), \boldsymbol{\varepsilon}(\boldsymbol{v}) - \boldsymbol{\varepsilon}(\dot{\boldsymbol{u}}_\rho(t)))_Q \\ + j_\rho(\boldsymbol{v}) - j_\rho(\dot{\boldsymbol{u}}_\rho(t)) \ge (\boldsymbol{f}(t), \boldsymbol{v} - \dot{\boldsymbol{u}}_\rho(t))_V \quad \forall\, \boldsymbol{v} \in V,\ t \in [0,T],\end{aligned} \tag{6.132}$$

$$\boldsymbol{u}_\rho(0) = \boldsymbol{u}_0. \tag{6.133}$$

We deduce from Theorem 6.29 that the Cauchy problem (6.132)–(6.133) has a unique solution $\boldsymbol{u}_\rho \in C^1([0,T];V)$, for each $\rho > 0$.

Assume that

$$F_\rho \to F \quad \text{in } L^2(\Gamma_3) \quad \text{as } \rho \to 0, \tag{6.134}$$

$$\mu_\rho \to \mu \quad \text{in } L^\infty(\Gamma_3) \quad \text{as } \rho \to 0. \tag{6.135}$$

Then the behavior of the solution $\boldsymbol{u}_\rho$ as ρ converges to zero is given in the following theorem.

Theorem 6.34 *Under the assumptions (6.121), (6.134) and (6.135), the solution $\boldsymbol{u}_\rho$ of Problem 6.33 converges to the solution $\boldsymbol{u}$ of Problem 6.27, i.e.*

$$\boldsymbol{u}_\rho \to \boldsymbol{u} \quad in\ C^1([0,T], V) \quad as \quad \rho \to 0. \tag{6.136}$$

Proof Let $\rho > 0$ and let $\boldsymbol{u}_1, \boldsymbol{u}_2 \in V$. We use (6.110), (6.131) and (5.124) to see that

$$
\begin{aligned}
& j_\rho(\boldsymbol{u}_1) - j_\rho(\boldsymbol{u}_2) + j(\boldsymbol{u}_2) - j(\boldsymbol{u}_1) \\
& \quad = \int_{\Gamma_3} (\mu_\rho F_\rho - \mu F)(\|\boldsymbol{u}_{1\tau}\| - \|\boldsymbol{u}_{2\tau}\|)\, da \\
& \quad \le \int_{\Gamma_3} |\mu_\rho F_\rho - \mu F|\, \|\boldsymbol{u}_1 - \boldsymbol{u}_2\|\, da \\
& \quad \le \int_{\Gamma_3} |\mu_\rho|\, |F_\rho - F|\, \|\boldsymbol{u}_1 - \boldsymbol{u}_2\|\, da + \int_{\Gamma_3} |\mu_\rho - \mu|\, |F|\, \|\boldsymbol{u}_1 - \boldsymbol{u}_2\|\, da
\end{aligned}
$$

and, therefore, by (4.13) it follows that

$$
\begin{aligned}
& j_\rho(\boldsymbol{u}_1) - j_\rho(\boldsymbol{u}_2) + j(\boldsymbol{u}_2) - j(\boldsymbol{u}_1) \\
& \quad \le c_0\, \|\mu_\rho\|_{L^\infty(\Gamma_3)} \|F_\rho - F\|_{L^2(\Gamma_3)} \|\boldsymbol{u}_1 - \boldsymbol{u}_2\|_V \\
& \quad + c_0\, \|\mu_\rho - \mu\|_{L^\infty(\Gamma_3)} \|F\|_{L^2(\Gamma_3)} \|\boldsymbol{u}_1 - \boldsymbol{u}_2\|_V. \qquad (6.137)
\end{aligned}
$$

Take

$$
G(\rho) = c_0\, \|\mu_\rho\|_{L^\infty(\Gamma_3)} \|F_\rho - F\|_{L^2(\Gamma_3)} + c_0\, \|\mu_\rho - \mu\|_{L^\infty(\Gamma_3)} \|F\|_{L^2(\Gamma_3)}.
$$

Then, by (6.134), (6.135) and (6.137) it follows that the functionals j_ρ and j satisfy condition (3.90) on the space $X = V$.

Moreover, recall that, as shown in the proof of Theorem 6.32, it follows that the corresponding operators B_ρ and B satisfy condition (3.82), too. Theorem 6.34 is now a consequence of Corollary 3.14, page 80. ■

Note that, besides the mathematical interest in the convergence results (6.123) and (6.136), they are of importance from a mechanical point of view, since they state that the solutions of the corresponding contact problems depend continuously both on the elasticity operator and the frictional contact conditions.

6.4 Viscoplastic frictionless contact problems

In this section we present two mathematical models of frictionless contact for viscoplastic materials under the assumption that there is no gap between the contact surface and the foundation. The main feature of these models consists of the fact that their variational formulation leads to a history-dependent quasivariational inequality for the displacement field.

6.4.1 Contact with normal compliance

The classical formulation of the first problem, in which the contact is modeled with normal compliance, is the following.

Problem 6.35 *Find a displacement field* $\boldsymbol{u} : \Omega \times [0,T] \to \mathbb{R}^d$ *and a stress field* $\boldsymbol{\sigma} : \Omega \times [0,T] \to \mathbb{S}^d$ *such that*

$$\dot{\boldsymbol{\sigma}} = \mathcal{E}\boldsymbol{\varepsilon}(\dot{\boldsymbol{u}}) + \mathcal{G}(\boldsymbol{\sigma}, \boldsymbol{\varepsilon}(\boldsymbol{u})) \quad \text{in} \quad \Omega \times (0,T), \tag{6.138}$$

$$\operatorname{Div} \boldsymbol{\sigma} + \boldsymbol{f}_0 = \boldsymbol{0} \quad \text{in} \quad \Omega \times (0,T), \tag{6.139}$$

$$\boldsymbol{u} = \boldsymbol{0} \quad \text{on} \quad \Gamma_1 \times (0,T), \tag{6.140}$$

$$\boldsymbol{\sigma\nu} = \boldsymbol{f}_2 \quad \text{on} \quad \Gamma_2 \times (0,T), \tag{6.141}$$

$$-\sigma_\nu = p(u_\nu) \quad \text{on} \quad \Gamma_3 \times (0,T), \tag{6.142}$$

$$\boldsymbol{\sigma}_\tau = \boldsymbol{0} \quad \text{on} \quad \Gamma_3 \times (0,T), \tag{6.143}$$

$$\boldsymbol{u}(0) = \boldsymbol{u}_0, \quad \boldsymbol{\sigma}(0) = \boldsymbol{\sigma}_0 \quad \text{in} \quad \Omega. \tag{6.144}$$

Equation (6.138) represents the viscoplastic constitutive law of the material introduced on page 98. Equation (6.139) is the equilibrium equation, (6.140) and (6.141) are the displacement and traction boundary conditions, while (6.142) and (6.143) represent the normal compliance and the frictionless contact conditions, respectively. Here and everywhere in this section we use the notation p for the normal compliance function, instead of p_ν, in order to simplify the notation. Finally, (6.144) represents the initial conditions in which $\boldsymbol{u}_0$ and $\boldsymbol{\sigma}_0$ denote the initial displacement and the initial stress field, respectively.

We now list the assumptions on the data and derive the variational formulation of Problem 6.35, then we state and prove its unique weak solvability. To this end we assume that the elasticity operator $\mathcal{E}$ and the nonlinear constitutive function $\mathcal{G}$ satisfy conditions (4.50) and (4.51), respectively. The normal compliance function satisfies condition (6.36). We also assume that the densities of body forces and surface tractions have the regularity (6.8) and, as usual, we use Riesz's representation theorem to define the operator $P : V \to V$ and the function $\boldsymbol{f} : [0,T] \to V$ by equalities (6.38) and (6.39), respectively. The initial data are such that

$$\boldsymbol{u}_0 \in V, \quad \boldsymbol{\sigma}_0 \in Q. \tag{6.145}$$

Alternatively, besides (6.145), we assume that the initial data satisfy the compatibility condition

$$(\boldsymbol{\sigma}_0, \boldsymbol{\varepsilon}(\boldsymbol{v}))_Q + (P\boldsymbol{u}_0, \boldsymbol{v})_V = (\boldsymbol{f}(0), \boldsymbol{v})_V \quad \forall\, \boldsymbol{v} \in V. \tag{6.146}$$

Note that this condition yields

$$\operatorname{Div} \boldsymbol{\sigma}_0 + \boldsymbol{f}_0(0) = \boldsymbol{0} \quad \text{in } \Omega.$$

Therefore, since $\boldsymbol{f}_0(0) \in L^2(\Omega)^d$ it follows that $\operatorname{Div} \boldsymbol{\sigma}_0 \in L^2(\Omega)^d$, which shows that $\boldsymbol{\sigma}_0 \in Q_1$.

Assume in what follows that $(\boldsymbol{u}, \boldsymbol{\sigma})$ are sufficiently regular functions which satisfy (6.138)–(6.144) and let $t \in [0,T]$ be given. We integrate the constitutive law (6.138) with the initial conditions (6.144) to obtain

$$\boldsymbol{\sigma}(t) = \mathcal{E}\boldsymbol{\varepsilon}(\boldsymbol{u}(t)) + \int_0^t \mathcal{G}(\boldsymbol{\sigma}(s), \boldsymbol{\varepsilon}(\boldsymbol{u}(s)))\,\mathrm{d}s + \boldsymbol{\sigma}_0 - \mathcal{E}\boldsymbol{\varepsilon}(\boldsymbol{u}_0). \tag{6.147}$$

Then we use Green's formula (4.22), the equilibrium equation (6.139), the boundary conditions (6.140)–(6.143) and the notation (6.38), (6.39) to see that

$$(\boldsymbol{\sigma}(t), \boldsymbol{\varepsilon}(\boldsymbol{v}))_Q + (P\boldsymbol{u}(t), \boldsymbol{v})_V = (\boldsymbol{f}(t), \boldsymbol{v})_V \quad \forall \boldsymbol{v} \in V. \tag{6.148}$$

We gather the two equalities above to obtain the following variational formulation of Problem 6.35.

Problem 6.36 *Find a displacement field $\boldsymbol{u} : [0,T] \to V$ and a stress field $\boldsymbol{\sigma} : [0,T] \to Q$ such that (6.147) and (6.148) hold, for all $t \in [0,T]$.*

In the study of Problem 6.36 we have the following existence and uniqueness result.

Theorem 6.37 *Assume (4.50), (4.51), (6.8), (6.36) and (6.145). Then:*

(1) *Problem* 6.36 *has a unique solution which satisfies*

$$\boldsymbol{u} \in C([0,T];V), \qquad \boldsymbol{\sigma} \in C([0,T];Q_1); \tag{6.149}$$

(2) *if, in addition, the initial data verify the compatibility condition (6.146) then*

$$\boldsymbol{u}(0) = \boldsymbol{u}_0, \qquad \boldsymbol{\sigma}(0) = \boldsymbol{\sigma}_0.$$

We turn now to the proof of the theorem. To this end we start with the following existence and uniqueness result.

Lemma 6.38 *Assume (4.50), (4.51) and (6.145). Then, for each function $\boldsymbol{u} \in C([0,T];V)$ there exists a unique function $\mathcal{S}\boldsymbol{u} \in C([0,T];Q)$ such that, for all $t \in [0,T]$,*

$$\mathcal{S}\boldsymbol{u}(t) = \int_0^t \mathcal{G}(\mathcal{S}\boldsymbol{u}(s) + \mathcal{E}\boldsymbol{\varepsilon}(\boldsymbol{u}(s)), \boldsymbol{\varepsilon}(\boldsymbol{u}(s)))\,\mathrm{d}s + \boldsymbol{\sigma}_0 - \mathcal{E}\boldsymbol{\varepsilon}(\boldsymbol{u}_0). \tag{6.150}$$

Moreover, the operator $\mathcal{S} : C([0,T];V) \to C([0,T];Q)$ is history-dependent, i.e. there exists $L_{\mathcal{S}} > 0$ such that

$$\|\mathcal{S}\boldsymbol{u}_1(t) - \mathcal{S}\boldsymbol{u}_2(t)\|_Q \le L_{\mathcal{S}} \int_0^t \|\boldsymbol{u}_1(s) - \boldsymbol{u}_2(s)\|_V\,\mathrm{d}s$$
$$\forall \boldsymbol{u}_1, \boldsymbol{u}_2 \in C([0,T];V),\ \forall t \in [0,T]. \tag{6.151}$$

Proof Let $\boldsymbol{u} \in C([0,T];V)$ and consider the operator $\Lambda : C([0,T];Q) \to C([0,T];Q)$ defined by

$$\Lambda\boldsymbol{\tau}(t) = \int_0^t \mathcal{G}(\boldsymbol{\tau}(s) + \mathcal{E}\boldsymbol{\varepsilon}(\boldsymbol{u}(s)), \boldsymbol{\varepsilon}(\boldsymbol{u}(s)))\mathrm{d}s + \boldsymbol{\sigma}_0 - \mathcal{E}\boldsymbol{\varepsilon}(\boldsymbol{u}_0)$$
$$\forall\, \boldsymbol{\tau} \in C([0,T];Q),\ \forall\, t \in [0,T]. \tag{6.152}$$

The operator Λ depends on $\boldsymbol{u}$ but, for simplicity, we do not indicate this dependence explicitly.

Let $\boldsymbol{\tau}_1$, $\boldsymbol{\tau}_2 \in C([0,T];Q)$ and let $t \in [0,T]$. Then, using (4.51) we have

$$\begin{aligned}
&\|\Lambda\boldsymbol{\tau}_1(t) - \Lambda\boldsymbol{\tau}_2(t)\|_Q \\
&= \Big\| \int_0^t \mathcal{G}(\boldsymbol{\tau}_1 + \mathcal{E}\boldsymbol{\varepsilon}(\boldsymbol{u}), \boldsymbol{\varepsilon}(\boldsymbol{u}))\mathrm{d}s - \int_0^t \mathcal{G}(\boldsymbol{\tau}_2 + \mathcal{E}\boldsymbol{\varepsilon}(\boldsymbol{u}), \boldsymbol{\varepsilon}(\boldsymbol{u}))\,\mathrm{d}s \Big\|_Q \\
&\le L_{\mathcal{G}} \int_0^t \|\boldsymbol{\tau}_1(s) - \boldsymbol{\tau}_2(s)\|_Q\,\mathrm{d}s.
\end{aligned}$$

Next, we use Proposition 3.1 on page 59 to see that Λ has a unique fixed point in $C([0,T];Q)$, denoted by $\mathcal{S}\boldsymbol{u}$. And, finally, we combine (6.152) with the equality $\mathcal{S}\boldsymbol{u} = \Lambda(\mathcal{S}\boldsymbol{u})$ to see that (6.150) holds.

Let $\boldsymbol{u}_1$, $\boldsymbol{u}_2 \in C([0,T];V)$ and let $t \in [0,T]$. Then, using (6.150), (4.50) and (4.51) we have

$$\begin{aligned}
&\|\mathcal{S}\boldsymbol{u}_1(t) - \mathcal{S}\boldsymbol{u}_2(t)\|_Q \\
&\le \mathcal{K}\Big(\int_0^t \|\boldsymbol{\varepsilon}(\boldsymbol{u}_1(s)) - \boldsymbol{\varepsilon}(\boldsymbol{u}_2(s))\|_Q\,\mathrm{d}s + \int_0^t \|\mathcal{S}\boldsymbol{u}_1(s) - \mathcal{S}\boldsymbol{u}_2(s)\|_Q\,\mathrm{d}s \Big) \\
&= \mathcal{K}\Big(\int_0^t \|\boldsymbol{u}_1(s) - \boldsymbol{u}_2(s)\|_V\,\mathrm{d}s + \int_0^t \|\mathcal{S}\boldsymbol{u}_1(s) - \mathcal{S}\boldsymbol{u}_2(s)\|_Q\,\mathrm{d}s \Big),
\end{aligned}$$

where $\mathcal{K}$ is a positive constant which depends on $\mathcal{G}$ and $\mathcal{E}$. Using now Gronwall's lemma we deduce that

$$\|\mathcal{S}\boldsymbol{u}_1(t) - \mathcal{S}\boldsymbol{u}_2(t)\|_Q \le \mathcal{K}\,\mathrm{e}^{T\mathcal{K}} \int_0^t \|\boldsymbol{u}_1(s) - \boldsymbol{u}_2(s)\|_V\,\mathrm{d}s.$$

This inequality shows that (6.151) holds with $L_{\mathcal{S}} = \mathcal{K}\,\mathrm{e}^{T\mathcal{K}}$. ■

Next, we use the operator $\mathcal{S} : C([0,T];V) \to C([0,T];Q)$ defined in Lemma 6.38 to obtain the following equivalence result.

Lemma 6.39 *Assume (4.50), (4.51), (6.8), (6.36) and (6.145), and let* $(\boldsymbol{u}, \boldsymbol{\sigma})$ *be a pair of functions which satisfies (6.149). Then* $(\boldsymbol{u}, \boldsymbol{\sigma})$ *is a* *solution of Problem 6.36 if and only if for all $t \in [0,T]$, the following equalities hold:*

$$\boldsymbol{\sigma}(t) = \mathcal{E}\boldsymbol{\varepsilon}(\boldsymbol{u}(t)) + \mathcal{S}\boldsymbol{u}(t), \tag{6.153}$$

$$\begin{aligned}(\mathcal{E}\boldsymbol{\varepsilon}(\boldsymbol{u}(t)), \boldsymbol{\varepsilon}(\boldsymbol{v}))_Q + (\mathcal{S}\boldsymbol{u}(t), \boldsymbol{\varepsilon}(\boldsymbol{v}))_Q \\ +(P\boldsymbol{u}(t), \boldsymbol{v})_V = (\boldsymbol{f}(t), \boldsymbol{v})_V \quad \forall \boldsymbol{v} \in V.\end{aligned} \tag{6.154}$$

Proof Assume that $(\boldsymbol{u}, \boldsymbol{\sigma})$ is a solution of Problem 6.36 and let $t \in [0,T]$. By (6.147) we have

$$\begin{aligned}&\boldsymbol{\sigma}(t) - \mathcal{E}\boldsymbol{\varepsilon}(\boldsymbol{u}(t)) \\ &\quad = \int_0^t \mathcal{G}(\boldsymbol{\sigma}(s) - \mathcal{E}\boldsymbol{\varepsilon}(\boldsymbol{u}(s)) + \mathcal{E}\boldsymbol{\varepsilon}(\boldsymbol{u}(s)), \boldsymbol{\varepsilon}(\boldsymbol{u}(s)))\,\mathrm{d}s + \boldsymbol{\sigma}_0 - \mathcal{E}\boldsymbol{\varepsilon}(\boldsymbol{u}_0)\end{aligned}$$

and, using the definition (6.150) of the operator $\mathcal{S}$, we obtain (6.153). Moreover, combining (6.148) and (6.153) we deduce (6.154).

Conversely, assume that $(\boldsymbol{u}, \boldsymbol{\sigma})$ satisfies (6.153) and (6.154) and let $t \in [0,T]$. Then by (6.153) and the definition (6.150) of the operator $\mathcal{S}$ we obtain (6.147). Moreover, using (6.153) and (6.154) we find that (6.148) holds, which concludes the proof. ∎

We are now in position to provide the proof of Theorem 6.37.

Proof (1) We define the operator $A : V \to V$ and the functional $\varphi : Q \times V \to \mathbb{R}$ by the equalities

$$(A\boldsymbol{u}, \boldsymbol{v})_V = (\mathcal{E}\boldsymbol{\varepsilon}(\boldsymbol{u}), \boldsymbol{\varepsilon}(\boldsymbol{v}))_Q + (P\boldsymbol{u}, \boldsymbol{v})_V \quad \forall \boldsymbol{u}, \boldsymbol{v} \in V, \tag{6.155}$$

$$\varphi(\boldsymbol{\tau}, \boldsymbol{v}) = (\boldsymbol{\tau}, \boldsymbol{\varepsilon}(\boldsymbol{v}))_Q \quad \forall \boldsymbol{\tau} \in Q,\ \boldsymbol{v} \in V. \tag{6.156}$$

Then we consider the problem of finding a function $\boldsymbol{u} : [0,T] \to V$ such that, for all $t \in [0,T]$, the following inequality holds:

$$\begin{aligned}(A\boldsymbol{u}(t), \boldsymbol{v} - \boldsymbol{u}(t))_V + \varphi(\mathcal{S}\boldsymbol{u}(t), \boldsymbol{v}) - \varphi(\mathcal{S}\boldsymbol{u}(t), \boldsymbol{u}(t)) \\ \geq (\boldsymbol{f}(t), \boldsymbol{v} - \boldsymbol{u}(t))_V \quad \forall \boldsymbol{v} \in V.\end{aligned} \tag{6.157}$$

To solve (6.157) we employ Theorem 3.6 with $X = K = V$, $Y = Q$ and $j \equiv 0$. We use (4.50), (6.36) and (4.13) to see that the operator A is strongly monotone and Lipschitz continuous, i.e. it verifies condition (3.37). In addition, we note that the functional φ satisfies condition (3.38) and we recall (6.151) which shows that (3.41) holds, too. Finally, using (6.8) and (6.39) we deduce that $\boldsymbol{f}$ has the regularity expressed in (3.42). It follows now from Theorem 3.6 that there exists a unique function $\boldsymbol{u} \in C([0,T];V)$ which solves the inequality (6.157), for all $t \in [0,T]$.

Next, we note that (6.157) is equivalent to the equation (6.154). Indeed, let $\boldsymbol{v}$ be an arbitrary element of V and let $t \in [0,T]$. We test in (6.157) with $\boldsymbol{u}(t) \pm \boldsymbol{v}$ and use the definitions (6.155)–(6.156) to obtain (6.154). Conversely, we test in (6.154) with $\boldsymbol{v} - \boldsymbol{u}(t)$, then we use the definitions (6.155)–(6.156) again to obtain (6.157).

Based on the results above we deduce the existence of a unique function $\boldsymbol{u} \in C([0,T];V)$ which satisfies (6.154) for any $t \in [0,T]$. Let $\boldsymbol{\sigma}$ be defined by (6.153) and note that, by arguments similar to those used on page 176, it follows that $\boldsymbol{\sigma} \in C([0,T];Q_1)$. Next, it follows that the pair $(\boldsymbol{u}, \boldsymbol{\sigma})$ is the unique pair of functions with regularity (6.149) which satisfies (6.153)–(6.154). Theorem 6.37 is now a consequence of Lemma 6.39.

(2) Assume the compatibility condition (6.146). We write (6.147) and (6.148) for $t = 0$ to obtain

$$\boldsymbol{\sigma}(0) = \mathcal{E}\boldsymbol{\varepsilon}(\boldsymbol{u}(0)) + \boldsymbol{\sigma}_0 - \mathcal{E}\boldsymbol{\varepsilon}(\boldsymbol{u}_0), \tag{6.158}$$

$$(\boldsymbol{\sigma}(0), \boldsymbol{\varepsilon}(\boldsymbol{v}))_Q + (P\boldsymbol{u}(0), \boldsymbol{v})_V = (\boldsymbol{f}(0), \boldsymbol{v})_V \quad \forall\, \boldsymbol{v} \in V. \tag{6.159}$$

We substitute (6.158) in (6.159), then we use (6.146) to find that

$$(\mathcal{E}\boldsymbol{\varepsilon}(\boldsymbol{u}(0)) - \mathcal{E}\boldsymbol{\varepsilon}(\boldsymbol{u}_0), \boldsymbol{\varepsilon}(\boldsymbol{v}))_Q + (P\boldsymbol{u}(0) - P\boldsymbol{u}_0, \boldsymbol{v})_V = 0 \quad \forall\, \boldsymbol{v} \in V.$$

We now take $\boldsymbol{v} = \boldsymbol{u}(0) - \boldsymbol{u}_0$ in this inequality and use the monotonicity of the operator P to see that

$$(\mathcal{E}\boldsymbol{\varepsilon}(\boldsymbol{u}(0)) - \mathcal{E}\boldsymbol{\varepsilon}(\boldsymbol{u}_0), \boldsymbol{\varepsilon}(\boldsymbol{u}(0)) - \boldsymbol{\varepsilon}(\boldsymbol{u}_0))_Q \le 0.$$

This inequality combined with assumption (4.50)(d) implies that

$$\|\boldsymbol{u}(0) - \boldsymbol{u}_0\|_V \le 0.$$

We conclude from this that $\boldsymbol{u}(0) = \boldsymbol{u}_0$. Therefore, from (6.158) we obtain $\boldsymbol{\sigma}(0) = \boldsymbol{\sigma}_0$, which concludes the proof. ■

6.4.2 Contact with unilateral constraint

In the second problem the contact is modelled with normal compliance and unilateral constraint. The classical formulation of the problem is the following.

Problem 6.40 *Find a displacement field* $\boldsymbol{u} : \Omega \times [0,T] \to \mathbb{R}^d$ *and a stress field* $\boldsymbol{\sigma} : \Omega \times [0,T] \to \mathbb{S}^d$ *such that*

$$\dot{\boldsymbol{\sigma}} = \mathcal{E}\boldsymbol{\varepsilon}(\dot{\boldsymbol{u}}) + \mathcal{G}(\boldsymbol{\sigma}, \boldsymbol{\varepsilon}(\boldsymbol{u})) \quad in \quad \Omega \times (0,T), \tag{6.160}$$

$$\operatorname{Div} \boldsymbol{\sigma} + \boldsymbol{f}_0 = \boldsymbol{0} \quad in \quad \Omega \times (0,T), \tag{6.161}$$

$$\boldsymbol{u} = \boldsymbol{0} \quad on \quad \Gamma_1 \times (0,T), \tag{6.162}$$

$$\boldsymbol{\sigma\nu} = \boldsymbol{f}_2 \quad on \quad \Gamma_2 \times (0,T), \tag{6.163}$$

$$\left.\begin{array}{l} u_\nu \le g, \quad \sigma_\nu + p(u_\nu) \le 0, \\ (u_\nu - g)(\sigma_\nu + p(u_\nu)) = 0 \end{array}\right\} \quad on \quad \Gamma_3 \times (0,T), \tag{6.164}$$

$$\boldsymbol{\sigma}_\tau = \boldsymbol{0} \quad on \quad \Gamma_3 \times (0,T), \tag{6.165}$$

$$\boldsymbol{u}(0) = \boldsymbol{u}_0, \ \boldsymbol{\sigma}(0) = \boldsymbol{\sigma}_0 \quad in \quad \Omega. \tag{6.166}$$

Note that the equations and conditions in Problem 6.40 have the same meaning as those in Problem 6.35. The difference arises from the fact that in Problem 6.40 we replaced the contact condition with normal compliance (6.142) with the contact condition with normal compliance and unilateral constraint (6.164) in which, recall, $g \ge 0$. Also, note that Problem 6.40 represents the viscoplastic version of the viscoelastic problem 6.7 on page 178.

In the study of Problem 6.40 we assume (4.50), (4.51), (6.8), (6.36) and (6.145). Alternatively, we assume that the initial data satisfy the compatibility conditions

$$\boldsymbol{u}_0 \in U, \tag{6.167}$$

$$(\boldsymbol{\sigma}_0, \boldsymbol{\varepsilon}(\boldsymbol{v}) - \boldsymbol{\varepsilon}(\boldsymbol{u}_0))_Q + (P\boldsymbol{u}_0, \boldsymbol{v} - \boldsymbol{u}_0)_V \ge (\boldsymbol{f}(0), \boldsymbol{v} - \boldsymbol{u}_0)_V \quad \boldsymbol{v} \in U, \tag{6.168}$$

where U denotes the set of admissible displacements defined by (6.37). Using similar arguments to those used on page 201 it follows that (6.168) implies $\boldsymbol{\sigma}_0 \in Q_1$.

Assume now that $(\boldsymbol{u}, \boldsymbol{\sigma})$ are sufficiently regular functions which satisfy (6.160)–(6.166) and, again, let $t \in [0,T]$ be given. Then (6.147) holds and, moreover,

$$\begin{aligned} &(\boldsymbol{\sigma}(t), \boldsymbol{\varepsilon}(\boldsymbol{v}) - \boldsymbol{\varepsilon}(\boldsymbol{u}(t)))_Q + (P\boldsymbol{u}(t), \boldsymbol{v} - \boldsymbol{u}(t))_V \\ &\qquad \ge (\boldsymbol{f}(t), \boldsymbol{v} - \boldsymbol{u}(t))_V \quad \forall\, \boldsymbol{v} \in U, \end{aligned} \tag{6.169}$$

where $\boldsymbol{f} : [0,T] \to V$ denotes the function defined by (6.39). In addition, we note that the first inequality in (6.164) and (6.37) imply that $\boldsymbol{u}(t) \in U$.

We deduce from the above the following variational formulation of Problem 6.40.

Problem 6.41 *Find a displacement field* $\boldsymbol{u} : [0, T] \to U$ *and a stress field* $\boldsymbol{\sigma} : [0, T] \to Q$ *such that (6.147) and (6.169) hold, for all* $t \in [0, T]$.

In the study of Problem 6.41 we have the following existence and uniqueness result.

Theorem 6.42 *Assume (4.50), (4.51), (6.8), (6.36) and (6.145). Then:*

(1) *Problem 6.41 has a unique solution which satisfies*

$$\boldsymbol{u} \in C([0, T]; U), \qquad \boldsymbol{\sigma} \in C([0, T]; Q_1); \tag{6.170}$$

(2) *if, in addition, the initial data verify the compatibility conditions (6.167), (6.168) then*

$$\boldsymbol{u}(0) = \boldsymbol{u}_0, \qquad \boldsymbol{\sigma}(0) = \boldsymbol{\sigma}_0.$$

In order to provide the proof of Theorem 6.42, besides the operator $\mathcal{S}$ defined in Lemma 6.38, we need the following version of Lemma 6.39.

Lemma 6.43 *Assume (4.50), (4.51), (6.8), (6.36), (6.145) and let* $(\boldsymbol{u}, \boldsymbol{\sigma})$ *be a pair of functions which satisfies (6.170). Then* $(\boldsymbol{u}, \boldsymbol{\sigma})$ *is a solution of Problem 6.41 if and only if for all* $t \in [0, T]$, *the equality and inequality below hold:*

$$\boldsymbol{\sigma}(t) = \mathcal{E}\boldsymbol{\varepsilon}(\boldsymbol{u}(t)) + \mathcal{S}\boldsymbol{u}(t), \tag{6.171}$$

$$\begin{aligned}(\mathcal{E}\boldsymbol{\varepsilon}(\boldsymbol{u}(t)), \boldsymbol{\varepsilon}(\boldsymbol{v}) - \boldsymbol{\varepsilon}(\boldsymbol{u}(t)))_Q + (\mathcal{S}\boldsymbol{u}(t), \boldsymbol{\varepsilon}(\boldsymbol{v}) - \boldsymbol{\varepsilon}(\boldsymbol{u}(t)))_Q \\ +(P\boldsymbol{u}(t), \boldsymbol{v} - \boldsymbol{u}(t))_V \geq (\boldsymbol{f}(t), \boldsymbol{v} - \boldsymbol{u}(t))_V \quad \forall \boldsymbol{v} \in U.\end{aligned} \tag{6.172}$$

The proof of Lemma 6.43 is similar to that of Lemma 6.39 and, therefore, we omit it. We are now in position to provide the proof of Theorem 6.42.

Proof (1) We use arguments similar to those used in the proof of Theorem 6.37. To avoid repetitions we skip the details and we restrict ourselves to the description of the main steps in the proof. First, we use notation (6.155)–(6.156) and Theorem 3.6 with $X = V$, $Y = Q$, $K = U$ and $j \equiv 0$ to prove the existence of a unique function $\boldsymbol{u} \in C([0, T]; U)$ which satisfies (6.172), for all $t \in [0, T]$. Then we define the function $\boldsymbol{\sigma}$ by (6.171) and we see that the pair $(\boldsymbol{u}, \boldsymbol{\sigma})$ satisfies (6.171)–(6.172) for all $t \in [0, T]$ and, moreover, it has the regularity (6.170). This concludes the existence part in Theorem 6.42. The uniqueness part follows from the uniqueness of the solution of the inequality (6.172), guaranteed by Theorem 3.6.

(2) We write (6.147) for $t = 0$ and substitute the result in (6.169) at $t = 0$ to obtain

$$\begin{aligned}(\mathcal{E}\boldsymbol{\varepsilon}(\boldsymbol{u}(0)) - \mathcal{E}\boldsymbol{\varepsilon}(\boldsymbol{u}_0), \boldsymbol{\varepsilon}(\boldsymbol{v}) - \boldsymbol{\varepsilon}(\boldsymbol{u}(0)))_Q + (\boldsymbol{\sigma}_0, \boldsymbol{\varepsilon}(\boldsymbol{v}) - \boldsymbol{\varepsilon}(\boldsymbol{u}(0)))_Q \\ +(P\boldsymbol{u}(0), \boldsymbol{v} - \boldsymbol{u}(0))_V \geq (\boldsymbol{f}(0), \boldsymbol{v} - \boldsymbol{u}(0))_V \quad \forall \boldsymbol{v} \in U.\end{aligned}$$

Then we take $\boldsymbol{v} = \boldsymbol{u}_0$ in this inequality and note that this choice is allowed by assumption (6.167). We obtain

$$\begin{aligned}&(\mathcal{E}\boldsymbol{\varepsilon}(\boldsymbol{u}(0)) - \mathcal{E}\boldsymbol{\varepsilon}(\boldsymbol{u}_0), \boldsymbol{\varepsilon}(\boldsymbol{u}_0) - \boldsymbol{\varepsilon}(\boldsymbol{u}(0)))_Q \\ &\quad + (\boldsymbol{\sigma}_0, \boldsymbol{\varepsilon}(\boldsymbol{u}_0) - \boldsymbol{\varepsilon}(\boldsymbol{u}(0)))_Q + (P\boldsymbol{u}(0), \boldsymbol{u}_0 - \boldsymbol{u}(0))_V \\ &\quad \geq (\boldsymbol{f}(0), \boldsymbol{u}_0 - \boldsymbol{u}(0))_V. \end{aligned} \tag{6.173}$$

Next, we take $\boldsymbol{v} = \boldsymbol{u}(0)$ in (6.168) and add the resulting inequality to (6.173) to see that

$$\begin{aligned}&(\mathcal{E}\boldsymbol{\varepsilon}(\boldsymbol{u}(0)) - \mathcal{E}\boldsymbol{\varepsilon}(\boldsymbol{u}_0), \boldsymbol{\varepsilon}(\boldsymbol{u}(0)) - \boldsymbol{\varepsilon}(\boldsymbol{u}_0))_Q \\ &\quad + (P\boldsymbol{u}(0) - P\boldsymbol{u}_0, \boldsymbol{u}(0) - \boldsymbol{u}_0)_V \leq 0. \end{aligned} \tag{6.174}$$

Finally, we use inequality (6.174), the monotonicity of P and the strong monotonicity of $\mathcal{E}$ to conclude the proof. ■

6.4.3 A convergence result

Everywhere in this section we assume that the function p does not depend on $\boldsymbol{x} \in \Gamma_3$ and, therefore, condition (6.36) reads

$$\left.\begin{array}{l}\text{(a) } p : \mathbb{R} \to \mathbb{R}_+; \\ \text{(b) there exists } L_p > 0 \text{ such that} \\ \qquad |p(r_1) - p(r_2)| \leq L_p|r_1 - r_2| \quad \forall\, r_1, r_2 \in \mathbb{R}; \\ \text{(c) } (p(r_1) - p(r_2))(r_1 - r_2) \geq 0 \quad \forall\, r_1, r_2 \in \mathbb{R}; \\ \text{(d) } p(r) = 0 \ \text{ for all } r \leq 0.\end{array}\right\} \tag{6.175}$$

Let q be a function which satisfies

$$\left.\begin{array}{l}\text{(a) } q : [g, +\infty[\to \mathbb{R}_+; \\ \text{(b) there exists } L_q > 0 \text{ such that} \\ \qquad |q(r_1) - q(r_2)| \leq L_q|r_1 - r_2| \quad \forall\, r_1, r_2 \geq g; \\ \text{(c) } (q(r_1) - q(r_2))(r_1 - r_2) > 0 \quad \forall\, r_1, r_2 \geq g,\ r_1 \neq r_2; \\ \text{(d) } q(g) = 0.\end{array}\right\} \tag{6.176}$$

Then for each $\mu > 0$ we consider the function p_μ defined by

$$p_\mu(r) = \begin{cases} p(r) & \text{if } \ r \leq g, \\ \frac{1}{\mu}\, q(r) + p(g) & \text{if } \ r > g. \end{cases} \tag{6.177}$$

Using assumptions (6.175) and (6.176) it follows that the function p_μ satisfies condition (6.175) too, that is,

$$\left.\begin{array}{l}\text{(a) } p_\mu : \mathbb{R} \to \mathbb{R}_+; \\ \text{(b) there exists } L_{p_\mu} > 0 \text{ such that} \\ \qquad |p_\mu(r_1) - p_\mu(r_2)| \leq L_{p_\mu}|r_1 - r_2| \quad \forall\, r_1, r_2 \in \mathbb{R}; \\ \text{(c) } (p_\mu(r_1) - p_\mu(r_2))(r_1 - r_2) \geq 0 \quad \forall\, r_1, r_2 \in \mathbb{R}; \\ \text{(d) } p_\mu(r) = 0 \ \text{ for all } r \leq 0.\end{array}\right\} \tag{6.178}$$

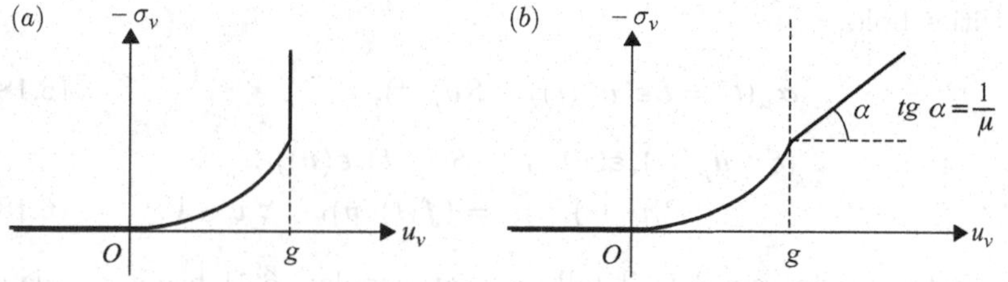

Figure 6.1 Representation of the contact conditions: (a) the contact condition (6.164); (b) the contact condition (6.180) with $q(r) = r - g$.

This allows us to consider the operator $P_\mu : V \to V$ defined by

$$(P_\mu \boldsymbol{u}, \boldsymbol{v})_V = \int_{\Gamma_3} p_\mu(u_\nu) v_\nu \, da \quad \forall \boldsymbol{u}, \boldsymbol{v} \in V \tag{6.179}$$

and, moreover, we note that P_μ is a monotone Lipschitz continuous operator.

We also consider Problem 6.35 in the case when the contact condition (6.142) is replaced with

$$-\sigma_\nu = p_\mu(u_\nu) \qquad \text{on } \Gamma_3 \times (0, T). \tag{6.180}$$

In this condition μ represents a penalization parameter which may be interpreted as a deformability of the foundation, and then $1/\mu$ is the surface stiffness coefficient, as explained on page 128. The graphical representation of the contact conditions (6.164) and (6.180), with p_μ given by (6.177) and $q(r) = r - g$, is presented in Figure 6.1.

Assume now that (4.50), (4.51), (6.8), (6.145), (6.175) and (6.176) hold. Then the variational formulation of the problem with normal compliance associated with the function p_μ is as follows.

Problem 6.44 *Find a displacement field $\boldsymbol{u}_\mu : [0, T] \to V$ and a stress field $\boldsymbol{\sigma}_\mu : [0, T] \to Q$ such that, for all $t \in [0, T]$, the following equalities hold:*

$$\boldsymbol{\sigma}_\mu(t) = \mathcal{E}\boldsymbol{\varepsilon}(\boldsymbol{u}_\mu(t)) + \int_0^t \mathcal{G}(\boldsymbol{\sigma}_\mu(s), \boldsymbol{\varepsilon}(\boldsymbol{u}_\mu(s))) \, ds + \boldsymbol{\sigma}_0 - \mathcal{E}\boldsymbol{\varepsilon}(\boldsymbol{u}_0),$$

$$(\boldsymbol{\sigma}_\mu(t), \boldsymbol{\varepsilon}(\boldsymbol{v}))_Q + (P_\mu \boldsymbol{u}_\mu(t), \boldsymbol{v})_V = (\boldsymbol{f}(t), \boldsymbol{v})_V \quad \forall \boldsymbol{v} \in V.$$

It follows from Theorem 6.37 that Problem 6.44 has a unique solution $(\boldsymbol{u}_\mu, \boldsymbol{\sigma}_\mu)$ which satisfies (6.149). Moreover, Lemma 6.39 shows that $(\boldsymbol{u}_\mu, \boldsymbol{\sigma}_\mu)$ is a solution of Problem 6.44 if and only if, for all $t \in [0, T]$, the following equalities hold:

$$\boldsymbol{\sigma}_\mu(t) = \mathcal{E}\boldsymbol{\varepsilon}(\boldsymbol{u}_\mu(t)) + \mathcal{S}\boldsymbol{u}_\mu(t), \tag{6.181}$$

$$(\mathcal{E}\boldsymbol{\varepsilon}(\boldsymbol{u}_\mu(t)), \boldsymbol{\varepsilon}(\boldsymbol{v}))_Q + (\mathcal{S}\boldsymbol{u}_\mu(t), \boldsymbol{\varepsilon}(\boldsymbol{v}))_Q \\ +(P_\mu \boldsymbol{u}_\mu(t), \boldsymbol{v})_V = (\boldsymbol{f}(t), \boldsymbol{v})_V \quad \forall\, \boldsymbol{v} \in V. \tag{6.182}$$

Finally, from Theorem 6.42 it follows that Problem 6.41 has a unique solution $(\boldsymbol{u}, \boldsymbol{\sigma})$ which satisfies (6.170). The behavior of the solution $(\boldsymbol{u}_\mu, \boldsymbol{\sigma}_\mu)$ as $\mu \to 0$ is given in the following result.

Theorem 6.45 *Assume (4.50), (4.51), (6.8), (6.145), (6.175) and (6.176). Then the solution $(\boldsymbol{u}_\mu, \boldsymbol{\sigma}_\mu)$ of Problem 6.44 converges to the solution $(\boldsymbol{u}, \boldsymbol{\sigma})$ of Problem 6.41, that is*

$$\|\boldsymbol{u}_\mu(t) - \boldsymbol{u}(t)\|_V + \|\boldsymbol{\sigma}_\mu(t) - \boldsymbol{\sigma}(t)\|_{Q_1} \to 0 \tag{6.183}$$

as $\mu \to 0$, for all $t \in [0, T]$.

The proof of Theorem 6.45 is carried out in several steps that we present in what follows. In the rest of this section we suppose that the assumptions of Theorem 6.45 hold and we denote by c a positive generic constant that may depend on time but does not depend on μ, and whose value may change from line to line.

Let $\mu > 0$. In the first step we consider the auxiliary problem of finding a displacement field $\widetilde{\boldsymbol{u}}_\mu : [0, T] \to V$ such that, for all $t \in [0, T]$,

$$(\mathcal{E}\boldsymbol{\varepsilon}(\widetilde{\boldsymbol{u}}_\mu(t)), \boldsymbol{\varepsilon}(\boldsymbol{v}))_Q + (\mathcal{S}\boldsymbol{u}(t), \boldsymbol{\varepsilon}(\boldsymbol{v}))_Q \\ +(P_\mu \widetilde{\boldsymbol{u}}_\mu(t), \boldsymbol{v})_V = (\boldsymbol{f}(t), \boldsymbol{v})_V \quad \forall\, \boldsymbol{v} \in V. \tag{6.184}$$

Note that the difference between the variational equations (6.182) and (6.184) arises from the fact that in (6.184) the operator $\mathcal{S}$ is applied to a known function.

We have the following existence and uniqueness result.

Lemma 6.46 *There exists a unique solution $\widetilde{\boldsymbol{u}}_\mu \in C([0, T]; V)$ which satisfies (6.184), for all $t \in [0, T]$.*

Proof We define the operator $A_\mu : V \to V$ and the function $\widetilde{\boldsymbol{f}} : [0, T] \to V$ by the equalities

$$(A_\mu \boldsymbol{u}, \boldsymbol{v})_V = (\mathcal{E}\boldsymbol{\varepsilon}(\boldsymbol{u}), \boldsymbol{\varepsilon}(\boldsymbol{v}))_Q + (P_\mu \boldsymbol{u}, \boldsymbol{v})_V \quad \forall\, \boldsymbol{u},\ \boldsymbol{v} \in V, \tag{6.185}$$

$$(\widetilde{\boldsymbol{f}}(t), \boldsymbol{v})_V = (\boldsymbol{f}(t), \boldsymbol{v})_V - (\mathcal{S}\boldsymbol{u}(t), \boldsymbol{\varepsilon}(\boldsymbol{v}))_Q \quad \forall\, \boldsymbol{v} \in V,\ t \in [0, T] \tag{6.186}$$

and note that (6.8), (6.39) and the regularity of $\mathcal{S}\boldsymbol{u}$ in Lemma 6.38 yield

$$\widetilde{\boldsymbol{f}} \in C([0, T]; V). \tag{6.187}$$

Let $t \in [0,T]$. Based on (6.185)–(6.186), it is easy to see that the variational equation (6.184) is equivalent to the nonlinear equation

$$A_\mu \widetilde{\boldsymbol{u}}_\mu(t) = \widetilde{\boldsymbol{f}}(t). \tag{6.188}$$

Next, by (4.50) and the properties of the operator P_μ it follows that A_μ is a strongly monotone and Lipschitz continuous operator. Therefore, combining Theorem 1.24 with Proposition 1.25 and regularity (6.187), we deduce the existence of a unique solution $\widetilde{\boldsymbol{u}}_\mu \in C([0,T];V)$ to the nonlinear equation (6.188), which concludes the proof. ■

We proceed with the following convergence result.

Lemma 6.47 *As* $\mu \to 0$,

$$\widetilde{\boldsymbol{u}}_\mu(t) \rightharpoonup \boldsymbol{u}(t) \quad \textit{in } V,$$

for all $t \in [0,T]$.

Proof Let $t \in [0,T]$. We take $\boldsymbol{v} = \widetilde{\boldsymbol{u}}_\mu(t)$ in (6.184) to obtain

$$\begin{aligned}(\mathcal{E}\boldsymbol{\varepsilon}(\widetilde{\boldsymbol{u}}_\mu(t)), \boldsymbol{\varepsilon}(\widetilde{\boldsymbol{u}}_\mu(t)))_Q + (\mathcal{S}\boldsymbol{u}(t), \boldsymbol{\varepsilon}(\widetilde{\boldsymbol{u}}_\mu(t)))_Q \\ +(P_\mu \widetilde{\boldsymbol{u}}_\mu(t), \widetilde{\boldsymbol{u}}_\mu(t))_V = (\boldsymbol{f}(t), \widetilde{\boldsymbol{u}}_\mu(t))_V.\end{aligned} \tag{6.189}$$

On the other hand, the properties (6.178) of the function p_μ yield

$$(P_\mu \widetilde{\boldsymbol{u}}_\mu(t), \widetilde{\boldsymbol{u}}_\mu(t))_V \geq 0. \tag{6.190}$$

We combine (6.189), (6.190) and use (4.50) to obtain that

$$\|\widetilde{\boldsymbol{u}}_\mu(t)\|_V \leq c(\|\boldsymbol{f}(t)\|_V + \|\mathcal{S}\boldsymbol{u}(t)\|_Q). \tag{6.191}$$

This inequality shows that the sequence $\{\widetilde{\boldsymbol{u}}_\mu(t)\} \subset V$ is bounded. Hence, there exists a subsequence of the sequence $\{\widetilde{\boldsymbol{u}}_\mu(t)\}$, still denoted $\{\widetilde{\boldsymbol{u}}_\mu(t)\}$, and an element $\widetilde{\boldsymbol{u}}(t) \in V$ such that

$$\widetilde{\boldsymbol{u}}_\mu(t) \rightharpoonup \widetilde{\boldsymbol{u}}(t) \quad \text{in } V \quad \text{as} \quad \mu \to 0. \tag{6.192}$$

It follows from (6.189) that

$$\begin{aligned}(P_\mu \widetilde{\boldsymbol{u}}_\mu(t), \widetilde{\boldsymbol{u}}_\mu(t))_V = (\boldsymbol{f}(t), \widetilde{\boldsymbol{u}}_\mu(t))_V - (\mathcal{E}\boldsymbol{\varepsilon}(\widetilde{\boldsymbol{u}}_\mu(t)), \boldsymbol{\varepsilon}(\widetilde{\boldsymbol{u}}_\mu(t)))_Q \\ -(\mathcal{S}\boldsymbol{u}(t), \boldsymbol{\varepsilon}(\widetilde{\boldsymbol{u}}_\mu(t)))_Q\end{aligned}$$

and, since $\{\widetilde{\boldsymbol{u}}_\mu(t)\}$ is a bounded sequence in V, we deduce that

$$(P_\mu \widetilde{\boldsymbol{u}}_\mu(t), \widetilde{\boldsymbol{u}}_\mu(t))_V \leq c.$$

This implies that

$$\int_{\Gamma_3} p_\mu(\widetilde{u}_{\mu\nu}(t))\widetilde{u}_{\mu\nu}(t)\, da \leq c$$

and, since $\int_{\Gamma_3} p_\mu(\widetilde{u}_{\mu\nu}(t))g\,\mathrm{d}a \geq 0$, it follows that

$$\int_{\Gamma_3} p_\mu(\widetilde{u}_{\mu\nu}(t))(\widetilde{u}_{\mu\nu}(t) - g)\,\mathrm{d}a \leq c. \tag{6.193}$$

We consider now the measurable subsets of Γ_3 defined by

$$\Gamma_{31} = \{\, \boldsymbol{x} \in \Gamma_3 : \widetilde{u}_{\mu\nu}(t)(\boldsymbol{x}) \leq g \,\}, \tag{6.194}$$

$$\Gamma_{32} = \{\, \boldsymbol{x} \in \Gamma_3 : \widetilde{u}_{\mu\nu}(t)(\boldsymbol{x}) > g \,\}. \tag{6.195}$$

Clearly, both Γ_{31} and Γ_{32} depend on t and μ but, for simplicity, we do not indicate this dependence explicitly. We use (6.193) to write

$$\int_{\Gamma_{31}} p_\mu(\widetilde{u}_{\mu\nu}(t))(\widetilde{u}_{\mu\nu}(t) - g)\,\mathrm{d}a + \int_{\Gamma_{32}} p_\mu(\widetilde{u}_{\mu\nu}(t))(\widetilde{u}_{\mu\nu}(t) - g)\,\mathrm{d}a \leq c$$

and, since

$$\int_{\Gamma_{31}} p_\mu(\widetilde{u}_{\mu\nu}(t))\widetilde{u}_{\mu\nu}(t)\,\mathrm{d}a \geq 0,$$

we obtain

$$\int_{\Gamma_{32}} p_\mu(\widetilde{u}_{\mu\nu}(t))(\widetilde{u}_{\mu\nu}(t) - g)\,\mathrm{d}a \leq \int_{\Gamma_{31}} p_\mu(\widetilde{u}_{\mu\nu}(t))g\,\mathrm{d}a + c.$$

Thus, taking into account that $p_\mu(r) = p(r)$ for $r \leq g$, by the monotonicity of the function p we can write

$$\int_{\Gamma_{32}} p_\mu(\widetilde{u}_{\mu\nu}(t))(\widetilde{u}_{\mu\nu}(t) - g)\,\mathrm{d}a \leq \int_{\Gamma_{31}} p(\widetilde{u}_{\mu\nu}(t))g\,\mathrm{d}a + c$$

$$\leq \int_{\Gamma_{31}} p(g)g\,\mathrm{d}a + c \leq \int_{\Gamma_3} p(g)g\,\mathrm{d}a + c.$$

Therefore, we deduce that

$$\int_{\Gamma_{32}} p_\mu(\widetilde{u}_{\mu\nu}(t))(\widetilde{u}_{\mu\nu}(t) - g)\,\mathrm{d}a \leq c. \tag{6.196}$$

We now use the definitions (6.177) and (6.195) to see that, a.e. on Γ_{32}, we have

$$p_\mu(\widetilde{u}_{\mu\nu}(t)) = \frac{1}{\mu}\,q(\widetilde{u}_{\mu\nu}(t)) + p(g), \quad p(g)(\widetilde{u}_{\mu\nu}(t) - g) \geq 0.$$

Consequently, the inequality (6.196) yields

$$\int_{\Gamma_{32}} q(\widetilde{u}_{\mu\nu}(t))(\widetilde{u}_{\mu\nu}(t) - g)\,\mathrm{d}a \leq c\mu. \tag{6.197}$$

Next, we consider the function $\widetilde{p}:\mathbb{R}\to\mathbb{R}_+$ defined by

$$\widetilde{p}(r)=\begin{cases}0 & \text{if } \ r\le g,\\ q(r) & \text{if } \ r>g\end{cases}$$

and we note that by (6.176) it follows that $\widetilde{p}$ is a Lipschitz continuous increasing function and, moreover,

$$\widetilde{p}(r)=0 \qquad \text{iff} \qquad r\le g. \tag{6.198}$$

We use (6.197), the equality $q(\widetilde{u}_{\mu\nu}(t))=\widetilde{p}(u_{\mu\nu}(t))$ a.e. on Γ_{32} and (6.194) to deduce that

$$\int_{\Gamma_3}\widetilde{p}(\widetilde{u}_{\mu\nu}(t))(\widetilde{u}_{\mu\nu}(t)-g)_+\mathrm{d}a\le c\mu,$$

where, as usual, $(\widetilde{u}_{\mu\nu}(t)-g)_+$ denotes the positive part of $\widetilde{u}_{\mu\nu}(t)-g$. Therefore, passing to the limit as $\mu\to 0$, by using (6.192) as well as compactness of the trace operator we find that

$$\int_{\Gamma_3}\widetilde{p}(\widetilde{u}_\nu(t))(\widetilde{u}_\nu(t)-g)_+\,\mathrm{d}a\le 0.$$

Since the integrand $\widetilde{p}(\widetilde{u}_\nu(t))(\widetilde{u}_\nu(t)-g)_+$ is positive a.e. on Γ_3, the last inequality yields

$$\widetilde{p}(\widetilde{u}_\nu(t))(\widetilde{u}_\nu(t)-g)_+=0\ \text{ a.e. on } \Gamma_3$$

and, using (6.198) and definition (6.37) we conclude that

$$\widetilde{\boldsymbol{u}}(t)\in U. \tag{6.199}$$

Next, we test in (6.184) with $\boldsymbol{v}-\widetilde{\boldsymbol{u}}_\mu(t)$, where $\boldsymbol{v}\in U$, to obtain

$$\begin{aligned}(\mathcal{E}\boldsymbol{\varepsilon}(\widetilde{\boldsymbol{u}}_\mu(t)),\boldsymbol{\varepsilon}(\boldsymbol{v})-\boldsymbol{\varepsilon}(\widetilde{\boldsymbol{u}}_\mu(t)))_Q+(\mathcal{S}\boldsymbol{u}(t),\boldsymbol{\varepsilon}(\boldsymbol{v})-\boldsymbol{\varepsilon}(\widetilde{\boldsymbol{u}}_\mu(t)))_Q\\ +(P_\mu\widetilde{\boldsymbol{u}}_\mu(t),\boldsymbol{v}-\widetilde{\boldsymbol{u}}_\mu(t))_V=(\boldsymbol{f}(t),\boldsymbol{v}-\widetilde{\boldsymbol{u}}_\mu(t))_V.\end{aligned} \tag{6.200}$$

Since $\boldsymbol{v}\in U$ we have $p_\mu(v_\nu)=p(v_\nu)$ a.e. on Γ_3. Taking into account this equality and the monotonicity of the function p_μ we have

$$p(v_\nu)(v_\nu-\widetilde{u}_{\mu\nu}(t))\ge p_\mu(\widetilde{u}_{\mu\nu}(t))(v_\nu-\widetilde{u}_{\mu\nu}(t))\ \text{ a.e. on } \Gamma_3$$

and, therefore, by using (6.179) we obtain

$$(P\boldsymbol{v},\boldsymbol{v}-\widetilde{\boldsymbol{u}}_\mu(t))_V\ge(P_\mu\widetilde{\boldsymbol{u}}_\mu(t),\boldsymbol{v}-\widetilde{\boldsymbol{u}}_\mu(t))_V. \tag{6.201}$$

Then, using (6.201) and (6.200) we find that

$$\begin{aligned}(\mathcal{E}\boldsymbol{\varepsilon}(\widetilde{\boldsymbol{u}}_\mu(t)),\boldsymbol{\varepsilon}(\boldsymbol{v})-\boldsymbol{\varepsilon}(\widetilde{\boldsymbol{u}}_\mu(t)))_Q+(\mathcal{S}\boldsymbol{u}(t),\boldsymbol{\varepsilon}(\boldsymbol{v})-\boldsymbol{\varepsilon}(\widetilde{\boldsymbol{u}}_\mu(t)))_Q\\ +(P\boldsymbol{v},\boldsymbol{v}-\widetilde{\boldsymbol{u}}_\mu(t))_V\ge(\boldsymbol{f}(t),\boldsymbol{v}-\widetilde{\boldsymbol{u}}_\mu(t))_V\quad\forall\,\boldsymbol{v}\in U.\end{aligned} \tag{6.202}$$

We pass to the lower limit in (6.202) and use (6.192) to obtain

$$(\mathcal{E}\varepsilon(\widetilde{\boldsymbol{u}}(t)), \varepsilon(\boldsymbol{v}) - \varepsilon(\widetilde{\boldsymbol{u}}(t)))_Q + (\mathcal{S}\boldsymbol{u}(t), \varepsilon(\boldsymbol{v}) - \varepsilon(\widetilde{\boldsymbol{u}}(t)))_Q \\ +(P\boldsymbol{v}, \boldsymbol{v} - \widetilde{\boldsymbol{u}}(t))_V \geq (\boldsymbol{f}(t), \boldsymbol{v} - \widetilde{\boldsymbol{u}}(t))_V \quad \forall \boldsymbol{v} \in U. \quad (6.203)$$

Next, we take $\boldsymbol{v} = \widetilde{\boldsymbol{u}}(t)$ in (6.172) and $\boldsymbol{v} = \boldsymbol{u}(t)$ in (6.203). Then, adding the resulting inequalities, we find that

$$(\mathcal{E}\varepsilon(\widetilde{\boldsymbol{u}}(t)) - \mathcal{E}\varepsilon(\boldsymbol{u}(t)), \varepsilon(\widetilde{\boldsymbol{u}}(t)) - \varepsilon(\boldsymbol{u}(t)))_Q \leq 0.$$

This inequality combined with (4.50) implies that

$$\widetilde{\boldsymbol{u}}(t) = \boldsymbol{u}(t).$$

It follows from this that the whole sequence $\{\widetilde{\boldsymbol{u}}_\mu(t)\}$ is weakly convergent to the element $\boldsymbol{u}(t) \in V$, which concludes the proof. ■

We proceed with the following strong convergence result.

Lemma 6.48 *As* $\mu \to 0$,

$$\|\widetilde{\boldsymbol{u}}_\mu(t) - \boldsymbol{u}(t)\|_V \to 0,$$

for all $t \in [0, T]$.

Proof Let $t \in [0, T]$. Using (4.50) we write

$$m_{\mathcal{E}}\|\widetilde{\boldsymbol{u}}_\mu(t) - \boldsymbol{u}(t)\|_V^2 \leq (\mathcal{E}\varepsilon(\widetilde{\boldsymbol{u}}_\mu(t)) - \mathcal{E}\varepsilon(\boldsymbol{u}(t)), \varepsilon(\widetilde{\boldsymbol{u}}_\mu(t)) - \varepsilon(\boldsymbol{u}(t)))_Q \\ = (\mathcal{E}\varepsilon(\boldsymbol{u}(t)), \varepsilon(\boldsymbol{u}(t)) - \varepsilon(\widetilde{\boldsymbol{u}}_\mu(t)))_Q \\ -(\mathcal{E}\varepsilon(\widetilde{\boldsymbol{u}}_\mu(t)), \varepsilon(\boldsymbol{u}(t)) - \varepsilon(\widetilde{\boldsymbol{u}}_\mu(t)))_Q.$$

Next, we take $\boldsymbol{v} = \boldsymbol{u}(t)$ in (6.202) to obtain

$$-(\mathcal{E}\varepsilon(\widetilde{\boldsymbol{u}}_\mu(t)), \varepsilon(\boldsymbol{u}(t)) - \varepsilon(\widetilde{\boldsymbol{u}}_\mu(t)))_Q \leq (\mathcal{S}\boldsymbol{u}(t), \varepsilon(\boldsymbol{u}(t)) - \varepsilon(\widetilde{\boldsymbol{u}}_\mu(t)))_Q \\ +(P\boldsymbol{u}(t), \boldsymbol{u}(t) - \widetilde{\boldsymbol{u}}_\mu(t))_V - (\boldsymbol{f}(t), \boldsymbol{u}(t) - \widetilde{\boldsymbol{u}}_\mu(t))_V$$

and, therefore, from the previous two inequalities we find that

$$m_{\mathcal{E}}\|\widetilde{\boldsymbol{u}}_\mu(t) - \boldsymbol{u}(t)\|_V^2 \leq (\mathcal{E}\varepsilon(\boldsymbol{u}(t)), \varepsilon(\boldsymbol{u}(t)) - \varepsilon(\widetilde{\boldsymbol{u}}_\mu(t)))_Q \\ +(\mathcal{S}\boldsymbol{u}(t), \varepsilon(\boldsymbol{u}(t)) - \varepsilon(\widetilde{\boldsymbol{u}}_\mu(t)))_Q \\ +(P\boldsymbol{u}(t), \boldsymbol{u}(t) - \widetilde{\boldsymbol{u}}_\mu(t))_V - (\boldsymbol{f}(t), \boldsymbol{u}(t) - \widetilde{\boldsymbol{u}}_\mu(t))_V.$$

We pass to the upper limit as $\mu \to 0$ in this inequality and use Lemma 6.47 to conclude the proof. ■

We are now in position to provide the proof of Theorem 6.45.

Proof Let $t \in [0,T]$ and $\mu > 0$. Then, testing with $\boldsymbol{v} = \boldsymbol{u}_\mu(t) - \widetilde{\boldsymbol{u}}_\mu(t)$ in (6.184) and (6.182), we have

$$\begin{aligned}(\mathcal{E}\boldsymbol{\varepsilon}(\widetilde{\boldsymbol{u}}_\mu(t)), \boldsymbol{\varepsilon}(\boldsymbol{u}_\mu(t)) - \boldsymbol{\varepsilon}(\widetilde{\boldsymbol{u}}_\mu(t)))_Q + (\mathcal{S}\boldsymbol{u}(t), \boldsymbol{\varepsilon}(\boldsymbol{u}_\mu(t)) - \boldsymbol{\varepsilon}(\widetilde{\boldsymbol{u}}_\mu(t)))_Q \\ +(P_\mu \widetilde{\boldsymbol{u}}_\mu(t), \boldsymbol{u}_\mu(t) - \widetilde{\boldsymbol{u}}_\mu(t))_V = (\boldsymbol{f}(t), \boldsymbol{u}_\mu(t) - \widetilde{\boldsymbol{u}}_\mu(t))_V,\end{aligned}$$

$$\begin{aligned}(\mathcal{E}\boldsymbol{\varepsilon}(\boldsymbol{u}_\mu(t)), \boldsymbol{\varepsilon}(\boldsymbol{u}_\mu(t)) - \boldsymbol{\varepsilon}(\widetilde{\boldsymbol{u}}_\mu(t)))_Q + (\mathcal{S}\boldsymbol{u}_\mu(t), \boldsymbol{\varepsilon}(\boldsymbol{u}_\mu(t)) - \boldsymbol{\varepsilon}(\widetilde{\boldsymbol{u}}_\mu(t)))_Q \\ +(P_\mu \boldsymbol{u}_\mu(t), \boldsymbol{u}_\mu(t) - \widetilde{\boldsymbol{u}}_\mu(t))_V = (\boldsymbol{f}(t), \boldsymbol{u}_\mu(t) - \widetilde{\boldsymbol{u}}_\mu(t))_V.\end{aligned}$$

We subtract the previous equalities and use the monotonicity of the operator P_μ to deduce that

$$\begin{aligned}(\mathcal{E}\boldsymbol{\varepsilon}(\boldsymbol{u}_\mu(t)) - \mathcal{E}\boldsymbol{\varepsilon}(\widetilde{\boldsymbol{u}}_\mu(t)), \boldsymbol{\varepsilon}(\boldsymbol{u}_\mu(t)) - \boldsymbol{\varepsilon}(\widetilde{\boldsymbol{u}}_\mu(t)))_Q \\ \le (\mathcal{S}\boldsymbol{u}(t) - \mathcal{S}\boldsymbol{u}_\mu(t), \boldsymbol{\varepsilon}(\boldsymbol{u}_\mu(t)) - \boldsymbol{\varepsilon}(\widetilde{\boldsymbol{u}}_\mu(t)))_Q\end{aligned}$$

and, therefore,

$$\|\boldsymbol{u}_\mu(t) - \widetilde{\boldsymbol{u}}_\mu(t)\|_V \le \frac{1}{m_\mathcal{E}} \|\mathcal{S}\boldsymbol{u}(t) - \mathcal{S}\boldsymbol{u}_\mu(t)\|_Q. \tag{6.204}$$

We combine (6.204) and (6.151) to find that

$$\|\boldsymbol{u}_\mu(t) - \widetilde{\boldsymbol{u}}_\mu(t)\|_V \le \frac{L_\mathcal{S}}{m_\mathcal{E}} \int_0^t \|\boldsymbol{u}(s) - \boldsymbol{u}_\mu(s)\|_V \, \mathrm{d}s,$$

where, recall, $L_\mathcal{S} > 0$. It follows from this that

$$\|\boldsymbol{u}_\mu(t) - \boldsymbol{u}(t)\|_V \le \|\widetilde{\boldsymbol{u}}_\mu(t) - \boldsymbol{u}(t)\|_V + \frac{L_\mathcal{S}}{m_\mathcal{E}} \int_0^t \|\boldsymbol{u}_\mu(s) - \boldsymbol{u}(s)\|_V \mathrm{d}s$$

and, using the Gronwall lemma, we obtain

$$\begin{aligned}\|\boldsymbol{u}_\mu(t) - \boldsymbol{u}(t)\|_V &\le \|\widetilde{\boldsymbol{u}}_\mu(t) - \boldsymbol{u}(t)\|_V \\ &+ \frac{L_\mathcal{S}}{m_\mathcal{E}} \int_0^t \mathrm{e}^{\frac{L_\mathcal{S}}{m_\mathcal{E}}(t-s)} \|\widetilde{\boldsymbol{u}}_\mu(s) - \boldsymbol{u}(s)\|_V \, \mathrm{d}s.\end{aligned} \tag{6.205}$$

Note that

$$\mathrm{e}^{\frac{L_\mathcal{S}}{m_\mathcal{E}}(t-s)} \le \mathrm{e}^{\frac{L_\mathcal{S}}{m_\mathcal{E}}t} \le \mathrm{e}^{\frac{L_\mathcal{S}T}{m_\mathcal{E}}}$$

for all $s \in [0,t]$ and, therefore, (6.205) yields

$$\begin{aligned}\|\boldsymbol{u}_\mu(t) - \boldsymbol{u}(t)\|_V &\le \|\widetilde{\boldsymbol{u}}_\mu(t) - \boldsymbol{u}(t)\|_V \\ &+ \frac{L_\mathcal{S}}{m_\mathcal{E}} \mathrm{e}^{\frac{L_\mathcal{S}T}{m_\mathcal{E}}} \int_0^t \|\widetilde{\boldsymbol{u}}_\mu(s) - \boldsymbol{u}(s)\|_V \, \mathrm{d}s.\end{aligned} \tag{6.206}$$

On the other hand, by estimate (6.191), Lemma 6.48 and Lebesgue's convergence theorem it follows that

$$\int_0^t \|\widetilde{\boldsymbol{u}}_\mu(s) - \boldsymbol{u}(s)\|_V \, ds \to 0 \qquad \text{as} \quad \mu \to 0. \tag{6.207}$$

We now use (6.206), (6.207) and Lemma 6.48 to see that

$$\|\boldsymbol{u}_\mu(t) - \boldsymbol{u}(t)\|_V \to 0 \qquad \text{as} \quad \mu \to 0. \tag{6.208}$$

Next, by (6.153) and (6.181) we obtain

$$\|\boldsymbol{\sigma}_\mu(t) - \boldsymbol{\sigma}(t)\|_Q \le \|\mathcal{E}\boldsymbol{\varepsilon}(\boldsymbol{u}_\mu(t)) - \mathcal{E}\boldsymbol{\varepsilon}(\boldsymbol{u}(t))\|_Q + \|\mathcal{S}\boldsymbol{u}_\mu(t) - \mathcal{S}\boldsymbol{u}(t)\|_Q$$

and, using (4.50) and (6.151) it follows that

$$\|\boldsymbol{\sigma}_\mu(t) - \boldsymbol{\sigma}(t)\|_Q \le c\,\|\boldsymbol{u}_\mu(t) - \boldsymbol{u}(t)\|_V + L_S \int_0^t \|\boldsymbol{u}_\mu(s) - \boldsymbol{u}(s)\|_V \, ds. \tag{6.209}$$

Moreover, taking $\boldsymbol{v} = \boldsymbol{u}_\mu(t)$ in (6.182) and using the monotonicity of P_μ and $\mathcal{E}$ we find that

$$\|\boldsymbol{u}_\mu(t)\|_V \le c\big(\|\boldsymbol{f}(t)\|_V + \|\mathcal{S}\boldsymbol{u}_\mu(t)\|_Q\big).$$

We use now the property (6.151) of the operator $\mathcal{S}$ and the Gronwall argument to see that

$$\|\boldsymbol{u}_\mu(t)\|_V \le c. \tag{6.210}$$

Then we use the convergence (6.208), estimate (6.210) and Lebesgue's theorem, again, and pass to the limit in (6.209). As a result we find that

$$\|\boldsymbol{\sigma}_\mu(t) - \boldsymbol{\sigma}(t)\|_Q \to 0 \qquad \text{as} \quad \mu \to 0. \tag{6.211}$$

Finally, by arguments similar to those used on page 176, it follows that

$$\text{Div}\,\boldsymbol{\sigma}_\mu(t) = \text{Div}\,\boldsymbol{\sigma}(t) = -\boldsymbol{f}_0(t)$$

and, therefore,

$$\|\boldsymbol{\sigma}_\mu(t) - \boldsymbol{\sigma}(t)\|_{Q_1} = \|\boldsymbol{\sigma}_\mu(t) - \boldsymbol{\sigma}(t)\|_Q. \tag{6.212}$$

Theorem 6.45 is now a consequence of (6.208), (6.211) and (6.212). ∎

In addition to the mathematical interest in the convergence (6.183), this result is important from the mechanical point of view, since it shows that the weak solution of the viscoplastic contact problem with normal compliance and unilateral constraint may be approached as closely as one wishes by the solution of a viscoplastic contact problem with normal compliance, with a sufficiently small deformability coefficient.

7
Analysis of piezoelectric contact problems

In this chapter we illustrate the use of the abstract results obtained in Chapters 2 and 3 in the study of three frictionless or frictional contact problems with piezoelectric bodies. We model the material's behavior with an electro-elastic, an electro-viscoelastic and an electro-viscoplastic constitutive law, respectively. The contact is either bilateral or modelled with the normal compliance condition, with or without unilateral constraint. The friction is modelled with versions of Coulomb's law. The foundation is assumed to be either an insulator or electrically conductive. For each problem we provide a variational formulation which is in the form of a nonlinear system in which the unknowns are the displacement field and the electric potential field. Then we use the abstract existence and uniqueness results presented in Chapters 2 and 3 to prove the unique weak solvability of the corresponding contact problems. For the electro-elastic problem we also provide a dual variational formulation in terms of the stress and electric displacement fields. Everywhere in this chapter we consider the physical setting and the notation presented in Section 4.5, as well as the function spaces introduced in Section 4.1.

7.1 An Electro-elastic frictional contact problem

In this section we consider a frictional contact problem for electro-elastic materials. The problem is static and, therefore, we investigate it by using the arguments of elliptic variational inequalities presented in Section 2.2.

7.1.1 Problem statement

We assume that the body is electro-elastic and the foundation is an insulator. The normal stress is prescribed on the contact surface and we use the static version of Coulomb's law of dry friction. The classical formulation of the problem is the following.

Problem 7.1 *Find a displacement field* $\boldsymbol{u} : \Omega \to \mathbb{R}^d$, *a stress field* $\boldsymbol{\sigma} : \Omega \to \mathbb{S}^d$, *an electric potential* $\varphi : \Omega \to \mathbb{R}$ *and an electric displacement field* $\boldsymbol{D} : \Omega \to \mathbb{R}^d$ *such that*

$$\boldsymbol{\sigma} = \mathcal{F}\boldsymbol{\varepsilon}(\boldsymbol{u}) - \mathcal{P}^\top \boldsymbol{E}(\varphi) \quad \text{in} \quad \Omega, \tag{7.1}$$

$$\boldsymbol{D} = \mathcal{P}\boldsymbol{\varepsilon}(\boldsymbol{u}) + \boldsymbol{\beta}\boldsymbol{E}(\varphi) \quad \text{in} \quad \Omega, \tag{7.2}$$

$$\operatorname{Div} \boldsymbol{\sigma} + \boldsymbol{f}_0 = \boldsymbol{0} \quad \text{in} \quad \Omega, \tag{7.3}$$

$$\operatorname{div} \boldsymbol{D} - q_0 = 0 \quad \text{in} \quad \Omega, \tag{7.4}$$

$$\boldsymbol{u} = \boldsymbol{0} \quad \text{on} \quad \Gamma_1, \tag{7.5}$$

$$\boldsymbol{\sigma\nu} = \boldsymbol{f}_2 \quad \text{on} \quad \Gamma_2, \tag{7.6}$$

$$\varphi = 0 \quad \text{on} \quad \Gamma_a, \tag{7.7}$$

$$\boldsymbol{D} \cdot \boldsymbol{\nu} = q_b \quad \text{on} \quad \Gamma_b, \tag{7.8}$$

$$-\sigma_\nu = F \quad \text{on} \quad \Gamma_3, \tag{7.9}$$

$$\left.\begin{array}{l} \|\boldsymbol{\sigma}_\tau\| \le \mu\,|\sigma_\nu|, \\ \boldsymbol{\sigma}_\tau = -\mu\,|\sigma_\nu|\,\dfrac{\boldsymbol{u}_\tau}{\|\boldsymbol{u}_\tau\|} \quad \text{if} \quad \boldsymbol{u}_\tau \neq \boldsymbol{0} \end{array}\right\} \quad \text{on} \quad \Gamma_3, \tag{7.10}$$

$$\boldsymbol{D} \cdot \boldsymbol{\nu} = 0 \quad \text{on} \quad \Gamma_3. \tag{7.11}$$

Note that Problem 7.1 represents a piezoelectric version of the elastic contact problem 5.17 studied in Section 5.2. Equations (7.1) and (7.2) represent the electro-elastic constitutive law of the material, (4.116) and (4.118), respectively. Recall that here $\mathcal{F}$ is a given nonlinear operator, $\boldsymbol{\varepsilon}(\boldsymbol{u})$ denotes the linearized strain tensor, $\boldsymbol{E}(\varphi) = -\nabla\varphi$ is the electric field, $\mathcal{P}$ represents the third order piezoelectric tensor, $\mathcal{P}^\top$ is its transpose and $\boldsymbol{\beta}$ denotes the electric permittivity tensor. Equations (7.3) and (7.4) represent the balance equations for the stress and electric displacement fields, respectively, in which "Div" and "div" denote the divergence operator for tensor and vector valued functions. Next, (7.5) and (7.6) are the displacement and traction boundary conditions, respectively, and (7.7), (7.8) represent the electric boundary conditions. Condition (7.9) is the contact condition (4.77) and condition (7.10) represents the friction law (4.86) associated with the friction bound (4.88). Finally, (7.11) represents the electrical contact condition (4.125) and we use it here since the foundation is assumed to be an insulator.

In the study of the piezoelectric contact problem 7.1 we assume that the elasticity operator, the piezoelectric tensor and the electric permittivity tensor satisfy the following conditions.

$$\left.\begin{array}{l}
\text{(a) } \mathcal{F} : \Omega \times \mathbb{S}^d \to \mathbb{S}^d. \\
\text{(b) There exists } L_{\mathcal{F}} > 0 \text{ such that} \\
\qquad \|\mathcal{F}(\boldsymbol{x}, \boldsymbol{\varepsilon}_1) - \mathcal{F}(\boldsymbol{x}, \boldsymbol{\varepsilon}_2)\| \le L_{\mathcal{F}} \|\boldsymbol{\varepsilon}_1 - \boldsymbol{\varepsilon}_2\| \\
\qquad \forall\, \boldsymbol{\varepsilon}_1, \boldsymbol{\varepsilon}_2 \in \mathbb{S}^d, \text{ a.e. } \boldsymbol{x} \in \Omega. \\
\text{(c) There exists } m_{\mathcal{F}} > 0 \text{ such that} \\
\qquad (\mathcal{F}(\boldsymbol{x}, \boldsymbol{\varepsilon}_1) - \mathcal{F}(\boldsymbol{x}, \boldsymbol{\varepsilon}_2)) \cdot (\boldsymbol{\varepsilon}_1 - \boldsymbol{\varepsilon}_2) \ge m_{\mathcal{F}}\, \|\boldsymbol{\varepsilon}_1 - \boldsymbol{\varepsilon}_2\|^2 \\
\qquad \forall\, \boldsymbol{\varepsilon}_1, \boldsymbol{\varepsilon}_2 \in \mathbb{S}^d, \text{ a.e. } \boldsymbol{x} \in \Omega. \\
\text{(d) The mapping } \boldsymbol{x} \mapsto \mathcal{F}(\boldsymbol{x}, \boldsymbol{\varepsilon}) \text{ is measurable on } \Omega, \\
\qquad \text{for any } \boldsymbol{\varepsilon} \in \mathbb{S}^d. \\
\text{(e) The mapping } \boldsymbol{x} \mapsto \mathcal{F}(\boldsymbol{x}, \boldsymbol{0}_{\mathbb{S}^d}) \text{ belongs to } Q.
\end{array}\right\} \tag{7.12}$$

$$\left.\begin{array}{l}
\text{(a) } \mathcal{P} : \Omega \times \mathbb{S}^d \to \mathbb{R}^d. \\
\text{(b) } \mathcal{P}(\boldsymbol{x}, \boldsymbol{\tau}) = (p_{ijk}(\boldsymbol{x})\tau_{jk})\ \forall\, \boldsymbol{\tau} = (\tau_{ij}) \in \mathbb{S}^d, \text{ a.e. } \boldsymbol{x} \in \Omega. \\
\text{(c) } p_{ijk} = p_{ikj} \in L^\infty(\Omega),\ 1 \le i, j, k \le d.
\end{array}\right\} \tag{7.13}$$

$$\left.\begin{array}{l}
\text{(a) } \boldsymbol{\beta} : \Omega \times \mathbb{R}^d \to \mathbb{R}^d. \\
\text{(b) } \boldsymbol{\beta}(\boldsymbol{x}, \boldsymbol{E}) = (\beta_{ij}(\boldsymbol{x})E_j)\ \forall\, \boldsymbol{E} = (E_i) \in \mathbb{R}^d, \text{ a.e. } \boldsymbol{x} \in \Omega. \\
\text{(c) } \beta_{ij} = \beta_{ji} \in L^\infty(\Omega),\ 1 \le i, j \le d. \\
\text{(d) There exists } m_\beta > 0 \text{ such that} \\
\qquad \beta(\boldsymbol{x}, \boldsymbol{E}) \cdot \boldsymbol{E} \ge m_\beta \|\boldsymbol{E}\|^2\ \forall\, \boldsymbol{E} \in \mathbb{R}^d, \text{ a.e. } \boldsymbol{x} \in \Omega.
\end{array}\right\} \tag{7.14}$$

The densities of body forces, surface tractions, volume and surface electric charges satisfy

$$\boldsymbol{f}_0 \in L^2(\Omega)^d, \quad \boldsymbol{f}_2 \in L^2(\Gamma_2)^d, \tag{7.15}$$

$$q_0 \in L^2(\Omega), \quad q_b \in L^2(\Gamma_b). \tag{7.16}$$

The prescribed normal stress is such that

$$F \in L^2(\Gamma_3) \text{ and } F(\boldsymbol{x}) \ge 0 \text{ a.e. } \boldsymbol{x} \in \Gamma_3 \tag{7.17}$$

and, finally, the coefficient of friction satisfies

$$\mu \in L^\infty(\Gamma_3) \text{ and } \mu(\boldsymbol{x}) \ge 0 \text{ a.e. } \boldsymbol{x} \in \Gamma_3. \tag{7.18}$$

We turn now to the variational formulation of Problem 7.1 and, to this end, besides the spaces V and W introduced in Section 4.1, we need further notation. Let $j : V \to \mathbb{R}$ be the functional

$$j(\boldsymbol{v}) = \int_{\Gamma_3} \mu\, F\, \|\boldsymbol{v}_\tau\|\, \mathrm{d}a \qquad \forall\, \boldsymbol{v} \in V \tag{7.19}$$

and, using Riesz's representation theorem, consider the elements $\boldsymbol{f} \in V$ and $q \in W$ given by

$$(\boldsymbol{f}, \boldsymbol{v})_V = \int_\Omega \boldsymbol{f}_0 \cdot \boldsymbol{v}\, \mathrm{d}x + \int_{\Gamma_2} \boldsymbol{f}_2 \cdot \boldsymbol{v}\, \mathrm{d}a \qquad \forall\, \boldsymbol{v} \in V,$$

$$(q, \psi)_W = \int_\Omega q_0 \psi\, \mathrm{d}x - \int_{\Gamma_b} q_b \psi\, \mathrm{d}a \qquad \forall\, \psi \in W.$$

Using the Green-type formulae (4.22) and (4.30) it is straightforward to see that if $(\boldsymbol{u}, \boldsymbol{\sigma}, \varphi, \boldsymbol{D})$ are sufficiently regular functions which satisfy (7.3)–(7.11) then

$$(\boldsymbol{\sigma}, \boldsymbol{\varepsilon}(\boldsymbol{v}) - \boldsymbol{\varepsilon}(\boldsymbol{u}))_Q + j(\boldsymbol{v}) - j(\boldsymbol{u}) \geq (\boldsymbol{f}, \boldsymbol{v} - \boldsymbol{u})_V \quad \forall\, \boldsymbol{v} \in V, \tag{7.20}$$

$$(\boldsymbol{D}, \nabla\psi)_{L^2(\Omega)^d} + (q, \psi)_W = 0 \quad \forall\, \psi \in W. \tag{7.21}$$

We substitute (7.1) in (7.20), (7.2) in (7.21) then we use the notation $\boldsymbol{E}(\varphi) = -\nabla\varphi$ to obtain the following variational formulation of Problem 7.1, in terms of displacement field and electric potential.

Problem 7.2 *Find a displacement field $\boldsymbol{u} \in V$ and an electric potential $\varphi \in W$ such that*

$$\begin{aligned}(\mathcal{F}\boldsymbol{\varepsilon}(\boldsymbol{u}), \boldsymbol{\varepsilon}(\boldsymbol{v}) - \boldsymbol{\varepsilon}(\boldsymbol{u}))_Q + (\mathcal{P}^\top \nabla\varphi, \boldsymbol{\varepsilon}(\boldsymbol{v}) - \boldsymbol{\varepsilon}(\boldsymbol{u}))_{L^2(\Omega)^d} \\ + j(\boldsymbol{v}) - j(\boldsymbol{u}) \geq (\boldsymbol{f}, \boldsymbol{v} - \boldsymbol{u})_V \quad \forall\, \boldsymbol{v} \in V,\end{aligned} \tag{7.22}$$

$$(\boldsymbol{\beta}\nabla\varphi, \nabla\psi)_{L^2(\Omega)^d} - (\mathcal{P}\boldsymbol{\varepsilon}(\boldsymbol{u}), \nabla\psi)_{L^2(\Omega)^d} = (q, \psi)_W \quad \forall\, \psi \in W. \tag{7.23}$$

Note that, unlike the variational formulation of the elastic contact problems studied in Chapter 5, the variational formulation of Problem 7.1 is given by a system coupling a variational inequality, (7.22), with a linear variational equation, (7.23).

7.1.2 Existence and uniqueness

Our main existence and uniqueness result in the study of Problem 7.2 is the following.

Theorem 7.3 *Assume (7.12)–(7.18). Then Problem 7.2 has a unique solution* $(\boldsymbol{u}, \varphi) \in V \times W$.

The proof of Theorem 7.3 is based on a preliminary result. To present it we consider the product space $X = V \times W$ together with the inner product

$$(x, y)_X = (\boldsymbol{u}, \boldsymbol{v})_V + (\varphi, \psi)_W \qquad \forall\, x = (\boldsymbol{u}, \varphi),\ y = (\boldsymbol{v}, \psi) \in X \tag{7.24}$$

and the associated norm $\|\cdot\|_X$. Everywhere below we assume that (7.12)–(7.18) hold. We introduce the operator $A : X \to X$ defined by

$$(Ax, y)_X = (\mathcal{F}\boldsymbol{\varepsilon}(\boldsymbol{u}), \boldsymbol{\varepsilon}(\boldsymbol{v}))_Q + (\boldsymbol{\beta}\nabla\varphi, \nabla\psi)_{L^2(\Omega)^d}$$

$$+(\mathcal{P}^\top\nabla\varphi, \boldsymbol{\varepsilon}(\boldsymbol{v}))_Q - (\mathcal{P}\boldsymbol{\varepsilon}(\boldsymbol{u}), \nabla\psi)_{L^2(\Omega)^d}$$

$$\forall\, x = (\boldsymbol{u}, \varphi),\ y = (\boldsymbol{v}, \psi) \in X \tag{7.25}$$

and we extend the functional j defined by (7.19) to a functional h defined on X, that is

$$h(x) = j(\boldsymbol{u}) \qquad \forall\, x = (\boldsymbol{u}, \varphi) \in X. \tag{7.26}$$

Finally, we consider the element $f \in X$ given by

$$f = (\boldsymbol{f}, q) \in X. \tag{7.27}$$

We have the following equivalence result.

Lemma 7.4 *The pair* $x = (\boldsymbol{u}, \varphi) \in X$ *is a solution to Problem 7.2 if and only if*

$$(Ax, y - x)_X + h(y) - h(x) \geq (f, y - x)_X \quad \forall\, y \in X. \tag{7.28}$$

Proof Let $x = (\boldsymbol{u}, \varphi) \in X$ be a solution to Problem 7.2 and let $y = (\boldsymbol{v}, \psi) \in X$. We test in (7.23) with $\psi - \varphi$, add the resulting equality to (7.22) and use (7.24)–(7.27) to obtain (7.28). Conversely, let $x = (\boldsymbol{u}, \varphi) \in X$ be a solution to the elliptic variational inequality (7.28). We take $y = (\boldsymbol{v}, \varphi)$ in (7.28), where $\boldsymbol{v}$ is an arbitrary element of V, and obtain (7.22); then we take successively $y = (\boldsymbol{u}, \varphi + \psi)$ and $y = (\boldsymbol{u}, \varphi - \psi)$ in (7.28), where ψ is an arbitrary element of W; as a result we obtain (7.23), which concludes the proof. ■

We are now in position to provide the proof of Theorem 7.3.

Proof First, we investigate the properties of the operator $A : X \to X$ given by (7.25). To this end, we consider two elements $x_1 = (\boldsymbol{u}_1, \varphi_1)$,

$x_2 = (\boldsymbol{u}_2, \varphi_2) \in X$. We have

$$\begin{aligned}(Ax_1 - Ax_2, x_1 - x_2)_X \\ = (\mathcal{F}\boldsymbol{\varepsilon}(\boldsymbol{u}_1) - \mathcal{F}\boldsymbol{\varepsilon}(\boldsymbol{u}_2), \boldsymbol{\varepsilon}(\boldsymbol{u}_1) - \boldsymbol{\varepsilon}(\boldsymbol{u}_2))_Q \\ +(\boldsymbol{\beta}\nabla\varphi_1 - \boldsymbol{\beta}\nabla\varphi_2, \nabla\varphi_1 - \nabla\varphi_2)_{L^2(\Omega)^d} \\ +(\mathcal{P}^\top\nabla\varphi_1 - \mathcal{P}^\top\nabla\varphi_2, \boldsymbol{\varepsilon}(\boldsymbol{u}_1) - \boldsymbol{\varepsilon}(\boldsymbol{u}_2))_Q \\ -(\mathcal{P}\boldsymbol{\varepsilon}(\boldsymbol{u}_1) - \mathcal{P}\boldsymbol{\varepsilon}(\boldsymbol{u}_2), \nabla\varphi_1 - \nabla\varphi_2)_{L^2(\Omega)^d}\end{aligned}$$

and, since (4.117) implies that $(\mathcal{P}^\top\nabla\varphi, \boldsymbol{\varepsilon}(\boldsymbol{u}))_Q = (\mathcal{P}\boldsymbol{\varepsilon}(\boldsymbol{u}), \nabla\varphi)_{L^2(\Omega)^d}$ for all $x = (\boldsymbol{u}, \varphi) \in X$, we find that

$$\begin{aligned}(Ax_1 - Ax_2, x_1 - x_2)_X \\ = (\mathcal{F}\boldsymbol{\varepsilon}(\boldsymbol{u}_1) - \mathcal{F}\boldsymbol{\varepsilon}(\boldsymbol{u}_2), \boldsymbol{\varepsilon}(\boldsymbol{u}_1) - \boldsymbol{\varepsilon}(\boldsymbol{u}_2))_Q \\ +(\boldsymbol{\beta}\nabla\varphi_1 - \boldsymbol{\beta}\nabla\varphi_2, \nabla\varphi_1 - \nabla\varphi_2)_{L^2(\Omega)^d}.\end{aligned}$$

We now use (7.12) and (7.14) to see that there exists $c_1 > 0$ which depends only on $\mathcal{F}$, $\boldsymbol{\beta}$, Ω such that

$$(Ax_1 - Ax_2, x_1 - x_2)_X \geq c_1(\|\boldsymbol{u}_1 - \boldsymbol{u}_2\|_V^2 + \|\varphi_1 - \varphi_2\|_W^2)$$

and, keeping in mind (7.24), we obtain

$$(Ax_1 - Ax_2, x_1 - x_2)_X \geq c_1 \|x_1 - x_2\|_X^2. \tag{7.29}$$

In the same way, using (7.12)–(7.14), after some algebra it follows that there exists $c_2 > 0$ which depends only on $\mathcal{F}$, $\boldsymbol{\beta}$ and $\mathcal{P}$ such that

$$\begin{aligned}(Ax_1 - Ax_2, y)_X \leq c_2(\|\boldsymbol{u}_1 - \boldsymbol{u}_2\|_V\|\boldsymbol{v}\|_V + \|\varphi_1 - \varphi_2\|_W\|\psi\|_W \\ +\|\varphi_1 - \varphi_2\|_W\|\boldsymbol{v}\|_V + \|\boldsymbol{u}_1 - \boldsymbol{u}_2\|_V\|\psi\|_W)\end{aligned}$$

for all $y = (\boldsymbol{v}, \psi) \in X$. We use (7.24) and the previous inequality to obtain

$$(Ax_1 - Ax_2, y)_X \leq 4c_2 \|x_1 - x_2\|_X\|y\|_X \qquad \forall\, y \in X$$

and, taking $y = Ax_1 - Ax_2 \in X$, we find

$$\|Ax_1 - Ax_2\|_X \leq 4c_2 \|x_1 - x_2\|_X. \tag{7.30}$$

Inequalities (7.29) and (7.30) show that the operator $A : X \to X$ is strongly monotone and Lipschitz continuous.

Next, we investigate the properties of the functional h given by (7.26), (7.19). To this end we recall that the functional j is a seminorm on V and, moreover, it satisfies inequality (5.116). This inequality, definition (7.26) and Proposition 1.34 show that h is a continuous seminorm on X.

The properties of A and h above allow us to use Theorem 2.8 on page 40. Therefore, we obtain that the variational inequality (7.28) has a unique solution $x = (\boldsymbol{u}, \varphi) \in X$ and, using Lemma 7.4, we deduce that the pair $(\boldsymbol{u}, \varphi) \in V \times W$ is the unique solution to Problem 7.2, which concludes the proof. ■

A quadruple of functions $(\boldsymbol{u}, \boldsymbol{\sigma}, \varphi, \boldsymbol{D})$ which satisfy (7.1), (7.2), (7.22) and (7.23) is called a *weak solution* of the piezoelectric contact problem (7.1)–(7.11). We conclude by Theorem 7.3 that, under the assumptions (7.12)–(7.18), the piezoelectric contact problem (7.1)–(7.11) has a unique weak solution $(\boldsymbol{u}, \boldsymbol{\sigma}, \varphi, \boldsymbol{D})$ such that $\boldsymbol{u} \in V$, $\varphi \in W$. Moreover, using arguments similar to those on page 127 it follows that

$$\operatorname{Div} \boldsymbol{\sigma} + \boldsymbol{f}_0 = \boldsymbol{0}, \quad \operatorname{div} \boldsymbol{D} = q_0 \quad \text{in } \Omega. \tag{7.31}$$

Using the regularity of $\boldsymbol{f}_0$ and q_0 in (7.15) and (7.16) it follows from (7.31) that $\operatorname{Div} \boldsymbol{\sigma} \in L^2(\Omega)^d$ and $\operatorname{div} \boldsymbol{D} \in L^2(\Omega)$. Therefore, recalling the definitions (4.20) and (4.29) we find that $\boldsymbol{\sigma} \in Q_1$ and $\boldsymbol{D} \in \mathcal{W}_1$.

7.1.3 Dual variational formulation

Problem 7.2 is called the *primal variational formulation* of the frictional contact problem 7.1. Our interest in this subsection lies in the study of a *dual formulation* of Problem 7.1, in terms of stress and electric displacement field. To this end we assume in what follows that (7.12)–(7.18) hold and we consider the product Hilbert space $Y = Q \times L^2(\Omega)^d$ endowed with the canonical inner product and the associated norm $\|\cdot\|_Y$, i.e.

$$(x, y)_Y = (\boldsymbol{\sigma}, \boldsymbol{\tau})_Q + (\boldsymbol{D}, \boldsymbol{E})_{L^2(\Omega)^d}, \quad \|y\|_Y = (y, y)_Y^{\frac{1}{2}}$$

for all $x = (\boldsymbol{\sigma}, \boldsymbol{D})$, $y = (\boldsymbol{\tau}, \boldsymbol{E}) \in Y$. Let $T : Y \to Y$ be the operator

$$Ty = (T_1 y, T_2 y) \quad \text{where} \quad T_1 y = \mathcal{F}\boldsymbol{\tau} + \mathcal{P}^\top \boldsymbol{E}, \; T_2 y = \mathcal{B}\boldsymbol{E} - \mathcal{P}\boldsymbol{\tau} \tag{7.32}$$

for all $y = (\boldsymbol{\tau}, \boldsymbol{E}) \in Y$.

Using (7.12)–(7.14) it is easy to see that T is a strongly monotone Lipschitz continuous operator on the space Y and, therefore, Proposition 1.25 implies that T is invertible and its inverse $T^{-1} : Y \to Y$ is again a strongly monotone Lipschitz continuous operator. Denote by T_1^{-1} and T_2^{-1} its components, i.e. $T^{-1}y = (T_1^{-1}y, T_2^{-1}y)$, for all $y \in Y$. We also introduce the set of admissible stress fields and the set of admissible electric displacement fields, respectively, defined by

$$\Sigma = \{\, \boldsymbol{\tau} \in Q \;:\; (\boldsymbol{\tau}, \boldsymbol{\varepsilon}(\boldsymbol{v}))_Q + j(\boldsymbol{v}) \ge (\boldsymbol{f}, \boldsymbol{v})_V \quad \forall\, \boldsymbol{v} \in V \,\}, \tag{7.33}$$

$$\Theta = \{\, \boldsymbol{E} \in L^2(\Omega)^d \;:\; (\boldsymbol{E}, \nabla\psi)_{L^2(\Omega)^d} = (q, \psi)_W \quad \forall\, \psi \in W \,\}. \tag{7.34}$$

Consider now the following variational problem.

Problem 7.5 *Find a stress field $\boldsymbol{\sigma}$ and an electric displacement field $\boldsymbol{D}$ such that*

$$\boldsymbol{\sigma} \in \Sigma, \qquad (T_1^{-1}(\boldsymbol{\sigma}, -\boldsymbol{D}), \boldsymbol{\tau} - \boldsymbol{\sigma})_Q \geq 0 \quad \forall \boldsymbol{\tau} \in \Sigma, \tag{7.35}$$

$$-\boldsymbol{D} \in \Theta, \qquad (T_2^{-1}(\boldsymbol{\sigma}, -\boldsymbol{D}), \boldsymbol{E} + \boldsymbol{D})_{L^2(\Omega)^d} = 0 \quad \forall \boldsymbol{E} \in \Theta. \tag{7.36}$$

Using arguments similar to those used on page 156 it can be proved that if $(\boldsymbol{u}, \boldsymbol{\sigma}, \varphi, \boldsymbol{D})$ are sufficiently regular functions which satisfy (7.1)–(7.11) then $(\boldsymbol{\sigma},\boldsymbol{D})$ is a solution to Problem 7.5. We conclude from this that Problem 7.5 represents a variational formulation of the contact problem 7.1 in terms of stress and electric displacement fields and, therefore, we call it a *dual variational formulation* of the frictional contact problem 7.1. Note that this formulation is in the form of a system coupling an elliptic variational inequality of the first kind, (7.35), with a linear variational equation, (7.36). The existence of a unique solution to Problem 7.5 can be obtained by using Theorem 2.1 in the product Hilbert space Y, with the convex set $K = \Sigma \times \Theta$ and the strongly monotone Lipschitz continuous operator $T^{-1} : Y \to Y$. The proof is based on arguments similar to those used in the proof of Theorem 7.3 and, for this reason, we skip it. Nevertheless, we remark that in both the statement and the solution of Problem 7.5 the real unknowns are, in fact, $\boldsymbol{\sigma}$ and $-\boldsymbol{D}$. This feature arises from the fact that the electric correspondent of the stress field $\boldsymbol{\sigma}$ is the opposite of the electric displacement field, $-\boldsymbol{D}$. Indeed, note that in both the balance equations (7.3)–(7.4) and the constitutive equations (7.37)–(7.38) below, $\boldsymbol{\sigma}$ and $-\boldsymbol{D}$ play a symmetric role.

The relation between the primal and dual problems 7.2 and 7.5 is given by the following result, which represents an extension of the result provided in Theorem 5.34 on page 158.

Theorem 7.6 *Assume (7.12)–(7.18).*

(1) *If $(\boldsymbol{u}, \varphi)$ is a solution to Problem 7.2 and $(\boldsymbol{\sigma}, \boldsymbol{D})$ are defined by*

$$\boldsymbol{\sigma} = \mathcal{F}\boldsymbol{\varepsilon}(\boldsymbol{u}) + \mathcal{P}^\top \nabla\varphi, \tag{7.37}$$

$$\boldsymbol{D} = \mathcal{P}\boldsymbol{\varepsilon}(\boldsymbol{u}) - \boldsymbol{\beta}\nabla\varphi, \tag{7.38}$$

then $(\boldsymbol{\sigma}, \boldsymbol{D})$ is a solution to Problem 7.5.

(2) *Conversely, let $(\boldsymbol{\sigma}, \boldsymbol{D})$ be a solution to Problem 7.5. Then there exists a unique pair $(\boldsymbol{u}, \varphi) \in V \times W$ such that (7.37) and (7.38) hold. Moreover, $(\boldsymbol{u}, \varphi)$ is a solution to Problem 7.2.*

Proof (1) Let $(\boldsymbol{u}, \varphi) \in V \times W$ be a solution to Problem 7.2 and let $(\boldsymbol{\sigma}, \boldsymbol{D})$ be given by (7.37)–(7.38). It follows that $(\boldsymbol{\sigma}, \boldsymbol{D}) \in Q \times L^2(\Omega)^d$ and using

(7.22)–(7.23) yields

$$(\boldsymbol{\sigma}, \boldsymbol{\varepsilon}(\boldsymbol{v}) - \boldsymbol{\varepsilon}(\boldsymbol{u}))_Q + j(\boldsymbol{v}) - j(\boldsymbol{u}) \geq (\boldsymbol{f}, \boldsymbol{v} - \boldsymbol{u})_V \quad \forall \boldsymbol{v} \in V, \tag{7.39}$$

$$(\boldsymbol{D}, \nabla\psi)_{L^2(\Omega)^d} + (q, \psi)_W = 0 \quad \forall \psi \in W. \tag{7.40}$$

We choose $\boldsymbol{v} = 2\boldsymbol{u}$ and $\boldsymbol{v} = \boldsymbol{0}_V$ in (7.39) to obtain

$$(\boldsymbol{\sigma}, \boldsymbol{\varepsilon}(\boldsymbol{u}))_Q + j(\boldsymbol{u}) = (\boldsymbol{f}, \boldsymbol{u})_V \tag{7.41}$$

and using (7.39), (7.41) and the definition of the set Σ we find that

$$\boldsymbol{\sigma} \in \Sigma. \tag{7.42}$$

Next, by using (7.41) and (7.33) we have

$$(\boldsymbol{\tau} - \boldsymbol{\sigma}, \boldsymbol{\varepsilon}(\boldsymbol{u}))_Q \geq 0 \qquad \forall \boldsymbol{\tau} \in \Sigma. \tag{7.43}$$

We now use equalities (7.37), (7.38) and the definition (7.32) of the operator T to see that

$$(\boldsymbol{\sigma}, -\boldsymbol{D}) = T(\boldsymbol{\varepsilon}(\boldsymbol{u}), \nabla\varphi),$$

which yields

$$\boldsymbol{\varepsilon}(\boldsymbol{u}) = T_1^{-1}(\boldsymbol{\sigma}, -\boldsymbol{D}), \quad \nabla\varphi = T_2^{-1}(\boldsymbol{\sigma}, -\boldsymbol{D}). \tag{7.44}$$

Inequality (7.35) is now a consequence of (7.42)–(7.44). Next, we use (7.40) and the definition (7.34) of the set Θ to find that

$$-\boldsymbol{D} \in \Theta. \tag{7.45}$$

Moreover, we take $\psi = \varphi$ in (7.40) and use the definition of the set Θ again to obtain

$$(\boldsymbol{D} + \boldsymbol{E}, \nabla\varphi)_{L^2(\Omega)^d} = 0 \qquad \forall \boldsymbol{E} \in \Theta. \tag{7.46}$$

We note that (7.36) is a consequence of (7.44)–(7.46), which concludes the first part of the proof.

(2) Let $(\boldsymbol{\sigma}, \boldsymbol{D})$ be a solution of Problem 7.5 and consider an element $\boldsymbol{z} \in Q$ such that $(\boldsymbol{z}, \boldsymbol{\varepsilon}(\boldsymbol{v}))_Q = 0$ for all $\boldsymbol{v} \in V$. It follows from the definition of Σ that $\boldsymbol{\sigma} \pm \boldsymbol{z} \in \Sigma$ and, by choosing $\boldsymbol{\tau} = \boldsymbol{\sigma} \pm \boldsymbol{z}$ in (7.35), we find that $(T_1^{-1}(\boldsymbol{\sigma}, -\boldsymbol{D}), \boldsymbol{z})_Q = 0$. This last inequality shows that $T_1^{-1}(\boldsymbol{\sigma}, -\boldsymbol{D}) \in \boldsymbol{\varepsilon}(V)^{\perp\perp}$, where $\perp$ represents the orthogonal complement in Q. It follows from Theorem 4.1 that $\boldsymbol{\varepsilon}(V)$ is a closed subspace of Q and, therefore, Theorem 1.17 implies that $\boldsymbol{\varepsilon}(V)^{\perp\perp} = \boldsymbol{\varepsilon}(V)$. We conclude that

$$T_1^{-1}(\boldsymbol{\sigma}, -\boldsymbol{D}) \in \boldsymbol{\varepsilon}(V)$$

and, therefore, there exists $\boldsymbol{u} \in V$ such that

$$T_1^{-1}(\boldsymbol{\sigma}, -\boldsymbol{D}) = \boldsymbol{\varepsilon}(\boldsymbol{u}). \tag{7.47}$$

Consider now an element $\boldsymbol{F} \in L^2(\Omega)^d$ such that $(\boldsymbol{F}, \nabla\psi)_{L^2(\Omega)^d} = 0$ for all $\psi \in W$. It follows from the definition of Θ that $-\boldsymbol{D} + \boldsymbol{F} \in \Theta$ and, by choosing $\boldsymbol{E} = -\boldsymbol{D} + \boldsymbol{F}$ in (7.36), we find that $(T_2^{-1}(\boldsymbol{\sigma}, -\boldsymbol{D}), \boldsymbol{F})_{L^2(\Omega)^d} = 0$. This last inequality shows that $T_2^{-1}(\boldsymbol{\sigma}, -\boldsymbol{D}) \in \nabla(W)^{\perp\perp}$, where $\nabla(W)$ is the space given by (4.28) and $\perp$ now represents the orthogonal complement in $L^2(\Omega)^d$. As a consequence of Theorems 4.3 and 1.17 it follows that $\nabla(W)^{\perp\perp} = \nabla(W)$ and thus $T_2^{-1}(\boldsymbol{\sigma}, -\boldsymbol{D}) \in \nabla(W)$. We conclude that there exists an element $\varphi \in W$ such that

$$T_2^{-1}(\boldsymbol{\sigma}, -\boldsymbol{D}) = \nabla\varphi. \tag{7.48}$$

We now combine equalities (7.47), (7.48) and use the definition (7.32) of the operator T to deduce the existence of a pair $(\boldsymbol{u}, \varphi) \in V \times W$ which satisfies (7.37)–(7.38). Its uniqueness follows from equalities (7.47) and (7.48) combined with Korn's and Friedrichs–Poincaré's inequalities, respectively.

We prove now that $(\boldsymbol{u}, \varphi)$ is a solution to Problem 7.2. To this end we consider the element $\boldsymbol{\tau}(\boldsymbol{u}) \in Q$ defined by (5.169) where, recall, $\boldsymbol{d}(\boldsymbol{u}) \in V$ is given by Lemma 5.33. We use (5.169), the definition (4.11) of the inner product on V and (5.160) to see that

$$(\boldsymbol{\tau}(\boldsymbol{u}), \boldsymbol{\varepsilon}(\boldsymbol{v}) - \boldsymbol{\varepsilon}(\boldsymbol{u}))_Q + j(\boldsymbol{v}) - j(\boldsymbol{u}) \geq (\boldsymbol{f}, \boldsymbol{v} - \boldsymbol{u})_V \quad \forall\, \boldsymbol{v} \in V. \tag{7.49}$$

We now choose $\boldsymbol{v} = 2\boldsymbol{u}$ and $\boldsymbol{v} = \boldsymbol{0}_V$ in (7.49) to obtain

$$(\boldsymbol{\tau}(\boldsymbol{u}), \boldsymbol{\varepsilon}(\boldsymbol{u}))_Q + j(\boldsymbol{u}) = (\boldsymbol{f}, \boldsymbol{u})_V. \tag{7.50}$$

As a consequence of (7.49) and (7.50) we find that

$$(\boldsymbol{\tau}(\boldsymbol{u}), \boldsymbol{\varepsilon}(\boldsymbol{v}))_Q + j(\boldsymbol{v}) \geq (\boldsymbol{f}, \boldsymbol{v})_V \quad \forall\, \boldsymbol{v} \in V,$$

which shows that $\boldsymbol{\tau}(\boldsymbol{u}) \in \Sigma$. Therefore, choosing $\boldsymbol{\tau} = \boldsymbol{\tau}(\boldsymbol{u})$ in (7.35) and using (7.47) it follows that

$$(\boldsymbol{\tau}(\boldsymbol{u}), \boldsymbol{\varepsilon}(\boldsymbol{u}))_Q \geq (\boldsymbol{\sigma}, \boldsymbol{\varepsilon}(\boldsymbol{u}))_Q. \tag{7.51}$$

We now combine (7.50) and (7.51) and obtain

$$(\boldsymbol{f}, \boldsymbol{u})_V \geq (\boldsymbol{\sigma}, \boldsymbol{\varepsilon}(\boldsymbol{u}))_Q + j(\boldsymbol{u}).$$

Since the converse inequality follows from the fact that $\boldsymbol{\sigma} \in \Sigma$ and $\boldsymbol{u} \in V$, we obtain (7.41) which, in turn, implies (7.39). Inequality (7.22) follows now from (7.37) and (7.39). On the other hand (7.36) implies that $-\boldsymbol{D} \in \Theta$ and, therefore, the definition (7.34) of the set Θ shows that (7.40) holds. Next, we combine (7.40) and (7.38) to obtain inequality (7.23), which concludes the proof. ■

The mechanical interpretation of the results in Theorem 7.6 is the following.

(1) If the pair $(\boldsymbol{u}, \varphi)$ is a solution of the primal variational formulation of the piezoelectric contact problem (7.1)–(7.11) (i.e. Problem 7.2), then the pair $(\boldsymbol{\sigma}, \boldsymbol{D})$ associated with $(\boldsymbol{u}, \varphi)$ by the electro-elastic constitutive law (7.37), (7.38) is a solution of the dual variational formulation of the piezoelectric contact problem (7.1)–(7.11) (i.e. Problem 7.5).

(2) If the pair $(\boldsymbol{\sigma}, \boldsymbol{D})$ is a solution of the dual variational formulation of the piezoelectric contact problem (7.1)–(7.11) (i.e. Problem 7.5), then there exists a unique pair $(\boldsymbol{u}, \varphi)$ associated with $(\boldsymbol{\sigma}, \boldsymbol{D})$ by the electro-elastic constitutive law and, moreover, $(\boldsymbol{u}, \varphi)$ is a solution of the primal variational formulation of the piezoelectric contact problem (7.1)–(7.11) (i.e. Problem 7.2).

Since Theorem 7.3 states that the primal Problem 7.2 has a unique solution $(\boldsymbol{u}, \varphi) \in V \times W$, it follows from Theorem 7.6 that the dual Problem 7.5 has a unique solution $(\boldsymbol{\sigma}, \boldsymbol{D}) \in Q \times L^2(\Omega)^d$. Moreover, the solutions of the two problems are connected by the electro-elastic constitutive law (7.37)–(7.38). For this reason, as mentioned on page 223, we refer to the quadruple $(\boldsymbol{u}, \boldsymbol{\sigma}, \varphi, \boldsymbol{D})$ given by Theorems 7.3 and 7.6 as a *weak solution* for the piezoelectric contact problem (7.1)–(7.11) and we conclude that this problem has a unique weak solution.

7.2 An electro-viscoelastic frictional contact problem

In this section we provide the analysis of a quasistatic frictional contact problem for electro-viscoelastic materials. To this end we use the results on evolutionary variational inequalities in Section 3.3.

7.2.1 Problem statement

We assume that the material is electro-viscoelastic and there is a gap between the body and the foundation. The contact is with normal compliance and it is associated with the Coulomb's law of dry friction. Also, the foundation is electrically conductive. With these assumptions, the classical model for the process is as follows.

Problem 7.7 *Find a displacement field* $\boldsymbol{u} : \Omega \times [0, T] \to \mathbb{R}^d$*, a stress field* $\boldsymbol{\sigma} : \Omega \times [0, T] \to \mathbb{S}^d$*, an electric potential* $\varphi : \Omega \times [0, T] \to \mathbb{R}$ *and an electric* *displacement field* $\boldsymbol{D} : \Omega \times [0,T] \to \mathbb{R}^d$ *such that*

$$\boldsymbol{\sigma} = \mathcal{A}\boldsymbol{\varepsilon}(\dot{\boldsymbol{u}}) + \mathcal{B}\boldsymbol{\varepsilon}(\boldsymbol{u}) - \mathcal{P}^\top \boldsymbol{E}(\varphi) \quad in \quad \Omega \times (0,T), \tag{7.52}$$

$$\boldsymbol{D} = \mathcal{P}\boldsymbol{\varepsilon}(\boldsymbol{u}) + \boldsymbol{\beta}\boldsymbol{E}(\varphi) \quad in \quad \Omega \times (0,T), \tag{7.53}$$

$$\mathrm{Div}\,\boldsymbol{\sigma} + \boldsymbol{f}_0 = \boldsymbol{0} \quad in \quad \Omega \times (0,T), \tag{7.54}$$

$$\mathrm{div}\,\boldsymbol{D} - q_0 = 0 \quad in \quad \Omega \times (0,T), \tag{7.55}$$

$$\boldsymbol{u} = \boldsymbol{0} \quad on \quad \Gamma_1 \times (0,T), \tag{7.56}$$

$$\boldsymbol{\sigma\nu} = \boldsymbol{f}_2 \quad on \quad \Gamma_2 \times (0,T), \tag{7.57}$$

$$\varphi = 0 \quad on \quad \Gamma_{\mathrm{a}} \times (0,T), \tag{7.58}$$

$$\boldsymbol{D} \cdot \boldsymbol{\nu} = q_{\mathrm{b}} \quad on \quad \Gamma_{\mathrm{b}} \times (0,T), \tag{7.59}$$

$$-\sigma_\nu = p_\nu(u_\nu - g_{\mathrm{a}}) \quad on \quad \Gamma_3 \times (0,T), \tag{7.60}$$

$$\left.\begin{aligned} &\|\boldsymbol{\sigma}_\tau\| \le p_\tau(u_\nu - g_{\mathrm{a}}), \\ &\boldsymbol{\sigma}_\tau = -p_\tau(u_\nu - g_{\mathrm{a}})\frac{\dot{\boldsymbol{u}}_\tau}{\|\dot{\boldsymbol{u}}_\tau\|} \quad \text{if} \quad \dot{\boldsymbol{u}}_\tau \neq \boldsymbol{0} \end{aligned}\right\} \quad on \quad \Gamma_3 \times (0,T), \tag{7.61}$$

$$\boldsymbol{D} \cdot \boldsymbol{\nu} = p_{\mathrm{e}}(u_\nu - g_{\mathrm{a}})h_{\mathrm{e}}(\varphi - \varphi_{\mathrm{F}}) \quad on \quad \Gamma_3 \times (0,T), \tag{7.62}$$

$$\boldsymbol{u}(0) = \boldsymbol{u}_0 \quad in \quad \Omega. \tag{7.63}$$

We now describe Problem 7.7 and provide explanation of the equations and the boundary conditions. First, equations (7.52) and (7.53) represent the electro-viscoelastic constitutive law, (4.119) and (4.118), respectively. Next, equations (7.54) and (7.55) are the balance equations for the stress and electric displacement fields, respectively. Conditions (7.56) and (7.57) are the displacement and traction boundary conditions, whereas (7.58) and (7.59) represent the electric boundary conditions. Conditions (7.60) and (7.61) represent the normal compliance condition and the associated friction law, already used in Section 6.3. Equality (7.62) represents the electrical condition on the potential contact surface, (4.130). Recall that here p_{e} and h_{e} are given functions, g_{a} represents the initial gap and φ_{F} denotes the electric potential of the foundation. Finally, (7.63) represents the initial condition for the displacement field in which $\boldsymbol{u}_0$ is given.

We list now the assumptions on the problem's data. The viscosity operator $\mathcal{A}$ and the elasticity operator $\mathcal{B}$ are assumed to satisfy the following conditions.

$$\left.\begin{array}{l}\text{(a) } \mathcal{A}:\Omega\times\mathbb{S}^d\to\mathbb{S}^d.\\ \text{(b) There exists } L_{\mathcal{A}}>0 \text{ such that}\\ \qquad \|\mathcal{A}(\boldsymbol{x},\boldsymbol{\varepsilon}_1)-\mathcal{A}(\boldsymbol{x},\boldsymbol{\varepsilon}_2)\|\le L_{\mathcal{A}}\|\boldsymbol{\varepsilon}_1-\boldsymbol{\varepsilon}_2\|\\ \qquad \forall\,\boldsymbol{\varepsilon}_1,\boldsymbol{\varepsilon}_2\in\mathbb{S}^d,\ \text{a.e. } \boldsymbol{x}\in\Omega.\\ \text{(c) There exists } m_{\mathcal{A}}>0 \text{ such that}\\ \qquad (\mathcal{A}(\boldsymbol{x},\boldsymbol{\varepsilon}_1)-\mathcal{A}(\boldsymbol{x},\boldsymbol{\varepsilon}_2))\cdot(\boldsymbol{\varepsilon}_1-\boldsymbol{\varepsilon}_2)\ge m_{\mathcal{A}}\,\|\boldsymbol{\varepsilon}_1-\boldsymbol{\varepsilon}_2\|^2\\ \qquad \forall\,\boldsymbol{\varepsilon}_1,\boldsymbol{\varepsilon}_2\in\mathbb{S}^d,\ \text{a.e. } \boldsymbol{x}\in\Omega.\\ \text{(d) The mapping } \boldsymbol{x}\mapsto\mathcal{A}(\boldsymbol{x},\boldsymbol{\varepsilon}) \text{ is measurable on } \Omega,\\ \qquad \text{for any } \boldsymbol{\varepsilon}\in\mathbb{S}^d.\\ \text{(e) The mapping } \boldsymbol{x}\mapsto\mathcal{A}(\boldsymbol{x},\boldsymbol{0}_{\mathbb{S}^d}) \text{ belongs to } Q.\end{array}\right\}\qquad(7.64)$$

$$\left.\begin{array}{l}\text{(a) } \mathcal{B}:\Omega\times\mathbb{S}^d\to\mathbb{S}^d.\\ \text{(b) There exists } L_{\mathcal{B}}>0 \text{ such that}\\ \qquad \|\mathcal{B}(\boldsymbol{x},\boldsymbol{\varepsilon}_1)-\mathcal{B}(\boldsymbol{x},\boldsymbol{\varepsilon}_2)\|\le L_{\mathcal{B}}\,\|\boldsymbol{\varepsilon}_1-\boldsymbol{\varepsilon}_2\|\\ \qquad \forall\,\boldsymbol{\varepsilon}_1,\boldsymbol{\varepsilon}_2\in\mathbb{S}^d,\ \text{a.e. } \boldsymbol{x}\in\Omega.\\ \text{(c) The mapping } \boldsymbol{x}\mapsto\mathcal{B}(\boldsymbol{x},\boldsymbol{\varepsilon}) \text{ is measurable on } \Omega,\\ \qquad \text{for any } \boldsymbol{\varepsilon}\in\mathbb{S}^d.\\ \text{(d) The mapping } \boldsymbol{x}\mapsto\mathcal{B}(\boldsymbol{x},\boldsymbol{0}_{\mathbb{S}^d}) \text{ belongs to } Q.\end{array}\right\}\qquad(7.65)$$

The piezoelectric tensor $\mathcal{P}$ and the electric permittivity tensor $\boldsymbol{\beta}$ satisfy conditions (7.13) and (7.14), respectively.

The normal compliance functions p_r $(r=\nu,\tau)$ satisfy

$$\left.\begin{array}{l}\text{(a) } p_r:\Gamma_3\times\mathbb{R}\to\mathbb{R}_+;\\ \text{(b) there exists } L_r>0 \text{ such that}\\ \qquad |p_r(\boldsymbol{x},u_1)-p_r(\boldsymbol{x},u_2)|\le L_r\,|u_1-u_2|\\ \qquad \forall\,u_1,\,u_2\in\mathbb{R},\ \text{a.e. } \boldsymbol{x}\in\Gamma_3;\\ \text{(c) the mapping } \boldsymbol{x}\mapsto p_r(\boldsymbol{x},u) \text{ is measurable on } \Gamma_3,\\ \qquad \text{for any } u\in\mathbb{R};\\ \text{(d) } p_r(\boldsymbol{x},u)=0 \text{ for all } u\le 0,\ \text{a.e. } \boldsymbol{x}\in\Gamma_3.\end{array}\right\}\qquad(7.66)$$

The functions $p_{\rm e}$ and h_e satisfy

$$\left.\begin{array}{l}\text{(a) } p_{\rm e}:\Gamma_3\times\mathbb{R}\to\mathbb{R};\\ \text{(b) there exists } L_{\rm e}>0 \text{ such that}\\ \qquad |p_{\rm e}(\boldsymbol{x},u_1)-p_{\rm e}(\boldsymbol{x},u_2)|\le L_{\rm e}|u_1-u_2|\\ \qquad \forall\,u_1,\,u_2\in\mathbb{R},\ \text{a.e. } \boldsymbol{x}\in\Gamma_3;\\ \text{(c) there exists } \overline{p}_{\rm e}>0 \text{ such that}\\ \qquad 0\le p_{\rm e}(\boldsymbol{x},u)\le\overline{p}_{\rm e}\ \forall\,u\in\mathbb{R},\ \text{a.e. } \boldsymbol{x}\in\Gamma_3;\\ \text{(d) the mapping } \boldsymbol{x}\mapsto p_{\rm e}(\boldsymbol{x},u) \text{ is measurable on } \Gamma_3,\\ \qquad \text{for any } u\in\mathbb{R};\\ \text{(e) } p_{\rm e}(\boldsymbol{x},u)=0 \text{ for all } u\le 0,\ \text{a.e. } \boldsymbol{x}\in\Gamma_3.\end{array}\right\}\qquad(7.67)$$

$$\left.\begin{array}{l}\text{(a) } h_{\mathrm{e}} : \Gamma_3 \times \mathbb{R} \to \mathbb{R};\\ \text{(b) there exists } l_{\mathrm{e}} > 0 \text{ such that}\\ \quad |h_{\mathrm{e}}(\boldsymbol{x},\varphi_1) - h_{\mathrm{e}}(\boldsymbol{x},\varphi_2)| \le l_{\mathrm{e}}|\varphi_1 - \varphi_2|\\ \quad \forall\, \varphi_1,\, \varphi_2 \in \mathbb{R}, \text{ a.e. } \boldsymbol{x} \in \Gamma_3;\\ \text{(c) there exists } \overline{h}_{\mathrm{e}} > 0 \text{ such that}\\ \quad |h_{\mathrm{e}}(\boldsymbol{x},\varphi)| \le \overline{h}_{\mathrm{e}}\ \forall\, \varphi \in \mathbb{R}, \text{ a.e. } \boldsymbol{x} \in \Gamma_3;\\ \text{(d) the mapping } \boldsymbol{x} \mapsto h_{\mathrm{e}}(\boldsymbol{x},u) \text{ is measurable on } \Gamma_3,\\ \quad \text{for any } \varphi \in \mathbb{R}.\end{array}\right\} \quad (7.68)$$

Moreover, the densities of body forces, surface tractions, volume and surface electric charges have the regularity

$$\boldsymbol{f}_0 \in C([0,T];L^2(\Omega)^d), \quad \boldsymbol{f}_2 \in C([0,T];L^2(\Gamma_2)^d), \tag{7.69}$$

$$q_0 \in C([0,T];L^2(\Omega)), \quad q_{\mathrm{b}} \in C([0,T];L^2(\Gamma_b)) \tag{7.70}$$

and, finally, we assume that the gap function, the electric potential of the foundation and the initial displacement satisfy

$$g_{\mathrm{a}} \in L^2(\Gamma_3), \quad g_{\mathrm{a}} \ge 0 \quad \text{a.e. on } \Gamma_3, \tag{7.71}$$

$$\varphi_{\mathrm{F}} \in L^2(\Gamma_3), \tag{7.72}$$

$$\boldsymbol{u}_0 \in V. \tag{7.73}$$

Next, we define the functionals $j : V \times V \to \mathbb{R}$, $J : V \times W \times W \to \mathbb{R}$ and the functions $\boldsymbol{f} : [0,T] \to V$ and $q : [0,T] \to W$, respectively, by

$$j(\boldsymbol{u},\boldsymbol{v}) = \int_{\Gamma_3} p_\nu(u_\nu - g_{\mathrm{a}})|v_\nu|\,\mathrm{d}a + \int_{\Gamma_3} p_\tau(u_\nu - g_{\mathrm{a}})\|\boldsymbol{v}_\tau\|\,\mathrm{d}a, \tag{7.74}$$

$$J(\boldsymbol{u},\varphi,\psi) = \int_{\Gamma_3} p_{\mathrm{e}}(u_\nu - g_{\mathrm{a}})h_{\mathrm{e}}(\varphi - \varphi_{\mathrm{F}})\,\psi\,\mathrm{d}a, \tag{7.75}$$

$$(\boldsymbol{f}(t),\boldsymbol{v})_V = \int_\Omega \boldsymbol{f}_0(t)\cdot\boldsymbol{v}\,\mathrm{d}x + \int_{\Gamma_2} \boldsymbol{f}_2(t)\cdot\boldsymbol{v}\,\mathrm{d}a, \tag{7.76}$$

$$(q(t),\psi)_W = \int_\Omega q_0(t)\psi\,\mathrm{d}x - \int_{\Gamma_{\mathrm{b}}} q_{\mathrm{b}}(t)\,\psi\,\mathrm{d}a, \tag{7.77}$$

for all $\boldsymbol{u}$, $\boldsymbol{v} \in V$, φ, $\psi \in W$ and $t \in [0,T]$. We note that the definitions of $\boldsymbol{f}$ and q are based on the Riesz representation theorem. Moreover, it follows from assumptions (7.66)–(7.72) that the integrals in (7.74)–(7.77) are well defined.

Using Green's formulae (4.22) and (4.30), it is straightforward to see that if $(\boldsymbol{u},\boldsymbol{\sigma},\varphi,\boldsymbol{D})$ are sufficiently regular functions which satisfy (7.54)–(7.62) then

$$\begin{aligned}(\boldsymbol{\sigma}(t),\boldsymbol{\varepsilon}(\boldsymbol{v}) - \boldsymbol{\varepsilon}(\dot{\boldsymbol{u}}(t))_Q + j(\boldsymbol{u}(t),\boldsymbol{v}) - j(\boldsymbol{u}(t),\dot{\boldsymbol{u}}(t))&\\ \ge (\boldsymbol{f}(t),\boldsymbol{v} - \dot{\boldsymbol{u}}(t))_V,&\end{aligned} \tag{7.78}$$

$$(\boldsymbol{D}(t),\nabla\psi)_{L^2(\Omega)^d} + (q(t),\psi)_W = J(\boldsymbol{u}(t),\varphi(t),\psi), \tag{7.79}$$

for all $\boldsymbol{v} \in V$, $\psi \in W$ and $t \in [0,T]$. We substitute (7.52) in (7.78), (7.53) in (7.79), then we use the notation $\boldsymbol{E}(\varphi) = -\nabla\varphi$ and the initial condition (7.63). As a result we obtain the following variational formulation of Problem 7.7, in terms of displacement and electric potential fields.

Problem 7.8 *Find a displacement field* $\boldsymbol{u} : [0,T] \to V$ *and an electric potential* $\varphi : [0,T] \to W$ *such that*

$$\begin{aligned}(\mathcal{A}\boldsymbol{\varepsilon}(\dot{\boldsymbol{u}}(t)), \boldsymbol{\varepsilon}(\boldsymbol{v}) - \boldsymbol{\varepsilon}(\dot{\boldsymbol{u}}(t)))_Q + (\mathcal{B}\boldsymbol{\varepsilon}(\boldsymbol{u}(t)), \boldsymbol{\varepsilon}(\boldsymbol{v}) - \boldsymbol{\varepsilon}(\dot{\boldsymbol{u}}(t)))_Q \\ +(\mathcal{P}^\top \nabla\varphi(t), \boldsymbol{\varepsilon}(\boldsymbol{v}) - \boldsymbol{\varepsilon}(\dot{\boldsymbol{u}}(t)))_Q + j(\boldsymbol{u}(t), \boldsymbol{v}) - j(\boldsymbol{u}(t), \dot{\boldsymbol{u}}(t)) \\ \geq (\boldsymbol{f}(t), \boldsymbol{v} - \dot{\boldsymbol{u}}(t))_V,\end{aligned} \tag{7.80}$$

for all $\boldsymbol{v} \in V$ *and* $t \in [0,T]$,

$$\begin{aligned}(\boldsymbol{\beta}\nabla\varphi(t), \nabla\psi)_{L^2(\Omega)^d} - (\mathcal{P}\boldsymbol{\varepsilon}(\boldsymbol{u}(t)), \nabla\psi)_{L^2(\Omega)^d} \\ +J(\boldsymbol{u}(t), \varphi(t), \psi) = (q(t), \psi)_W,\end{aligned} \tag{7.81}$$

for all $\psi \in W$ *and* $t \in [0,T]$,

$$\boldsymbol{u}(0) = \boldsymbol{u}_0. \tag{7.82}$$

To study Problem 7.8 we consider the smallness assumption

$$\overline{p}_e l_e < \frac{m_\beta}{\widetilde{c}_0^2}, \tag{7.83}$$

where $\overline{p}_e$, l_e, m_β and $\widetilde{c}_0$ are the constants in (7.67), (7.68), (7.14) and (4.27), respectively. We note that only the trace constant, the coercivity constant of $\boldsymbol{\beta}$, the bound of p_e and the Lipschitz constant of h_e are involved in (7.83); therefore, this smallness assumption involves only the geometry and the electrical data of the problem, and does not depend on the mechanical data of the problem. Moreover, (7.83) is satisfied when the obstacle is insulated, since then $h_e \equiv 0$ and so $l_e = 0$. Removing this smallness assumption remains a task for future research, since it is made for mathematical reasons, and does not seem to relate to any inherent physical constraints of the problem.

7.2.2 *Existence and uniqueness*

Our main existence and uniqueness result in the study of Problem 7.8 is the following.

Theorem 7.9 *Assume (7.13), (7.14), (7.64)–(7.73) and (7.83). Then there exists a unique solution of Problem 7.8. Moreover, the solution satisfies*

$$\boldsymbol{u} \in C^1([0,T]; V), \quad \varphi \in C([0,T]; W). \tag{7.84}$$

The proof of Theorem 7.9 is carried out in several steps. We assume in what follows that (7.13), (7.14), (7.64)–(7.73) and (7.83) hold and, everywhere below, we denote by c various positive constants which are independent of time and whose value may change from line to line.

Let $\boldsymbol{\eta} \in C([0,T],Q)$ be given. In the first step consider the following intermediate variational problem for the displacement field.

Problem 7.10 *Find a displacement field* $\boldsymbol{u}_\eta : [0,T] \to V$ *such that*

$$(\mathcal{A}\boldsymbol{\varepsilon}(\dot{\boldsymbol{u}}_\eta(t)), \boldsymbol{\varepsilon}(\boldsymbol{v}) - \boldsymbol{\varepsilon}(\dot{\boldsymbol{u}}_\eta(t)))_Q + (\mathcal{B}\boldsymbol{\varepsilon}(\boldsymbol{u}_\eta(t)), \boldsymbol{\varepsilon}(\boldsymbol{v}) - \boldsymbol{\varepsilon}(\dot{\boldsymbol{u}}_\eta(t)))_Q$$
$$+(\boldsymbol{\eta}(t), \boldsymbol{\varepsilon}(\boldsymbol{v}) - \boldsymbol{\varepsilon}(\dot{\boldsymbol{u}}_\eta(t)))_Q + j(\boldsymbol{u}_\eta(t), \boldsymbol{v}) - j(\boldsymbol{u}_\eta(t), \dot{\boldsymbol{u}}_\eta(t))$$
$$\geq (\boldsymbol{f}(t), \boldsymbol{v} - \dot{\boldsymbol{u}}_\eta(t))_V \quad \forall \boldsymbol{v} \in V,\ t \in [0,T], \tag{7.85}$$

$$\boldsymbol{u}_\eta(0) = \boldsymbol{u}_0. \tag{7.86}$$

We have the following result.

Lemma 7.11 *Problem 7.10 has a unique solution* $\boldsymbol{u}_\eta \in C^1([0,T];V)$. *Moreover, if* $\boldsymbol{u}_1$ *and* $\boldsymbol{u}_2$ *are the solutions of (7.85), (7.86) corresponding to the data* $\boldsymbol{\eta}_1$, $\boldsymbol{\eta}_2 \in C([0,T];Q)$, *then there exists* $c > 0$ *such that*

$$\|\dot{\boldsymbol{u}}_1(t) - \dot{\boldsymbol{u}}_2(t)\|_V \leq c\,(\|\boldsymbol{\eta}_1(t) - \boldsymbol{\eta}_2(t)\|_Q + \|\boldsymbol{u}_1(t) - \boldsymbol{u}_2(t)\|_V) \tag{7.87}$$

for all $t \in [0,T]$.

Proof We apply Theorem 3.10 on the space $X = V$. To this end we use the Riesz representation theorem to define the operators $A : V \to V$, $B : V \to V$ and the function $\boldsymbol{f}_\eta : [0,T] \to V$ by the equalities

$$(A\boldsymbol{u}, \boldsymbol{v})_V = (\mathcal{A}\boldsymbol{\varepsilon}(\boldsymbol{u}), \boldsymbol{\varepsilon}(\boldsymbol{v}))_Q, \tag{7.88}$$
$$(B\boldsymbol{u}, \boldsymbol{v})_V = (\mathcal{B}\boldsymbol{\varepsilon}(\boldsymbol{u}), \boldsymbol{\varepsilon}(\boldsymbol{v}))_Q, \tag{7.89}$$
$$(\boldsymbol{f}_\eta(t), \boldsymbol{v})_V = (\boldsymbol{f}(t), \boldsymbol{v})_V - (\boldsymbol{\eta}(t), \boldsymbol{\varepsilon}(\boldsymbol{v}))_Q, \tag{7.90}$$

for all $\boldsymbol{u}$, $\boldsymbol{v} \in V$ and $t \in [0,T]$. Assumptions (7.64) and (7.65) imply that the operators A and B satisfy conditions (3.11)–(3.13).

Next, it follows from assumption (7.66) that the functional j, (7.74), satisfies condition (3.72)(a); moreover

$$j(\boldsymbol{u}_1, \boldsymbol{v}_2) - j(\boldsymbol{u}_1, \boldsymbol{v}_1) + j(\boldsymbol{u}_2, \boldsymbol{v}_1) - j(\boldsymbol{u}_2, \boldsymbol{v}_2)$$
$$\leq c_0^2(L_\nu + L_\tau)\|\boldsymbol{u}_1 - \boldsymbol{u}_2\|_V \|\boldsymbol{v}_1 - \boldsymbol{v}_2\|_V, \tag{7.91}$$

for all $\boldsymbol{u}_1$, $\boldsymbol{u}_2$, $\boldsymbol{v}_1$, $\boldsymbol{v}_2 \in V$, which shows that the functional j satisfies condition (3.72)(b) on $X = V$. In addition, using (7.69) it is easy to see that the function $\boldsymbol{f}$ defined by (7.76) satisfies $\boldsymbol{f} \in C([0,T];V)$ and, keeping in mind that $\boldsymbol{\eta} \in C([0,T];Q)$, we deduce from (7.90) that $\boldsymbol{f}_\eta \in C([0,T];V)$, i.e.,

$\boldsymbol{f}_\eta$ satisfies (3.42). Also, we note that (7.73) shows that condition (3.73) is satisfied, too.

Using now (7.88)–(7.90) we find that the existence and uniqueness part in Lemma 7.11 is a direct consequence of Theorem 3.10. Finally, the estimate (7.87) follows from standard arguments, based on the properties of the operators $\mathcal{A}$ and $\mathcal{B}$ and the inequality (7.91) on the functional j. ■

In the next step we use the solution $\boldsymbol{u}_\eta \in C^1([0,T],V)$, obtained in Lemma 7.11, to construct the following variational problem for the electric potential.

Problem 7.12 *Find an electric potential* $\varphi_\eta : [0,T] \to W$ *such that*

$$(\boldsymbol{\beta}\nabla\varphi_\eta(t), \nabla\psi)_{L^2(\Omega)^d} - (\mathcal{P}\boldsymbol{\varepsilon}(\boldsymbol{u}_\eta(t)), \nabla\psi)_{L^2(\Omega)^d} \\ +J(\boldsymbol{u}_\eta(t), \varphi_\eta(t), \psi) = (q(t), \psi)_W, \tag{7.92}$$

for all $\psi \in W$, $t \in [0,T]$.

The well-posedness of Problem 7.12 follows.

Lemma 7.13 *There exists a unique solution* $\varphi_\eta \in C([0,T];W)$ *which satisfies (7.92). Moreover, if* φ_{η_1} *and* φ_{η_2} *are the solutions of (7.92) corresponding to* $\boldsymbol{\eta}_1$, $\boldsymbol{\eta}_2 \in C([0,T];Q)$, *then there exists* $c > 0$ *such that*

$$\|\varphi_{\eta_1}(t) - \varphi_{\eta_2}(t)\|_W \le c\,\|\boldsymbol{u}_{\eta_1}(t) - \boldsymbol{u}_{\eta_2}(t)\|_V \quad \forall\, t \in [0,T]. \tag{7.93}$$

Proof Let $t \in [0,T]$. We use the Riesz representation theorem to define the operator $A_\eta(t) : W \to W$ by

$$(A_\eta(t)\varphi, \psi)_W = (\boldsymbol{\beta}\nabla\varphi, \nabla\psi)_{L^2(\Omega)^d} \\ -(\mathcal{P}\boldsymbol{\varepsilon}(\boldsymbol{u}_\eta(t)), \nabla\psi)_{L^2(\Omega)^d} + J(\boldsymbol{u}_\eta(t), \varphi, \psi), \tag{7.94}$$

for all φ, $\psi \in W$. Let φ_1, $\varphi_2 \in W$. Then, assumption (7.14)(d) and notation (7.75) imply

$$(A_\eta(t)\varphi_1 - A_\eta(t)\varphi_2, \varphi_1 - \varphi_2)_W \ge m_\beta\,\|\varphi_1 - \varphi_2\|_W^2 \\ +\int_{\Gamma_3} p_{\rm e}(u_{\eta\nu}(t) - g_{\rm a})\big(h_{\rm e}(\varphi_1 - \varphi_{\rm F}) - h_{\rm e}(\varphi_2 - \varphi_{\rm F})\big)(\varphi_1 - \varphi_2)\,da.$$

Therefore, using the bound (7.67)(c), the Lipschitz continuity (7.68)(b) of $h_{\rm e}$ and the trace theorem (4.27) we obtain

$$(A_\eta(t)\varphi_1 - A_\eta(t)\varphi_2, \varphi_1 - \varphi_2)_W \\ \ge m_\beta\,\|\varphi_1 - \varphi_2\|_W^2 - \overline{p}_{\rm e} l_{\rm e} \widetilde{c}_0^2 \|\varphi_1 - \varphi_2\|_W^2. \tag{7.95}$$

It follows from inequality (7.95) and the smallness assumption (7.83) that there exists $c_1 > 0$ such that

$$(A_\eta(t)\varphi_1 - A_\eta(t)\varphi_2, \varphi_1 - \varphi_2)_W \geq c_1 \|\varphi_1 - \varphi_2\|_W^2. \tag{7.96}$$

On the other hand, using (7.14), (7.67) and (7.75) we have

$$(A_\eta(t)\varphi_1 - A_\eta(t)\varphi_2, \psi)_W \leq c_\beta \|\varphi_1 - \varphi_2\|_W \|\psi\|_W$$
$$+ \overline{p}_e\, l_e \int_{\Gamma_3} |\varphi_1 - \varphi_2|\, |\psi|\, da \qquad \forall\, \psi \in W, \tag{7.97}$$

where c_β is a positive constant which depends on $\boldsymbol{\beta}$. It follows from (7.97) and (4.27) that

$$(A_\eta(t)\varphi_1 - A_\eta(t)\varphi_2, \psi)_W \leq (c_\beta + \overline{p}_e l_e \widetilde{c}_0^2) \|\varphi_1 - \varphi_2\|_W \|\psi\|_W \quad \forall\, \psi \in W,$$

thus

$$\|A_\eta(t)\varphi_1 - A_\eta(t)\varphi_2\|_W \leq (c_\beta + \overline{p}_e l_e \widetilde{c}_0^2) \|\varphi_1 - \varphi_2\|_W. \tag{7.98}$$

Inequalities (7.96) and (7.98) show that the operator $A_\eta(t)$ is a strongly monotone Lipschitz continuous operator on W and, therefore, Theorem 1.24 implies that there exists a unique element $\varphi_\eta(t) \in W$ such that

$$A_\eta(t)\varphi_\eta(t) = q(t). \tag{7.99}$$

We now combine (7.94) and (7.99) to find that $\varphi_\eta(t) \in W$ is the unique solution of the nonlinear variational equation (7.92).

We show next that $\varphi_\eta \in C([0,T]; W)$. To this end, let t_1, $t_2 \in [0,T]$ and, for the sake of simplicity, we write $\varphi_\eta(t_i) = \varphi_i$, $\boldsymbol{u}_\eta(t_i) = \boldsymbol{u}_i$, $q_b(t_i) = q_i$, for $i = 1, 2$. Using (7.92), (7.13), (7.14) and (7.75) we find

$$m_\beta \|\varphi_1 - \varphi_2\|_W^2$$
$$\leq c_{\mathcal{P}} \|\boldsymbol{u}_1 - \boldsymbol{u}_2\|_V \|\varphi_1 - \varphi_2\|_W + \|q_1 - q_2\|_W \|\varphi_1 - \varphi_2\|_W$$
$$+ \int_{\Gamma_3} |p_e(u_{1\nu} - g_a) h_e(\varphi_1 - \varphi_F) - p_e(u_{2\nu} - g_a) h_e(\varphi_2 - \varphi_F)|\, |\varphi_1 - \varphi_2|\, da, \tag{7.100}$$

where $c_{\mathcal{P}}$ is a positive constant which depends on the piezoelectric tensor $\mathcal{P}$.

We use the bounds $|p_e(u_{i\nu} - g_a)| \leq \overline{p}_e$, $|h_e(\varphi_i - \varphi_F)| \leq \overline{h}_e$, the Lipschitz continuity of the functions p_e, h_e, and inequalities (4.13), (4.27) to obtain

$$\int_{\Gamma_3} |p_e(u_{1\nu} - g_a) h_e(\varphi_1 - \varphi_F) - p_e(u_{2\nu} - g_a) h_e(\varphi_2 - \varphi_F)|\, |\varphi_1 - \varphi_2|\, da$$
$$\leq \overline{p}_e l_e \int_{\Gamma_3} |\varphi_1 - \varphi_2|^2\, da + L_e \overline{h}_e \int_{\Gamma_3} |u_{1\nu} - u_{2\nu}|\, |\varphi_1 - \varphi_2|\, da$$
$$\leq \overline{p}_e l_e \widetilde{c}_0^2 \|\varphi_1 - \varphi_2\|_W^2 + L_e\, \overline{h}_e c_0 \widetilde{c}_0 \|\boldsymbol{u}_1 - \boldsymbol{u}_2\|_V \|\varphi_1 - \varphi_2\|_W.$$

Substituting this last inequality in (7.100) yields

$$m_\beta \|\varphi_1 - \varphi_2\|_W \le (c_{\mathcal{P}} + L_e \overline{h}_e c_0 \widetilde{c}_0) \|\boldsymbol{u}_1 - \boldsymbol{u}_2\|_V + \|q_1 - q_2\|_W + \overline{p}_e l_e \widetilde{c}_0^2 \|\varphi_1 - \varphi_2\|_W. \tag{7.101}$$

It follows from inequality (7.101) and the smallness assumption (7.83) that there exists $c_2 > 0$ such that

$$\|\varphi_1 - \varphi_2\|_W \le c_2 (\|\boldsymbol{u}_1 - \boldsymbol{u}_2\|_V + \|q_1 - q_2\|_W). \tag{7.102}$$

We also note that assumption (7.70) combined with definition (7.77) implies that $q \in C([0,T];W)$. Since $\boldsymbol{u}_\eta \in C^1([0,T];V)$, inequality (7.102) implies that $\varphi_\eta \in C([0,T];W)$.

Finally, let $\boldsymbol{\eta}_1, \boldsymbol{\eta}_2 \in C([0,T];Q)$ and let $\varphi_{\eta_i} = \varphi_i$, $\boldsymbol{u}_{\eta_i} = \boldsymbol{u}_i$, for $i = 1, 2$. We use (7.92) and arguments similar to those used in the proof of (7.101) to obtain

$$m_\beta \|\varphi_1(t) - \varphi_2(t)\|_W \le (c_{\mathcal{P}} + L_e \overline{h}_e c_0 \widetilde{c}_0) \|\boldsymbol{u}_1(t) - \boldsymbol{u}_2(t)\|_V + \overline{p}_e l_e \widetilde{c}_0^2 \|\varphi_1(t) - \varphi_2(t)\|_W \quad \forall\, t \in [0,T].$$

This inequality, combined with assumption (7.83) leads to (7.93), which concludes the proof. ∎

Next, we consider the operator $\Lambda : C([0,T];Q) \to C([0,T];Q)$ defined by

$$\Lambda \boldsymbol{\eta}(t) = \mathcal{P}^\top \nabla \varphi_\eta(t) \qquad \forall\, \boldsymbol{\eta} \in C([0,T];Q),\ t \in [0,T]. \tag{7.103}$$

We have the following fixed point result.

Lemma 7.14 *There exists a unique $\boldsymbol{\eta}^* \in C([0,T];Q)$ such that $\Lambda \boldsymbol{\eta}^* = \boldsymbol{\eta}^*$.*

Proof Let $\boldsymbol{\eta}_1, \boldsymbol{\eta}_2 \in C([0,T];Q)$ and denote by $\boldsymbol{u}_i$ and φ_i the functions $\boldsymbol{u}_{\eta_i}$ and φ_{η_i} obtained in Lemmas 7.11 and 7.13, for $i = 1, 2$. Let $t \in [0,T]$. Using (7.103) and (7.13) we obtain

$$\|\Lambda \boldsymbol{\eta}_1(t) - \Lambda \boldsymbol{\eta}_2(t)\|_Q \le c \|\varphi_1(t) - \varphi_2(t)\|_W,$$

and, keeping in mind (7.93), we find

$$\|\Lambda \boldsymbol{\eta}_1(t) - \Lambda \boldsymbol{\eta}_2(t)\|_Q \le c \|\boldsymbol{u}_1(t) - \boldsymbol{u}_2(t)\|_V. \tag{7.104}$$

On the other hand, since $\boldsymbol{u}_i(t) = \boldsymbol{u}_0 + \int_0^t \dot{\boldsymbol{u}}_i(s)\, \mathrm{d}s$, we have

$$\|\boldsymbol{u}_1(t) - \boldsymbol{u}_2(t)\|_V \le \int_0^t \|\dot{\boldsymbol{u}}_1(s) - \dot{\boldsymbol{u}}_2(s)\|_V\, \mathrm{d}s, \tag{7.105}$$

and using this inequality in (7.87) yields

$$\|\dot{\boldsymbol{u}}_1(t) - \dot{\boldsymbol{u}}_2(t)\|_V \le c\Big(\|\boldsymbol{\eta}_1(t) - \boldsymbol{\eta}_2(t)\|_Q + \int_0^t \|\dot{\boldsymbol{u}}_1(s) - \dot{\boldsymbol{u}}_2(s)\|_V \,\mathrm{d}s\Big).$$

This inequality implies that

$$\int_0^t \|\dot{\boldsymbol{u}}_1(s) - \dot{\boldsymbol{u}}_2(s)\|_V \,\mathrm{d}s \le c\int_0^t \|\boldsymbol{\eta}_1(s) - \boldsymbol{\eta}_2(s)\|_Q \,\mathrm{d}s$$
$$+ c\int_0^t \int_0^s \|\dot{\boldsymbol{u}}_1(r) - \dot{\boldsymbol{u}}_2(r)\|_V \,\mathrm{d}r\,\mathrm{d}s.$$

It follows now from the Gronwall inequality that

$$\int_0^t \|\dot{\boldsymbol{u}}_1(s) - \dot{\boldsymbol{u}}_2(s)\|_V \,\mathrm{d}s \le c\int_0^t \|\boldsymbol{\eta}_1(s) - \boldsymbol{\eta}_2(s)\|_Q \,\mathrm{d}s. \qquad (7.106)$$

Combining (7.104)–(7.106) leads to

$$\|\Lambda\boldsymbol{\eta}_1(t) - \Lambda\boldsymbol{\eta}_2(t)\|_Q \le c\int_0^t \|\boldsymbol{\eta}_1(s) - \boldsymbol{\eta}_2(s)\|_Q \,\mathrm{d}s.$$

Lemma 7.14 is now a direct consequence of Proposition 3.1. ■

We now have all the ingredients to prove Theorem 7.9.

Proof *Existence* Let $\boldsymbol{\eta}^* \in C([0,T];Q)$ be the fixed point of the operator Λ and let $\boldsymbol{u}_{\eta^*}$, φ_{η^*} be the solutions of Problem 7.10 and Problem 7.12, respectively, for $\boldsymbol{\eta} = \boldsymbol{\eta}^*$. It follows from (7.103) that $\mathcal{P}^\top \nabla\varphi_{\eta^*} = \boldsymbol{\eta}^*$ and, therefore, (7.85), (7.86) and (7.92) imply that $(\boldsymbol{u}_{\eta^*}, \varphi_{\eta^*})$ is a solution of Problem 7.8. Regularity (7.84) follows from Lemmas 7.11 and 7.13.

Uniqueness The uniqueness of the solution follows from the uniqueness of the fixed point of the operator Λ combined with the unique solvability of the Problems 7.10 and 7.12, guaranteed by Lemmas 7.11 and 7.13, respectively. It can also be obtained by using arguments similar to those used in [125], based on a direct computation and the Gronwall inequality. ■

A quadruple of functions ($\boldsymbol{u}$, $\boldsymbol{\sigma}$, φ, $\boldsymbol{D}$) which satisfies (7.52), (7.53), (7.80)–(7.82) is called a *weak solution* of the piezoelectric contact problem 7.7. It follows from Theorem 7.9 that, under the assumptions (7.13), (7.14), (7.64)–(7.73), (7.83), there exists a unique weak solution of Problem 7.7. Moreover, the weak solution satisfies (7.84) and, in addition, $\boldsymbol{\sigma} \in C([0,T];Q_1)$, $\boldsymbol{D} \in C([0,T];\mathcal{W}_1)$.

7.3 An electro-viscoplastic frictionless contact problem

In this section we provide the analysis of a frictionless contact problem for electro-viscoplastic materials. The unique weak solvability of the problem is obtained by using arguments of history-dependent quasivariational inequalities.

7.3.1 Problem statement

We assume that the material is electro-viscoplastic and there is no gap between the body and the foundation. The contact is frictionless and it is modelled with a version of the normal compliance condition with unilateral constraint, which takes into account the conductivity of the foundation. The classical formulation of the problem is as follows.

Problem 7.15 *Find a displacement field $\boldsymbol{u} : \Omega \times [0,T] \to \mathbb{R}^d$, a stress field $\boldsymbol{\sigma} : \Omega \times [0,T] \to \mathbb{S}^d$, an electric potential $\varphi : \Omega \times [0,T] \to \mathbb{R}$ and an electric displacement field $\boldsymbol{D} : \Omega \times [0,T] \to \mathbb{R}^d$ such that*

$$\dot{\boldsymbol{\sigma}} = \mathcal{E}\boldsymbol{\varepsilon}(\dot{\boldsymbol{u}}) - \mathcal{P}^\top \boldsymbol{E}(\dot{\varphi}) + \mathcal{G}(\boldsymbol{\sigma}, \boldsymbol{\varepsilon}(\boldsymbol{u}), \boldsymbol{D}, \boldsymbol{E}(\varphi)) \quad \text{in} \quad \Omega, \tag{7.107}$$

$$\dot{\boldsymbol{D}} = \beta \boldsymbol{E}(\dot{\varphi}) + \mathcal{P}\boldsymbol{\varepsilon}(\dot{\boldsymbol{u}}) + G(\boldsymbol{D}, \boldsymbol{E}(\varphi), \boldsymbol{\sigma}, \boldsymbol{\varepsilon}(\boldsymbol{u})) \quad \text{in} \quad \Omega, \tag{7.108}$$

$$\operatorname{Div} \boldsymbol{\sigma} + \boldsymbol{f}_0 = \boldsymbol{0} \quad \text{in} \quad \Omega, \tag{7.109}$$

$$\operatorname{div} \boldsymbol{D} - q_0 = 0 \quad \text{in} \quad \Omega, \tag{7.110}$$

$$\boldsymbol{u} = \boldsymbol{0} \quad \text{on} \quad \Gamma_1, \tag{7.111}$$

$$\boldsymbol{\sigma\nu} = \boldsymbol{f}_2 \quad \text{on} \quad \Gamma_2, \tag{7.112}$$

$$\varphi = 0 \quad \text{on} \quad \Gamma_a, \tag{7.113}$$

$$\boldsymbol{D} \cdot \boldsymbol{\nu} = q_b \quad \text{on} \quad \Gamma_b, \tag{7.114}$$

$$\left. \begin{array}{l} u_\nu \le g, \quad \sigma_\nu + h_\nu(\varphi - \varphi_F) p_\nu(u_\nu) \le 0, \\ (u_\nu - g)(\sigma_\nu + h_\nu(\varphi - \varphi_F) p_\nu(u_\nu)) = 0, \end{array} \right\} \quad \text{on} \quad \Gamma_3, \tag{7.115}$$

$$\boldsymbol{\sigma}_\tau = \boldsymbol{0} \quad \text{on} \quad \Gamma_3, \tag{7.116}$$

$$\boldsymbol{D} \cdot \boldsymbol{\nu} = p_e(u_\nu) h_e(\varphi - \varphi_F) \quad \text{on} \quad \Gamma_3, \tag{7.117}$$

$$\boldsymbol{u}(0) = \boldsymbol{u}_0, \ \boldsymbol{\sigma}(0) = \boldsymbol{\sigma}_0, \ \varphi(0) = \varphi_0, \ \boldsymbol{D}(0) = \boldsymbol{D}_0 \quad \text{in} \quad \Omega. \tag{7.118}$$

Note that Problem 7.15 represents a piezoelectric version of the contact problem 6.40 studied in Section 6.4. Equations (7.107) and (7.108) represent the electro-viscoplastic constitutive law of the material introduced in Section 4.5, (4.123) and (4.124), respectively. Equations (7.109) and (7.110) are the balance equations for the stress and the electric displacement fields, respectively. Conditions (7.111) and (7.112) are the displacement and traction boundary conditions, and conditions (7.113)–(7.114) represent the electric boundary conditions. The boundary conditions (7.115)–(7.117) describe the mechanical and electrical conditions on the contact surface Γ_3 in which, as usual, φ_{F} denotes the electric potential of the foundation and g is a given bound. First, (7.115) represents a version of the normal compliance contact condition with unilateral constraint, (4.127). Next, (7.116) shows that the tangential stress on the contact surface vanishes. We use it here since we assume that the contact process is frictionless. An important extension of the results in this section would take into consideration frictional conditions on the contact surface Γ_3. Finally, (7.117) is the regularized electrical contact condition (4.130) with $g_{\mathrm{a}} = 0$, introduced in Section 4.5. Also, (7.118) represents the initial conditions in which $\boldsymbol{u}_0$, $\boldsymbol{\sigma}_0$, φ_0, and $\boldsymbol{D}_0$ denote the initial displacement, the initial stress, the initial electric potential field and the initial electric displacement field, respectively.

Note that in (7.107)–(7.118) the coupling between the mechanical unknowns $(\boldsymbol{u}, \boldsymbol{\sigma})$ and the electrical unknowns $(\varphi, \boldsymbol{D})$ arises both in the constitutive equations (7.107)–(7.108) and the contact conditions (7.115) and (7.117). This feature of Problem 7.15 represents one of the differences with respect to other models of piezoelectric contact treated in the literature, and leads to additional mathematical difficulties.

In the study of the piezoelectric contact problem 7.15 we assume that the elasticity tensor is symmetric and positively definite, i.e.

$$\left.\begin{array}{l}
\text{(a) } \mathcal{E} : \Omega \times \mathbb{S}^d \to \mathbb{S}^d;\\
\text{(b) } \mathcal{E}(\boldsymbol{x}, \boldsymbol{\varepsilon}) = (e_{ijkl}(\boldsymbol{x})\varepsilon_{kl}) \text{ for all } \boldsymbol{\varepsilon} = (\varepsilon_{ij}) \in \mathbb{S}^d, \text{ a.e. } \boldsymbol{x} \in \Omega;\\
\text{(c) } e_{ijkl} = e_{jikl} = e_{klij} \in L^\infty(\Omega),\ 1 \le i, j, k, l \le d;\\
\text{(d) there exists } m_{\mathcal{E}} > 0 \text{ such that}\\
\qquad \mathcal{E}(\boldsymbol{x}, \boldsymbol{\tau}) \cdot \boldsymbol{\tau} \ge m_{\mathcal{E}} \|\boldsymbol{\tau}\|^2\ \forall \boldsymbol{\tau} \in \mathbb{S}^d, \text{ a.e. } \boldsymbol{x} \text{ in } \Omega.
\end{array}\right\} \quad (7.119)$$

The piezoelectric tensor $\mathcal{P}$ and the electric permittivity tensor $\boldsymbol{\beta}$ satisfy the conditions (7.13) and (7.14), respectively, and the constitutive functions $\mathcal{G}$ and G verify the following conditions.

$$\left.\begin{array}{l}
\text{(a) } \mathcal{G} : \Omega \times \mathbb{S}^d \times \mathbb{S}^d \times \mathbb{R}^d \times \mathbb{R}^d \to \mathbb{S}^d.\\
\text{(b) There exists } L_{\mathcal{G}} > 0 \text{ such that}\\
\qquad \|\mathcal{G}(\boldsymbol{x}, \boldsymbol{\sigma}_1, \boldsymbol{\varepsilon}_1, \boldsymbol{D}_1, \boldsymbol{E}_1) - \mathcal{G}(\boldsymbol{x}, \boldsymbol{\sigma}_2, \boldsymbol{\varepsilon}_2, \boldsymbol{D}_2, \boldsymbol{E}_2)\|\\
\qquad \le L_{\mathcal{G}}\,(\|\boldsymbol{\sigma}_1 - \boldsymbol{\sigma}_2\| + \|\boldsymbol{\varepsilon}_1 - \boldsymbol{\varepsilon}_2\|)\\
\qquad + L_{\mathcal{G}}\,(\|\boldsymbol{D}_1 - \boldsymbol{D}_2\| + \|\boldsymbol{E}_1 - \boldsymbol{E}_2\|)\\
\qquad \forall \boldsymbol{\sigma}_1, \boldsymbol{\sigma}_2, \boldsymbol{\varepsilon}_1, \boldsymbol{\varepsilon}_2 \in \mathbb{S}^d,\ \boldsymbol{D}_1, \boldsymbol{D}_2, \boldsymbol{E}_1, \boldsymbol{E}_2 \in \mathbb{R}^d, \text{ a.e. } \boldsymbol{x} \in \Omega.\\
\text{(c) The mapping } \boldsymbol{x} \mapsto \mathcal{G}(\boldsymbol{x}, \boldsymbol{\sigma}, \boldsymbol{\varepsilon}, \boldsymbol{D}, \boldsymbol{E}) \text{ is measurable on } \Omega,\\
\qquad \text{for any } \boldsymbol{\sigma}, \boldsymbol{\varepsilon} \in \mathbb{S}^d \text{ and } \boldsymbol{D}, \boldsymbol{E} \in \mathbb{R}^d.\\
\text{(d) The mapping } \boldsymbol{x} \mapsto \mathcal{G}(\boldsymbol{x}, \boldsymbol{0}_{\mathbb{S}^d}, \boldsymbol{0}_{\mathbb{S}^d}, \boldsymbol{0}_{\mathbb{R}^d}, \boldsymbol{0}_{\mathbb{R}^d}) \text{ belongs to } Q.
\end{array}\right\} \quad (7.120)$$

$$\left.\begin{array}{l}\text{(a) } G:\Omega\times\mathbb{R}^d\times\mathbb{R}^d\times\mathbb{S}^d\times\mathbb{S}^d\to\mathbb{R}^d.\\ \text{(b) There exists } L_G>0 \text{ such that}\\ \quad \|G(\boldsymbol{x},\boldsymbol{D}_1,\boldsymbol{E}_1,\boldsymbol{\sigma}_1,\boldsymbol{\varepsilon}_1)-G(\boldsymbol{x},\boldsymbol{D}_2,\boldsymbol{E}_2,\boldsymbol{\sigma}_2,\boldsymbol{\varepsilon}_2)\|\\ \quad \le L_G\,(\|\boldsymbol{D}_1-\boldsymbol{D}_2\|+\|\boldsymbol{E}_1-\boldsymbol{E}_2\|)\\ \quad +L_G(\|\boldsymbol{\sigma}_1-\boldsymbol{\sigma}_2\|+\|\boldsymbol{\varepsilon}_1-\boldsymbol{\varepsilon}_2\|)\\ \quad \forall\,\boldsymbol{D}_1,\boldsymbol{D}_2,\boldsymbol{E}_1,\boldsymbol{E}_2\in\mathbb{R}^d,\ \boldsymbol{\sigma}_1,\boldsymbol{\sigma}_2,\boldsymbol{\varepsilon}_1,\boldsymbol{\varepsilon}_2\in\mathbb{S}^d,\ \text{a.e. } \boldsymbol{x}\in\Omega.\\ \text{(c) The mapping } \boldsymbol{x}\mapsto G(\boldsymbol{x},\boldsymbol{D},\boldsymbol{E},\boldsymbol{\sigma},\boldsymbol{\varepsilon}) \text{ is measurable on } \Omega,\\ \quad \text{for any } \boldsymbol{D},\boldsymbol{E}\in\mathbb{R}^d \text{ and } \boldsymbol{\sigma},\boldsymbol{\varepsilon}\in\mathbb{S}^d.\\ \text{(d) The mapping } \boldsymbol{x}\mapsto G(\boldsymbol{x},\boldsymbol{0}_{\mathbb{R}^d},\boldsymbol{0}_{\mathbb{R}^d},\boldsymbol{0}_{\mathbb{S}^d},\boldsymbol{0}_{\mathbb{S}^d})\\ \quad \text{belongs to } L^2(\Omega)^d.\end{array}\right\}\quad(7.121)$$

The functions p_r and h_r (for $r=\nu$, e) satisfy the following conditions.

$$\left.\begin{array}{l}\text{(a) } p_r:\Gamma_3\times\mathbb{R}\to\mathbb{R}.\\ \text{(b) There exists } L_r>0 \text{ such that}\\ \quad |p_r(\boldsymbol{x},u_1)-p_r(\boldsymbol{x},u_2)|\le L_r|u_1-u_2|\ \forall\,u_1,\,u_2\in\mathbb{R},\\ \quad\ \text{a.e. } \boldsymbol{x}\in\Gamma_3.\\ \text{(c) There exists } \overline{p}_r>0 \text{ such that}\\ \quad 0\le p_r(\boldsymbol{x},u)\le\overline{p}_r\ \forall\,u\in\mathbb{R},\ \text{a.e. } \boldsymbol{x}\in\Gamma_3.\\ \text{(d) The mapping } \boldsymbol{x}\mapsto p_r(\boldsymbol{x},u) \text{ is measurable on } \Gamma_3,\\ \quad \text{for any } u\in\mathbb{R}.\\ \text{(e) } p_r(\boldsymbol{x},u)=0\quad\forall\,u\le 0,\ \text{a.e. } \boldsymbol{x}\in\Gamma_3.\end{array}\right\}\quad(7.122)$$

$$\left.\begin{array}{l}\text{(a) } h_r:\Gamma_3\times\mathbb{R}\to\mathbb{R}.\\ \text{(b) There exists } l_r>0 \text{ such that}\\ \quad |h_r(\boldsymbol{x},\varphi_1)-h_r(\boldsymbol{x},\varphi_2)|\le l_r|\varphi_1-\varphi_2|\\ \quad\ \forall\,\varphi_1,\,\varphi_2\in\mathbb{R},\ \text{a.e. } \boldsymbol{x}\in\Gamma_3.\\ \text{(c) There exists } \overline{h}_\nu>0 \text{ such that}\\ \quad 0\le h_\nu(\boldsymbol{x},\varphi)\le\overline{h}_\nu\ \forall\,\varphi\in\mathbb{R},\ \text{a.e. } \boldsymbol{x}\in\Gamma_3.\\ \text{(d) There exists } \overline{h}_{\rm e}>0 \text{ such that}\\ \quad |h_{\rm e}(\boldsymbol{x},\varphi)|\le\overline{h}_{\rm e}\ \forall\,\varphi\in\mathbb{R},\ \text{a.e. } \boldsymbol{x}\in\Gamma_3.\\ \text{(e) The mapping } \boldsymbol{x}\mapsto h_r(\boldsymbol{x},u) \text{ is measurable on } \Gamma_3,\\ \quad \text{for any } \varphi\in\mathbb{R}.\end{array}\right\}\quad(7.123)$$

We also assume that the bound of the normal displacement and the electric potential of the foundation satisfy

$$g\in L^2(\Gamma_3),\qquad g\ge 0\quad \text{a.e. on } \Gamma_3, \tag{7.124}$$

$$\varphi_{\rm F}\in L^2(\Gamma_3). \tag{7.125}$$

Moreover, the densities of body forces, surface tractions, volume and surface electric charges have the regularity (7.69) and (7.70), respectively, and, finally, the initial data satisfy

$$\boldsymbol{u}_0\in V,\quad \boldsymbol{\sigma}_0\in Q, \tag{7.126}$$

$$\varphi_0\in W,\quad \boldsymbol{D}_0\in\mathcal{W}. \tag{7.127}$$

Alternatively, besides (7.126)–(7.127), we assume that the initial data satisfy the compatibility conditions

$$\boldsymbol{u}_0 \in U, \quad (\boldsymbol{\sigma}_0, \boldsymbol{\varepsilon}(\boldsymbol{v}) - \boldsymbol{\varepsilon}(\boldsymbol{u}_0))_Q + J_\nu(\varphi_0, \boldsymbol{u}_0, \boldsymbol{v} - \boldsymbol{u}_0) \\ \geq (\boldsymbol{f}(0), \boldsymbol{v} - \boldsymbol{u}_0)_V \quad \forall\, \boldsymbol{v} \in U, \tag{7.128}$$

$$(\boldsymbol{D}_0, \nabla\psi)_{L^2(\Omega)^d} + (q(0), \psi)_W = J_e(\boldsymbol{u}_0, \varphi_0, \psi) \quad \forall\, \psi \in W. \tag{7.129}$$

Here and below U denotes the set of admissible displacements defined by

$$U = \{\, \boldsymbol{v} \in V : v_\nu \leq g \text{ a.e. on } \Gamma_3 \,\} \tag{7.130}$$

and $J_\nu : W \times V \times V \to \mathbb{R}$, $J_e : V \times W \times W \to \mathbb{R}$ denote the functionals given by

$$J_\nu(\varphi, \boldsymbol{u}, \boldsymbol{v}) = \int_{\Gamma_3} h_\nu(\varphi - \varphi_F) p_\nu(u_\nu) v_\nu \, da, \tag{7.131}$$

$$J_e(\boldsymbol{u}, \varphi, \psi) = \int_{\Gamma_3} p_e(u_\nu) h_e(\varphi - \varphi_F) \psi \, da, \tag{7.132}$$

for all $\boldsymbol{u}, \boldsymbol{v} \in V$, $\varphi, \psi \in W$. Similar arguments to those used on page 201 show that the compatibility conditions (7.128) and (7.129) imply that $\boldsymbol{\sigma}_0 \in Q_1$ and $\boldsymbol{D} \in \mathcal{W}_1$, respectively.

We turn now to the variational formulation of Problem 7.15 and, to this end, we consider the functions $\boldsymbol{f} : [0,T] \to V$ and $q : [0,T] \to W$ defined by (7.76) and (7.77), respectively. We recall that the definitions of $\boldsymbol{f}$ and q are based on the Riesz representation theorem and, moreover,

$$\boldsymbol{f} \in C([0,T]; V), \tag{7.133}$$

$$q \in C([0,T]; W). \tag{7.134}$$

Assume in what follows that $(\boldsymbol{u}, \boldsymbol{\sigma}, \varphi, \boldsymbol{D})$ are sufficiently regular functions which satisfy (7.107)–(7.118) and let $t \in [0,T]$ be given. We integrate equations (7.107) and (7.108) with the initial conditions (7.118) to obtain

$$\boldsymbol{\sigma}(t) = \mathcal{E}\boldsymbol{\varepsilon}(\boldsymbol{u}(t)) - \mathcal{P}^\top \boldsymbol{E}(\varphi(t)) + \int_0^t \mathcal{G}(\boldsymbol{\sigma}(s), \boldsymbol{\varepsilon}(\boldsymbol{u}(s)), \boldsymbol{D}(s), \boldsymbol{E}(\varphi(s)))\, ds \\ + \boldsymbol{\sigma}_0 - \mathcal{E}\boldsymbol{\varepsilon}(\boldsymbol{u}_0) + \mathcal{P}^\top \boldsymbol{E}(\varphi_0), \tag{7.135}$$

$$\boldsymbol{D}(t) = \boldsymbol{\beta}\boldsymbol{E}(\varphi(t)) + \mathcal{P}\boldsymbol{\varepsilon}(\boldsymbol{u}(t)) + \int_0^t G(\boldsymbol{D}(s), \boldsymbol{E}(\varphi(s)), \boldsymbol{\sigma}(s), \boldsymbol{\varepsilon}(\boldsymbol{u}(s)))\, ds \\ + \boldsymbol{D}_0 - \boldsymbol{\beta}\boldsymbol{E}(\varphi_0) - \mathcal{P}\boldsymbol{\varepsilon}(\boldsymbol{u}_0). \tag{7.136}$$

Also, we use the Green formulae (4.22) and (4.30), the balance equations (7.109)–(7.110), the boundary conditions (7.111)–(7.117) and notation (7.131), (7.132), (7.76), (7.77) to see that

$$(\boldsymbol{\sigma}(t), \boldsymbol{\varepsilon}(\boldsymbol{v}) - \boldsymbol{\varepsilon}(\boldsymbol{u}(t)))_Q + J_\nu(\varphi(t), \boldsymbol{u}(t), \boldsymbol{v})$$
$$- J_\nu(\varphi(t), \boldsymbol{u}(t), \boldsymbol{u}(t)) \geq (\boldsymbol{f}(t), \boldsymbol{v} - \boldsymbol{u}(t))_V \quad \forall \boldsymbol{v} \in U, \tag{7.137}$$
$$(\boldsymbol{D}(t), \nabla\psi)_{L^2(\Omega)^d} + (q(t), \psi)_W = J_e(\boldsymbol{u}(t), \varphi(t), \psi) \quad \forall \psi \in W. \tag{7.138}$$

And, finally, we note that the first inequality in (7.115) and (7.130) imply that $\boldsymbol{u}(t) \in U$.

We gather the results above to obtain the following variational formulation of Problem 7.15.

Problem 7.16 *Find a displacement field $\boldsymbol{u} : [0,T] \to U$, a stress field $\boldsymbol{\sigma} : [0,T] \to Q$, an electric potential $\varphi : [0,T] \to W$ and an electric displacement field $\boldsymbol{D} : [0,T] \to L^2(\Omega)^d$ such that (7.135)–(7.138) hold, for all $t \in [0,T]$.*

A solution $(\boldsymbol{u}, \boldsymbol{\sigma}, \varphi, \boldsymbol{D})$ of Problem 7.16 is called a *weak solution* to the piezoelectric contact Problem 7.15. Moreover, recall that in (7.135), (7.136) and below, $\boldsymbol{E}(\varphi)$ represents the electric field, i.e. $\boldsymbol{E}(\varphi) = -\nabla\varphi$.

7.3.2 Existence and uniqueness

In the study of Problem 7.16 we have the following result, which provides the unique weak solvability of Problem 7.15.

Theorem 7.17 *Assume (7.13), (7.14), (7.69), (7.70), (7.119)–(7.127). Then:*

(1) *There exists $L_0 > 0$ which depends on Ω, Γ_1, Γ_3, $\mathcal{E}$ and $\boldsymbol{\beta}$ such that Problem 7.16 has a unique solution, if*

$$\overline{h}_\nu L_\nu + \overline{h}_e L_e + \overline{p}_\nu l_\nu + \overline{p}_e l_e < L_0. \tag{7.139}$$

Moreover, the solution satisfies

$$\boldsymbol{u} \in C([0,T]; U), \qquad \boldsymbol{\sigma} \in C([0,T]; Q_1), \tag{7.140}$$
$$\varphi \in C([0,T]; W), \qquad \boldsymbol{D} \in C([0,T]; \mathcal{W}_1). \tag{7.141}$$

(2) *If (7.139) holds and, in addition, the initial data verify the compatibility conditions (7.128) and (7.129), then the solution satisfies*

$$\boldsymbol{u}(0) = \boldsymbol{u}_0, \quad \boldsymbol{\sigma}(0) = \boldsymbol{\sigma}_0, \quad \varphi(0) = \varphi_0, \quad \boldsymbol{D}(0) = \boldsymbol{D}_0. \tag{7.142}$$

Note that condition (7.139) represents a smallness assumption on the functions involved in the boundary conditions of Problem 7.16. It is satisfied if, for instance, either the quantities $\overline{p}_\nu$, $\overline{h}_\nu$, $\overline{p}_e$, $\overline{h}_e$, or the quantities L_ν, L_e, l_ν, l_e are small enough. And, this means that either the range of the functions p_ν, p_e, h_ν, h_e, or the range of their derivatives with respect to the second variable (which exists, a.e.), is small enough. We conclude that the result in Theorem 7.17 works in the case when either the normal compliance function, the stiffness coefficient, the electric conductivity coefficient and the electric charge function are small enough, or their variation is small enough.

The proof of Theorem 7.17 will be carried out in several steps. Everywhere below we assume that (7.13), (7.14), (7.69), (7.70), (7.119)–(7.127) hold and we denote by c a positive generic constant which does not depend on time and whose value may change from place to place. We consider the spaces $X = V \times W$, $Y = Q \times L^2(\Omega)^d$, together with the canonical inner products $(\cdot,\cdot)_X$, $(\cdot,\cdot)_Y$ and the associated norms $\|\cdot\|_X$, $\|\cdot\|_Y$, respectively. In addition, for the convenience of the reader we shall use the short hand notation

$$\begin{aligned}&\widetilde{\mathcal{G}}(\boldsymbol{u},\varphi,\boldsymbol{\sigma},\boldsymbol{D})\\&\quad=\mathcal{G}\left(\mathcal{E}\boldsymbol{\varepsilon}(\boldsymbol{u})-\mathcal{P}^\top\boldsymbol{E}(\varphi)+\boldsymbol{\sigma},\boldsymbol{\varepsilon}(\boldsymbol{u}),\boldsymbol{\beta}\boldsymbol{E}(\varphi)+\mathcal{P}\boldsymbol{\varepsilon}(\boldsymbol{u})+\boldsymbol{D},\boldsymbol{E}(\varphi)\right),\end{aligned} \tag{7.143}$$

$$\begin{aligned}&\widetilde{G}(\boldsymbol{u},\varphi,\boldsymbol{\sigma},\boldsymbol{D})\\&\quad=G\left(\boldsymbol{\beta}\boldsymbol{E}(\varphi)+\mathcal{P}\boldsymbol{\varepsilon}(\boldsymbol{u})+\boldsymbol{D},\boldsymbol{E}(\varphi),\mathcal{E}\boldsymbol{\varepsilon}(\boldsymbol{u})-\mathcal{P}^\top\boldsymbol{E}(\varphi)+\boldsymbol{\sigma},\boldsymbol{\varepsilon}(\boldsymbol{u})\right),\end{aligned} \tag{7.144}$$

for all $\boldsymbol{u} \in V$, $\boldsymbol{\sigma} \in Q$, $\varphi \in W$, $\boldsymbol{D} \in L^2(\Omega)^d$.

The first step of the proof is given by the following existence and uniqueness result.

Lemma 7.18 *For all $(\boldsymbol{u},\varphi) \in C([0,T];X)$ there exists a unique pair of functions $(\boldsymbol{\sigma}^I(\boldsymbol{u},\varphi), \boldsymbol{D}^I(\boldsymbol{u},\varphi)) \in C^1([0,T];Y)$ such that, for all $t \in [0,T]$, the following equalities hold:*

$$\begin{aligned}\boldsymbol{\sigma}^I(\boldsymbol{u},\varphi)(t) &= \int_0^t \widetilde{\mathcal{G}}(\boldsymbol{u}(s),\varphi(s),\boldsymbol{\sigma}^I(\boldsymbol{u},\varphi)(s),\boldsymbol{D}^I(\boldsymbol{u},\varphi)(s))\,\mathrm{d}s\\&\quad+\boldsymbol{\sigma}_0-\mathcal{E}\boldsymbol{\varepsilon}(\boldsymbol{u}_0)+\mathcal{P}^\top\boldsymbol{E}(\varphi_0),\end{aligned} \tag{7.145}$$

$$\begin{aligned}\boldsymbol{D}^I(\boldsymbol{u},\varphi)(t) &= \int_0^t \widetilde{G}(\boldsymbol{u}(s),\varphi(s),\boldsymbol{\sigma}^I(\boldsymbol{u},\varphi)(s),\boldsymbol{D}^I(\boldsymbol{u},\varphi)(s))\,\mathrm{d}s\\&\quad+\boldsymbol{D}_0-\boldsymbol{\beta}\boldsymbol{E}(\varphi_0)-\mathcal{P}\boldsymbol{\varepsilon}(\boldsymbol{u}_0).\end{aligned} \tag{7.146}$$

Proof Let $(\boldsymbol{u},\varphi) \in C([0,T];X)$ be given. We introduce the operator $\Lambda : C([0,T];Y) \to C([0,T];Y)$ defined by

$$\Lambda(\boldsymbol{\sigma},\boldsymbol{D})(t) = (\Lambda_1(\boldsymbol{\sigma},\boldsymbol{D})(t), \Lambda_2(\boldsymbol{\sigma},\boldsymbol{D})(t)), \tag{7.147}$$

where

$$\Lambda_1(\boldsymbol{\sigma}, \boldsymbol{D})(t) = \int_0^t \widetilde{\mathcal{G}}(\boldsymbol{u}(s), \varphi(s), \boldsymbol{\sigma}(s), \boldsymbol{D}(s))\,\mathrm{d}s$$
$$+ \boldsymbol{\sigma}_0 - \mathcal{E}\boldsymbol{\varepsilon}(\boldsymbol{u}_0) + \mathcal{P}^\top \boldsymbol{E}(\varphi_0), \qquad (7.148)$$

$$\Lambda_2(\boldsymbol{\sigma}, \boldsymbol{D})(t) = \int_0^t \widetilde{G}(\boldsymbol{u}(s), \varphi(s), \boldsymbol{\sigma}(s), \boldsymbol{D}(s))\,\mathrm{d}s$$
$$+ \boldsymbol{D}_0 - \beta \boldsymbol{E}(\varphi_0) - \mathcal{P}\boldsymbol{\varepsilon}(\boldsymbol{u}_0) \qquad (7.149)$$

for all $(\boldsymbol{\sigma}, \boldsymbol{D}) \in C([0,T]; Y)$ and $t \in [0,T]$. The operator Λ depends on the pair $(\boldsymbol{u}, \varphi)$ but, for simplicity, we do not indicate this dependence explicitly.

Let $(\boldsymbol{\sigma}_1, \boldsymbol{D}_1)$, $(\boldsymbol{\sigma}_2, \boldsymbol{D}_2) \in C([0,T]; Y)$ and let $t \in 0, T]$. Then, using (7.147)–(7.149), the definitions of the functions $\widetilde{\mathcal{G}}$ and $\widetilde{G}$ and the properties (7.119)–(7.121) of the constitutive functions it follows that

$$\|\Lambda(\boldsymbol{\sigma}_1, \boldsymbol{D}_1)(t) - \Lambda(\boldsymbol{\sigma}_2, \boldsymbol{D}_2)(t)\|_Y$$
$$\le \left\| \int_0^t \widetilde{\mathcal{G}}(\boldsymbol{u}, \varphi, \boldsymbol{\sigma}_1, \boldsymbol{D}_1)\,\mathrm{d}s - \int_0^t \widetilde{\mathcal{G}}(\boldsymbol{u}, \varphi, \boldsymbol{\sigma}_2, \boldsymbol{D}_2)\,\mathrm{d}s \right\|_Q$$
$$+ \left\| \int_0^t \widetilde{G}(\boldsymbol{u}, \varphi, \boldsymbol{\sigma}_1, \boldsymbol{D}_1)\,\mathrm{d}s - \int_0^t \widetilde{G}(\boldsymbol{u}, \varphi, \boldsymbol{\sigma}_2, \boldsymbol{D}_2)\,\mathrm{d}s \right\|_{L^2(\Omega)^d}$$
$$\le c \int_0^t \big(\|\boldsymbol{\sigma}_1(s) - \boldsymbol{\sigma}_2(s)\|_Q + \|\boldsymbol{D}_1(s) - \boldsymbol{D}_2(s)\|_{L^2(\Omega)^d}\big)\,\mathrm{d}s.$$

Therefore, we obtain that

$$\|\Lambda(\boldsymbol{\sigma}_1, \boldsymbol{D}_1)(t) - \Lambda(\boldsymbol{\sigma}_2, \boldsymbol{D}_2)(t)\|_Y$$
$$\le c \int_0^t \|(\boldsymbol{\sigma}_1(s), \boldsymbol{D}_1(s)) - (\boldsymbol{\sigma}_2(s), \boldsymbol{D}_2(s))\|_Y\,\mathrm{d}s.$$

We now use Proposition 3.1 to deduce that the operator Λ has a unique fixed point, denoted by $(\boldsymbol{\sigma}^I(\boldsymbol{u}, \varphi), \boldsymbol{D}^I(\boldsymbol{u}, \varphi))$. We have $(\boldsymbol{\sigma}^I(\boldsymbol{u}, \varphi), \boldsymbol{D}^I(\boldsymbol{u}, \varphi)) \in C([0,T]; Y)$ and, using (7.147)–(7.149), we deduce that equalities (7.145)–(7.146) hold. Moreover, note that the mappings

$$t \mapsto \widetilde{\mathcal{G}}(\boldsymbol{u}(t), \varphi(t), \boldsymbol{\sigma}^I(\boldsymbol{u}, \varphi)(t), \boldsymbol{D}^I(\boldsymbol{u}, \varphi)(t)),$$
$$t \mapsto \widetilde{G}(\boldsymbol{u}(t), \varphi(t), \boldsymbol{\sigma}^I(\boldsymbol{u}, \varphi)(t), \boldsymbol{D}^I(\boldsymbol{u}, \varphi)(t))$$

are continuous from $[0,T]$ with values in Q and $L^2(\Omega)^d$, respectively. Therefore, (7.145) and (7.146) imply the regularity

$$(\boldsymbol{\sigma}^I(\boldsymbol{u}, \varphi), \boldsymbol{D}^I(\boldsymbol{u}, \varphi)) \in C^1([0,T]; Y),$$

which concludes the existence part of the lemma. The uniqueness part is a consequence of the uniqueness of the fixed point of the operator Λ, guaranteed by Proposition 3.1. ■

Lemma 7.18 allows us to consider the operator $\mathcal{S} : C([0,T],X) \to C^1([0,T],Y)$ defined by

$$\mathcal{S}(x) = \big(\boldsymbol{\sigma}^I(\boldsymbol{u},\varphi), -\boldsymbol{D}^I(\boldsymbol{u},\varphi)\big) \quad \forall x = (\boldsymbol{u},\varphi) \in C([0,T],X). \tag{7.150}$$

Moreover, it leads to the following equivalence result.

Lemma 7.19 *A quadruple of functions $(\boldsymbol{u},\boldsymbol{\sigma},\boldsymbol{D},\varphi)$ which satisfy (7.140) and (7.141) is a solution to Problem 7.16 if and only if*

$$\boldsymbol{\sigma}(t) = \mathcal{E}\boldsymbol{\varepsilon}(\boldsymbol{u}(t)) + \mathcal{P}^\top\nabla\varphi(t) + \boldsymbol{\sigma}^I(\boldsymbol{u},\varphi)(t), \tag{7.151}$$

$$\boldsymbol{D}(t) = -\boldsymbol{\beta}\nabla\varphi(t) + \mathcal{P}\boldsymbol{\varepsilon}(\boldsymbol{u}(t)) + \boldsymbol{D}^I(\boldsymbol{u},\varphi)(t), \tag{7.152}$$

$$\begin{aligned}&(\mathcal{E}\boldsymbol{\varepsilon}(\boldsymbol{u}(t)),\boldsymbol{\varepsilon}(\boldsymbol{v}) - \boldsymbol{\varepsilon}(\boldsymbol{u}(t)))_Q + (\mathcal{P}^\top\nabla\varphi(t),\boldsymbol{\varepsilon}(\boldsymbol{v}) - \boldsymbol{\varepsilon}(\boldsymbol{u}(t)))_Q\\ &\quad+(\boldsymbol{\sigma}^I(\boldsymbol{u},\varphi)(t),\boldsymbol{\varepsilon}(\boldsymbol{v}) - \boldsymbol{\varepsilon}(\boldsymbol{u}(t)))_Q + J_\nu(\varphi(t),\boldsymbol{u}(t),\boldsymbol{v})\\ &\quad-J_\nu(\varphi(t),\boldsymbol{u}(t),\boldsymbol{u}(t)) \ge (\boldsymbol{f}(t),\boldsymbol{v} - \boldsymbol{u}(t))_V \quad \forall \boldsymbol{v} \in U,\end{aligned} \tag{7.153}$$

$$\begin{aligned}&(\boldsymbol{\beta}\nabla\varphi(t),\nabla\psi)_{L^2(\Omega)^d} - (\mathcal{P}\boldsymbol{\varepsilon}(\boldsymbol{u}(t)),\nabla\psi)_{L^2(\Omega)^d} - (\boldsymbol{D}^I(\boldsymbol{u},\varphi)(t),\nabla\psi)_{L^2(\Omega)^d}\\ &\quad+J_e(\boldsymbol{u}(t),\varphi(t),\psi) = (q(t),\psi)_W \quad \forall \psi \in W,\end{aligned} \tag{7.154}$$

for all $t \in [0,T]$.

Proof Lemma 7.19 is a direct consequence of the definition of the functions $\boldsymbol{\sigma}^I$ and $\boldsymbol{D}^I$ introduced in Lemma 7.18, combined with the equality $\boldsymbol{E}(\varphi) = -\nabla\varphi$. ■

To proceed, we consider the set $K = U \times W \subset X$, the operator $A : K \to X$, the functions $\phi : Y \times K \to \mathbb{R}$, $j : K \times K \to \mathbb{R}$ and $f : [0,T] \to X$ defined by

$$\begin{aligned}(Ax,y)_X = {}&(\mathcal{E}\boldsymbol{\varepsilon}(\boldsymbol{u}),\boldsymbol{\varepsilon}(\boldsymbol{v}))_Q + (\mathcal{P}^\top\nabla\varphi,\boldsymbol{\varepsilon}(\boldsymbol{v}))_Q\\ &-(\mathcal{P}\boldsymbol{\varepsilon}(\boldsymbol{u}),\nabla\psi)_{L^2(\Omega)^d} + (\boldsymbol{\beta}\nabla\varphi,\nabla\psi)_{L^2(\Omega)^d},\end{aligned} \tag{7.155}$$

$$\phi(z,x) = (\boldsymbol{\sigma},\boldsymbol{\varepsilon}(\boldsymbol{u}))_Q + (\boldsymbol{D},\nabla\varphi)_{L^2(\Omega)^d}, \tag{7.156}$$

$$j(x,y) = J_\nu(\varphi,\boldsymbol{u},\boldsymbol{v}) + J_e(\boldsymbol{u},\varphi,\psi), \tag{7.157}$$

$$f(t) = (\boldsymbol{f}(t),q(t)), \tag{7.158}$$

for all $x = (\boldsymbol{u},\varphi) \in K$, $y = (\boldsymbol{v},\psi) \in K$, $z = (\boldsymbol{\sigma},\boldsymbol{D}) \in Y$ and $t \in [0,T]$. We recall that, again, the definition of the operator A follows by using Riesz's representation theorem.

The next step is provided by the following result.

Lemma 7.20 *Let* $t \in [0,T]$, $\boldsymbol{u} \in C([0,T],U)$, $\varphi \in C([0,T],W)$ *and denote* $x = (\boldsymbol{u},\varphi) \in C([0,T],K)$. *Then (7.153)–(7.154) hold if and only if* $x(t)$ *satisfies the inequality*

$$\begin{aligned}(Ax(t), y - x(t))_X &+ \phi(\mathcal{S}x(t), y) - \phi(\mathcal{S}x(t), x(t)) \\ &+ j(x(t), y) - j(x(t), x(t)) \geq (f(t), x(t) - y)_X \quad \forall\, y \in K. \qquad (7.159)\end{aligned}$$

Proof Assume that (7.153)–(7.154) hold and let $y = (\boldsymbol{v},\psi) \in K$. We note that (7.154) yields

$$\begin{aligned}&(\boldsymbol{\beta}\nabla\varphi(t), \nabla\psi - \nabla\varphi(t))_{L^2(\Omega)^d} - (\mathcal{P}\boldsymbol{\varepsilon}(\boldsymbol{u}(t)), \nabla\psi - \nabla\varphi(t))_{L^2(\Omega)^d} \\ &\quad -(\boldsymbol{D}^I(\boldsymbol{u},\varphi)(t), \nabla\psi - \nabla\varphi(t))_{L^2(\Omega)^d} \\ &\quad\quad + J_{\mathrm{e}}(\boldsymbol{u}(t), \varphi(t), \psi) - J_{\mathrm{e}}(\boldsymbol{u}(t), \varphi(t), \varphi(t)) \\ &\quad\quad = (q(t), \psi - \varphi(t))_W \qquad (7.160)\end{aligned}$$

and, moreover, combining (7.156) and (7.150) we have

$$\phi(\mathcal{S}x(t), y) = (\boldsymbol{\sigma}^I(\boldsymbol{u},\varphi)(t), \boldsymbol{\varepsilon}(\boldsymbol{v}))_Q - (\boldsymbol{D}^I(\boldsymbol{u},\varphi)(t), \nabla\psi)_{L^2(\Omega)^d}, \quad (7.161)$$

$$\phi(\mathcal{S}x(t), x(t)) = (\boldsymbol{\sigma}^I(\boldsymbol{u},\varphi)(t), \boldsymbol{\varepsilon}(\boldsymbol{u}(t)))_Q - (\boldsymbol{D}^I(\boldsymbol{u},\varphi)(t), \nabla\varphi(t))_{L^2(\Omega)^d}. \quad (7.162)$$

We add equality (7.160) and inequality (7.153) then we use (7.161)–(7.162) and the definition (7.155) of A, the definition (7.157) of j as well as the definition (7.158) of f. As a result we deduce that $x(t)$ satisfies the inequality (7.159).

Conversely, assume that (7.159) holds and let $\boldsymbol{v} \in U$, $\psi \in W$. We test in (7.159) with $y = (\boldsymbol{v}, \varphi(t)) \in K$, then with $y = (\boldsymbol{u}(t), \varphi(t) \pm \psi) \in K$ and use (7.161)–(7.162) together with the definitions of A, j and f. As a result we obtain (7.153) and (7.154), which concludes the proof. ■

We continue with the following existence and uniqueness result.

Lemma 7.21 *There exists* $L_0 > 0$ *which depends on* Ω, Γ_1, Γ_3, $\mathcal{E}$ *and* $\boldsymbol{\beta}$ *such that there exists a unique function* $x \in C([0,T],K)$ *which satisfies the inequality (7.159) for all* $t \in [0,T]$, *if (7.139) holds.*

Proof We apply Theorem 3.6 with $X = V \times W$ and $Y = Q \times L^2(\Omega)^d$. To this end, we first note that $K = U \times W$ is a convex nonempty subset of the space X and, therefore, (3.36) holds.

Next, recall that the definition of the transpose tensor implies

$$(\mathcal{P}^\top\nabla\varphi, \boldsymbol{\varepsilon}(\boldsymbol{u}))_Q = (\mathcal{P}\boldsymbol{\varepsilon}(\boldsymbol{u}), \nabla\varphi)_{L^2(\Omega)^d} \qquad \forall\, \boldsymbol{u} \in V,\ \varphi \in W. \qquad (7.163)$$

Using (7.163) and assumptions (7.119), (7.13), (7.14) on the tensors $\mathcal{E}$, $\mathcal{P}$ and $\boldsymbol{\beta}$ we find that the operator A defined by (7.155) satisfies condition (3.37) with

$$m = \min\{\, m_{\mathcal{E}}, m_{\boldsymbol{\beta}} \,\}. \qquad (7.164)$$

Also, it is easy to see that the function $\phi : Y \times K \to \mathbb{R}$ defined by (7.156) satisfies condition (3.38).

Next, using (7.131) and (7.132) it follows that the function $j : K \times K \to \mathbb{R}$ defined by (7.157) satisfies condition (3.39)(a). Consider now four elements $x_1, x_2, y_1, y_2 \in K$, $x_i = (\boldsymbol{u}_i, \varphi_i)$, $y_i = (\boldsymbol{v}_i, \psi_i)$, $i = 1, 2$. Then

$$
\begin{aligned}
& j(x_1, y_2) - j(x_1, y_1) + j(x_2, y_1) - j(x_2, y_2) \\
& \quad = J_\nu(\varphi_1, \boldsymbol{u}_1, \boldsymbol{v}_2) - J_\nu(\varphi_1, \boldsymbol{u}_1, \boldsymbol{v}_1) + J_\nu(\varphi_2, \boldsymbol{u}_2, \boldsymbol{v}_1) - J_\nu(\varphi_2, \boldsymbol{u}_2, \boldsymbol{v}_2) \\
& \qquad + J_{\rm e}(\boldsymbol{u}_1, \varphi_1, \psi_2) - J_{\rm e}(\boldsymbol{u}_1, \varphi_1, \psi_1) + J_{\rm e}(\boldsymbol{u}_2, \varphi_2, \psi_1) - J_{\rm e}(\boldsymbol{u}_2, \varphi_2, \psi_2).
\end{aligned}
\tag{7.165}
$$

On the other hand, by (7.131) we find that

$$
\begin{aligned}
& J_\nu(\varphi_1, \boldsymbol{u}_1, \boldsymbol{v}_2) - J_\nu(\varphi_1, \boldsymbol{u}_1, \boldsymbol{v}_1) + J_\nu(\varphi_2, \boldsymbol{u}_2, \boldsymbol{v}_1) - J_\nu(\varphi_2, \boldsymbol{u}_2, \boldsymbol{v}_2) \\
& \quad = \int_{\Gamma_3} (h_\nu(\varphi_1 - \varphi_{\rm F}) p_\nu(u_{1\nu}) - h_\nu(\varphi_2 - \varphi_{\rm F}) p_\nu(u_{2\nu}))\,(v_{2\nu} - v_{1\nu})\, da
\end{aligned}
$$

and, using (7.122)–(7.123), we obtain

$$
\begin{aligned}
& J_\nu(\varphi_1, \boldsymbol{u}_1, \boldsymbol{v}_2) - J_\nu(\varphi_1, \boldsymbol{u}_1, \boldsymbol{v}_1) + J_\nu(\varphi_2, \boldsymbol{u}_2, \boldsymbol{v}_1) - J_\nu(\varphi_2, \boldsymbol{u}_2, \boldsymbol{v}_2) \\
& \quad \le \int_{\Gamma_3} \big(\overline{h}_\nu L_\nu |u_{1\nu} - u_{2\nu}| + \overline{p}_\nu l_\nu |\varphi_1 - \varphi_2|\big)\, |v_{1\nu} - v_{2\nu}|\, da \\
& \quad \le \int_{\Gamma_3} \big(\overline{h}_\nu L_\nu \|\boldsymbol{u}_1 - \boldsymbol{u}_2\| + \overline{p}_\nu l_\nu |\varphi_1 - \varphi_2|\big)\, \|\boldsymbol{v}_1 - \boldsymbol{v}_2\|\, da.
\end{aligned}
$$

We now use inequalities (4.13) and (4.27) to see that

$$
\begin{aligned}
& J_\nu(\varphi_1, \boldsymbol{u}_1, \boldsymbol{v}_2) - J_\nu(\varphi_1, \boldsymbol{u}_1, \boldsymbol{v}_1) \\
& \quad + J_\nu(\varphi_2, \boldsymbol{u}_2, \boldsymbol{v}_1) - J_\nu(\varphi_2, \boldsymbol{u}_2, \boldsymbol{v}_2) \\
& \qquad \le \max\{\, c_0^2, \widetilde{c}_0^2 \,\}\, (\overline{h}_\nu L_\nu + \overline{p}_\nu l_\nu)\, \|x_1 - x_2\|_X \|y_1 - y_2\|_X.
\end{aligned}
\tag{7.166}
$$

Moreover, (7.132) and similar arguments yield

$$
\begin{aligned}
& J_{\rm e}(\boldsymbol{u}_1, \varphi_1, \psi_2) - J_{\rm e}(\boldsymbol{u}_1, \varphi_1, \psi_1) \\
& \quad + J_{\rm e}(\boldsymbol{u}_2, \varphi_2, \psi_1) - J_{\rm e}(\boldsymbol{u}_2, \varphi_2, \psi_2) \\
& \qquad \le \max\{\, c_0^2, \widetilde{c}_0^2 \,\}\, (\overline{h}_{\rm e} L_{\rm e} + \overline{p}_{\rm e} l_{\rm e})\, \|x_1 - x_2\|_X \|y_1 - y_2\|_X.
\end{aligned}
\tag{7.167}
$$

We add inequalities (7.166) and (7.167) and then we use (7.165) to infer that the functional j satisfies condition (3.39)(b) with

$$
\alpha = \max\{\, c_0^2, \widetilde{c}_0^2 \,\}\, (\overline{h}_\nu L_\nu + \overline{p}_\nu l_\nu + \overline{h}_{\rm e} L_{\rm e} + \overline{p}_{\rm e} l_{\rm e}). \tag{7.168}
$$

Consider now two elements $x_i = (\boldsymbol{u}_i, \varphi_i) \in C([0,T], X)$ for $i = 1, 2$ and let $t \in [0, T]$. Then by the definition (7.150) we have

$$\|\mathcal{S}x_1(t) - \mathcal{S}x_2(t)\|_Y \le \|\boldsymbol{\sigma}^I(\boldsymbol{u}_1, \varphi_1)(t) - \boldsymbol{\sigma}^I(\boldsymbol{u}_2, \varphi_2)(t)\|_Q$$
$$+ \|\boldsymbol{D}^I(\boldsymbol{u}_1, \varphi_1)(t) - \boldsymbol{D}^I(\boldsymbol{u}_2, \varphi_2)(t)\|_{L^2(\Omega)^d}.$$

Next, we use equalities (7.145)–(7.146), the definitions (7.143)–(7.144) of the functions $\widetilde{\mathcal{G}}$ and $\widetilde{G}$ as well as assumptions (7.13)–(7.14), (7.119)–(7.121) to see that

$$\|\mathcal{S}x_1(t) - \mathcal{S}x_2(t)\|_Y \le c \int_0^t \big(\|\boldsymbol{u}_1(s) - \boldsymbol{u}_2(s)\|_V + \|\varphi_1(s) - \varphi_2(s)\|_W\big)\,\mathrm{d}s$$
$$+ c \int_0^t \|\boldsymbol{\sigma}^I(\boldsymbol{u}_1, \varphi_1)(s) - \boldsymbol{\sigma}^I(\boldsymbol{u}_2, \varphi_2)(s)\|_Q\,\mathrm{d}s$$
$$+ c \int_0^t \|\boldsymbol{D}^I(\boldsymbol{u}_1, \varphi_1)(s) - \boldsymbol{D}^I(\boldsymbol{u}_2, \varphi_2)(s)\|_{L^2(\Omega)^d}\,\mathrm{d}s.$$

This inequality combined with the definition (7.150) of the operator $\mathcal{S}$ and the notation $x_i = (\boldsymbol{u}_i, \varphi_i)$ shows that

$$\|\mathcal{S}x_1(t) - \mathcal{S}x_2(t)\|_Y \le c \int_0^t \|x_1(s) - x_2(s)\|_X\,\mathrm{d}s + c \int_0^t \|\mathcal{S}x_1(t) - \mathcal{S}x_2(t)\|_Y\,\mathrm{d}s.$$

Using now the Gronwall argument it follows that

$$\|\mathcal{S}x_1(t) - \mathcal{S}x_2(t)\|_Y \le c\,\mathrm{e}^{ct} \int_0^t \|x_1(s) - x_2(s)\|_X\,\mathrm{d}s.$$

This inequality shows that

$$\|\mathcal{S}x_1(t) - \mathcal{S}x_2(t)\|_Y \le L_{\mathcal{S}} \int_0^t \|x_1(s) - x_2(s)\|_V\,\mathrm{d}s$$

with $L_{\mathcal{S}} = c\,\mathrm{e}^{cT}$. We conclude from this that the operator $\mathcal{S}$ verifies condition (3.41).

Finally, we note that the regularity (7.133)–(7.134) combined with (7.158) implies that (3.42) holds, too.

Let L_0 be defined by

$$L_0 = \frac{\min\{m_{\mathcal{E}}, m_{\beta}\}}{\max\{c_0^2, \widetilde{c}_0^{\,2}\}} \tag{7.169}$$

and note that, clearly, L_0 depends on Ω, Γ_1, Γ_3, $\mathcal{E}$ and $\boldsymbol{\beta}$. Assume now that (7.139) holds. Then, using (7.169), (7.168) and (7.164) if follows that $m > \alpha$, i.e. the smallness assumption (3.40) holds. Lemma 7.21 is now a consequence of Theorem 3.6. ∎

We turn now to the proof of Theorem 7.17.

Proof (1) Let L_0 be defined as in Lemma 7.21 and assume that (7.139) holds. Then by Lemma 7.21 we obtain the existence of a unique function $x = (\boldsymbol{u}, \varphi) \in C([0,T], K)$ which satisfies (7.159), for all $t \in [0,T]$. It follows that $\boldsymbol{u} \in C([0,T]; U)$, $\varphi \in C([0,T]; W)$. Moreover, by Lemma 7.20 we deduce that the pair of functions $(\boldsymbol{u}, \varphi)$ is the unique pair of functions with this regularity which satisfies the system (7.153)–(7.154), for all $t \in [0,T]$.

Consider now the functions $\boldsymbol{\sigma}$ and $\boldsymbol{D}$ defined by equalities (7.151)–(7.152), for all $t \in [0,T]$. It follows that

$$\boldsymbol{\sigma} \in C([0,T]; Q), \quad \boldsymbol{D} \in C([0,T]; L^2(\Omega)^d)$$

and, moreover, (7.137)–(7.138) hold, for all $t \in [0,T]$. We test in (7.137) with $\boldsymbol{u}(t) \pm \boldsymbol{v}$ where $\boldsymbol{v} \in C_0^\infty(\Omega)^d$, then we take $\psi \in C_0^\infty(\Omega)$ in (7.138) to obtain that

$$\mathrm{Div}\,\boldsymbol{\sigma}(t) + \boldsymbol{f}_0(t) = \boldsymbol{0}, \quad \mathrm{div}\,\boldsymbol{D}(t) = q_0(t) \quad \text{a.e. in } \Omega, \ \forall t \in [0,T]. \tag{7.170}$$

Next, we use assumptions (7.69)–(7.70) to deduce that

$$\mathrm{Div}\,\boldsymbol{\sigma} \in C([0,T]; L^2(\Omega)^d), \quad \mathrm{div}\,\boldsymbol{D} \in C([0,T]; L^2(\Omega))$$

and, therefore,

$$\boldsymbol{\sigma} \in C([0,T]; Q_1), \quad \boldsymbol{D} \in C([0,T]; \mathcal{W}_1).$$

We conclude from the above that the quadruple of functions $(\boldsymbol{u}, \boldsymbol{\sigma}, \boldsymbol{D}, \varphi)$ satisfies (7.140)–(7.141) and is a solution of the system (7.151)–(7.154), for all $t \in [0,T]$. Then Lemma 7.19 shows that $(\boldsymbol{u}, \boldsymbol{\sigma}, \boldsymbol{D}, \varphi)$ represents a solution of Problem 7.16 which concludes the existence part of the theorem. The uniqueness part is a consequence of Lemma 7.19 combined with the unique solvability of the system (7.153)–(7.154), proved above.

(2) Assume that (7.139) holds and, in addition, assume the compatibility conditions (7.128)–(7.129). We write (7.135)–(7.138) for $t = 0$ to obtain

$$\boldsymbol{\sigma}(0) = \mathcal{E}\boldsymbol{\varepsilon}(\boldsymbol{u}(0)) - \mathcal{P}^\top \boldsymbol{E}(\varphi(0)) + \boldsymbol{\sigma}_0 - \mathcal{E}\boldsymbol{\varepsilon}(\boldsymbol{u}_0) + \mathcal{P}^\top \boldsymbol{E}(\varphi_0), \tag{7.171}$$

$$\boldsymbol{D}(0) = \boldsymbol{\beta}\boldsymbol{E}(\varphi(0)) + \mathcal{P}\boldsymbol{\varepsilon}(\boldsymbol{u}(0)) + \boldsymbol{D}_0 - \boldsymbol{\beta}\boldsymbol{E}(\varphi_0) - \mathcal{P}\boldsymbol{\varepsilon}(\boldsymbol{u}_0), \tag{7.172}$$

$$\begin{aligned}&(\boldsymbol{\sigma}(0), \boldsymbol{\varepsilon}(\boldsymbol{v}) - \boldsymbol{\varepsilon}(\boldsymbol{u}(0)))_Q + J_\nu(\varphi(0), \boldsymbol{u}(0), \boldsymbol{v})\\ &\quad - J_\nu(\varphi(0), \boldsymbol{u}(0), \boldsymbol{u}(0)) \geq (\boldsymbol{f}(0), \boldsymbol{v} - \boldsymbol{u}(0))_V \quad \forall \boldsymbol{v} \in U,\end{aligned} \tag{7.173}$$

$$\begin{aligned}&(\boldsymbol{D}(0), \nabla\psi)_{L^2(\Omega)^d} + (q(0), \psi)_W = J_e(\boldsymbol{u}(0), \varphi(0), \psi)\\ &\quad \forall \psi \in W.\end{aligned} \tag{7.174}$$

We substitute now (7.171) in (7.173), then (7.172) in (7.174) to find that

$$(\mathcal{E}\boldsymbol{\varepsilon}(\boldsymbol{u}(0)) - \mathcal{P}^\top \boldsymbol{E}(\varphi(0)), \boldsymbol{\varepsilon}(\boldsymbol{v}) - \boldsymbol{\varepsilon}(\boldsymbol{u}(0)))_Q$$
$$+(\boldsymbol{\sigma}_0 - \mathcal{E}\boldsymbol{\varepsilon}(\boldsymbol{u}_0) + \mathcal{P}^\top \boldsymbol{E}(\varphi_0), \boldsymbol{\varepsilon}(\boldsymbol{v}) - \boldsymbol{\varepsilon}(\boldsymbol{u}(0)))_Q$$
$$+J_\nu(\varphi(0), \boldsymbol{u}(0), \boldsymbol{v}) - J_\nu(\varphi(0), \boldsymbol{u}(0), \boldsymbol{u}(0))$$
$$\geq (\boldsymbol{f}(0), \boldsymbol{v} - \boldsymbol{u}(0))_V \quad \forall \boldsymbol{v} \in U, \tag{7.175}$$
$$(\boldsymbol{\beta}\boldsymbol{E}(\varphi(0)) + \mathcal{P}\boldsymbol{\varepsilon}(\boldsymbol{u}(0)) + \boldsymbol{D}_0 - \boldsymbol{\beta}\boldsymbol{E}(\varphi_0) - \mathcal{P}\boldsymbol{\varepsilon}(\boldsymbol{u}_0), \nabla\psi)_{L^2(\Omega)^d}$$
$$+(q(0), \psi)_W = J_e(\boldsymbol{u}(0), \varphi(0), \psi) \quad \forall \psi \in W. \tag{7.176}$$

We now take $\boldsymbol{v} = \boldsymbol{u}_0$ in (7.175) and $\psi = \varphi_0 - \varphi(0)$ in (7.176). Then, using (7.128) with $\boldsymbol{v} = \boldsymbol{u}(0)$ and (7.129) with $\psi = \varphi_0 - \varphi(0)$, it follows that

$$(\mathcal{E}\boldsymbol{\varepsilon}(\boldsymbol{u}(0)) - \mathcal{P}^\top \boldsymbol{E}(\varphi(0)) - \mathcal{E}\boldsymbol{\varepsilon}(\boldsymbol{u}_0) + \mathcal{P}^\top \boldsymbol{E}(\varphi_0), \boldsymbol{\varepsilon}(\boldsymbol{u}_0) - \boldsymbol{\varepsilon}(\boldsymbol{u}(0)))_Q$$
$$+J_\nu(\varphi(0), \boldsymbol{u}(0), \boldsymbol{u}_0 - \boldsymbol{u}(0)) - J_\nu(\varphi_0, \boldsymbol{u}_0, \boldsymbol{u}_0 - \boldsymbol{u}(0)) \geq 0, \tag{7.177}$$
$$(\boldsymbol{\beta}\boldsymbol{E}(\varphi(0)) + \mathcal{P}\boldsymbol{\varepsilon}(\boldsymbol{u}(0)) - \boldsymbol{\beta}\boldsymbol{E}(\varphi_0) - \mathcal{P}\boldsymbol{\varepsilon}(\boldsymbol{u}_0), \nabla\varphi_0 - \nabla\varphi(0))_{L^2(\Omega)^d}$$
$$= J_e(\boldsymbol{u}(0), \varphi(0), \varphi_0 - \varphi(0)) - J_e(\boldsymbol{u}_0, \varphi_0, \varphi_0 - \varphi(0)). \tag{7.178}$$

We subtract (7.178) from (7.177) and then we use the equalities $\boldsymbol{E}(\varphi_0) = -\nabla\varphi_0$, $\boldsymbol{E}(\varphi(0)) = -\nabla\varphi(0)$ and (7.163) to obtain

$$(\mathcal{E}\boldsymbol{\varepsilon}(\boldsymbol{u}(0)) - \mathcal{E}\boldsymbol{\varepsilon}(\boldsymbol{u}_0), \boldsymbol{\varepsilon}(\boldsymbol{u}(0)) - \boldsymbol{\varepsilon}(\boldsymbol{u}_0))_Q$$
$$+(\boldsymbol{\beta}\nabla\varphi(0) - \boldsymbol{\beta}\nabla\varphi_0, \nabla\varphi(0) - \nabla\varphi_0)_{L^2(\Omega)^d}$$
$$\leq J_\nu(\varphi(0), \boldsymbol{u}(0), \boldsymbol{u}_0 - \boldsymbol{u}(0)) - J_\nu(\varphi_0, \boldsymbol{u}_0, \boldsymbol{u}_0 - \boldsymbol{u}(0))$$
$$+J_e(\boldsymbol{u}(0), \varphi(0), \varphi_0 - \varphi(0)) - J_e(\boldsymbol{u}_0, \varphi_0, \varphi_0 - \varphi(0)).$$

We now use (7.119), (7.14) and arguments similar to those used to obtain the estimates (7.166) and (7.167). As a result we find that

$$\min\{m_{\mathcal{E}}, m_\beta\}(\|\boldsymbol{u}(0) - \boldsymbol{u}_0\|_V^2 + \|\varphi(0) - \varphi_0\|_W^2)$$
$$\leq \max\{c_0^2, \widetilde{c}_0^2\}(\overline{h}_\nu L_\nu + \overline{h}_e L_e + \overline{p}_e l_e + \overline{p}_\nu l_\nu)(\|\boldsymbol{u}(0) - \boldsymbol{u}_0\|_V^2 + \|\varphi(0) - \varphi_0\|_W^2).$$

Then we use (7.169) and (7.139) to deduce that $\boldsymbol{u}(0) = \boldsymbol{u}_0$ and $\varphi(0) = \varphi_0$. Moreover, by (7.171) and (7.172) we obtain that $\boldsymbol{\sigma}(0) = \boldsymbol{\sigma}_0$ and $\boldsymbol{D}(0) = \boldsymbol{D}_0$, which concludes the proof. ■

Bibliographical notes

Part I

The material presented in Chapter 1 is standard and can be found in many books on functional analysis. For more information in the field we refer the reader to the books [4, 22, 33, 34, 155, 156, 160]. A comprehensive treatment of the fixed point theory for Lipschitz-type mappings with various applications in pure and applied mathematics is presented in [2]. The proof of Proposition 1.23 was written by using the ideas in [85, p. 179]. Details on pseudomonotone operators can be found in [19, 74, 158, 159]. References on existence, uniqueness and regularity results for nonlinear equations with monotone operators in Hilbert and Banach spaces include [19, 21, 85]. A complete treatment of the general theory of convex functions as well as proofs of various results on convex analysis can be found in the works [14, 33, 43, 58, 79], for instance. A self-contained reference on the main results on convex analysis and monotone operators theory in the context of Hilbert spaces is the recent book [16]. A unitary framework for abstract linear variational equations with applications to boundary value problems for partial differential equations was provided in [130].

The literature in the study of the elliptic variational inequalities presented in Chapter 2 is extensive. It contains the pioneering papers [19, 45, 87] and the survey [85], as well. A basic reference in the field is [40], where many problems in mechanics and physics are formulated and solved in the framework of variational inequalities. Various applications to free-boundary problems arising in the flow of fluids through porous media can be found in [6] and [74]. More recent references in the mathematical analysis of variational inequalities include [5, 13, 34, 46, 76, 119, 139]. Details concerning their numerical analysis can be found in [48, 49, 57, 59, 75]. A reference in the study of mathematical and numerical analysis of variational inequalities arising in hardening plasticity is [52].

The results presented in Sections 2.1–2.3 are standard and can be found in many books and surveys. In the proof of Theorem 2.1 we followed arguments similar to those used in [43, 74]. The proof of Theorem 2.8, based on the properties of the proximity operators, follows the ideas in [19]. The proof of Theorem 2.19, based on the Banach fixed point argument, follows [27]; there, a general existence and uniqueness result for elliptic quasivariational inequalities involving strongly monotone Lipschitz continuous operators was presented. The proof of Theorem 2.22, based on the Schauder fixed point argument, was written using some ideas in [114]; there, a general result on quasivariational inequalities of the form (2.58) is obtained in the case when $K = X$; the existence of the solution is proved using the properties of the directional derivative of the function j and topological degree arguments; moreover, an application in the study of a static elastic contact problem with normal compliance and Coulomb's friction law is considered.

Background on spaces of functions defined on a time interval $[0, T]$ with values in a Banach or Hilbert space can be found in [12, 14, 21, 24, 79]. An extension of the result presented in Proposition 3.1 was obtained in [135], in the case of operators on spaces of continuous functions defined on the unbounded interval $\mathbb{R}_+ = [0, \infty)$. The rest of the results presented in Sections 3.1–3.2 are based on some ideas in [140]. There, a version of the existence and uniqueness result in Theorem 3.6 was presented, together with a regularity result for the solution of history-dependent quasivariational inequalities. These results have been used in the study of several representative frictional and frictionless contact problems with elastic and viscoelastic materials.

The material presented in Section 3.3 in the study of evolutionary equations and variational inequalities is based on some results obtained in [53, 54]; there, the unique solvability of quasivariational inequalities of the form (3.66)–(3.67) and (3.68)–(3.69) is proved together with their numerical analysis, based on the study of semi-discrete and fully discrete schemes; error estimates and convergence results are also obtained; moreover, applications in the study of three-dimensional frictional contact problems with viscoelastic materials are provided.

Various results on evolutionary equations with maximal monotone operators can be found in [12, 14, 22]. Evolutionary variational inequalities of the form (3.70)–(3.71) in which $A = 0$ and B is a linear operator have been studied by many authors, under various assumptions on the function j, see e.g. [20, 52, 55]. The results in [20] are based on arguments of nonlinear equations with maximal monotone operators. The results in [52] and [55] use a time discretization method, under the assumption that j is a positively homogeneous function or a seminorm, respectively. These results were extended in [113] in the case of evolutionary quasivariational inequalities.

The existence and uniqueness result presented in Theorem 3.6 represents an important tool in the study of a large number of quasistatic contact problems with viscoelastic and viscoplastic materials, as was shown in Chapters 6 and 7 of this book. Nevertheless, there exist several mathematical models of contact which, in a variational formulation, are not cast in the general framework (3.35). This is the case of quasistatic frictional contact problems for elastic materials or the Signorini frictionless contact problem for viscoelastic materials with short memory, for instance. The analysis of the corresponding variational inequalities was carried out in the literature, in an abstract framework, using functional arguments which are different from those presented in Chapter 3. They include the time-discretization method, the penalization method, the regularization method via a viscosity term and the use of abstract arguments of evolutionary variational inequalities with monotone operators, as well. To keep this book to a reasonable length we made the choice to not present these methods here. Nevertheless, we send the reader to [12, 21, 139] for details and information on this topic.

Some of the abstract results presented in Chapters 2 and 3 of this book could be naturally extended to nonlinear inclusions, subdifferential inclusions and hemivariational inequalities. For more details on these topics we send the reader to the references [33, 34, 120] and to the recent monograph [111]. Also, we mention that a version of Theorem 3.6 is obtained in [110], in the study of a general class of history-dependent hemivariational inequalities.

Part II

For a comprehensive treatment of L^p and Sobolev spaces we refer the reader to [1, 22, 44, 79, 86, 115, 124]. For the basic aspects of solid mechanics the reader is referred to [3, 35, 47, 51, 73, 83, 91, 116, 117, 148] and to [25] for an in-depth mathematical treatment of three-dimensional elasticity. Details on the function spaces in Contact Mechanics introduced in Section 4.1 can be found in [55, 66, 75], for instance.

More information concerning the elastic and viscoelastic constitutive laws presented in Section 4.2 can be found in [38, 55, 133, 139]. Rate-type viscoplastic models of the form (4.49) were considered in the engineering literature in order to describe the behavior of real materials like rubbers, metals, pastes, rocks and so on. Various results and mechanical interpretation concerning models of this form can be found in [31] and the references therein. Existence and uniqueness results for initial and boundary value problems with rate-type viscoplastic materials of the form (4.49) can be found in [66]; there, the case of displacement-traction boundary conditions is analyzed.

Experimental background and elements of surface physics which justify some of the contact and frictional boundary conditions presented in Sections 4.3 and 4.4 can be found in [50, 72, 75, 118, 122]. In particular, the relationship between the Coulomb and Tresca friction laws is discussed in [122] and, there, a natural transition from the classical Coulomb to the Tresca friction laws is justified.

Currently, there is a considerable interest in the study of contact problems involving piezoelectric materials and, therefore, there exist a large number of references which can be used to complete the contents of Section 4.5. General models for electro-elastic materials can be found in [15, 60, 121]. Problems involving piezoelectric contact arising in smart structures and various device applications can be found in [152, 154] and the references therein. The relative motion of two bodies may be detected by a piezoelectric sensor in frictional contact with them, as stated in [17], and vibration of elastic plates may be obtained by contact with a piezoelectric actuator under electric voltage, see [153] for details. Also, the contact of a read/write piezoelectric head on a hard disk is based on the mechanical deformation generated by the inverse piezoelectric effect, see for instance [7]; there, error estimates and numerical simulations in the study of the corresponding piezoelectric contact problem are provided.

Contact problems in elasticity have been studied by many authors. The famous Signorini problem was formulated in [134] to model the frictionless unilateral contact between a linearly elastic body and a rigid foundation. Mathematical analysis of the Signorini problem is provided in [45] and its numerical analysis is carried out in [75]. The analysis of the static Signorini frictional contact problem for elastic materials is provided in [27, 39]. References concerning the Signorini problem for elastic and inelastic materials include [147], as well.

In writing Section 5.1 we used some of the references above and, for part of the results, we provided our own proofs. In particular, Subsection 5.1.4, including the statement and the proof of Theorem 5.13, was written following [37]. The material in Section 5.2 was written based on our own proofs, too. For instance, Subsection 5.2.5, including the statement and the proof of Theorem 5.34, was written following some ideas in [36]; there, the analysis of a dual variational formulation of the Signorini frictional contact problem for elastic materials, including existence, uniqueness and equivalence results, is provided. The existence and uniqueness results presented in Section 5.3 were written using some ideas in [27] and [114]. The convergence result in Theorem 5.41 was written based on [55]. A slip-dependent frictional contact problem for nonlinear elastic materials which leads to an elliptic quasivariational inequality of the form (5.187) is considered in [26]; there, the existence and uniqueness of the solution is proved by using the abstract result for elliptic quasivariational inequalities proved in [114].

Static frictional contact problems with linearly elastic materials are studied in [89] within the framework of hemivariational inequalities, see also [111] for a survey. Quasistatic frictional contact problems with linearly elastic materials are studied in [28, 29, 113]. In [28, 29] the contact is modelled with the Signorini condition and in [113] it is modelled with normal compliance. References on dynamic contact problems with elastic materials include [64, 65].

Chapter 6 was written using our original results. Thus, besides the introductive problems in Section 6.1, in the study of the contact problems with viscoelastic materials with long memory presented in Section 6.2 we used some ideas from [140]. Other references on quasistatic contact problems with such materials include [109, 127, 128, 129, 142]. The analysis of contact problems in Section 6.3 follows [125, 126] and [140], as well. In [125] the contact is modelled with normal compliance and in [126] it is described with normal damped response; the variational formulation of the problems is derived and the unique weak solvability of the models is proved using arguments of evolutionary variational inequalities and fixed point. The numerical analysis of the corresponding contact problems can be found in [55]. In [140] some of these problems are revisited and their unique solvability is proved using arguments of history-dependent quasivariational inequalities. And, finally, Section 6.4 follows our recent paper [8]; there, besides the existence and uniqueness results in Theorems 6.37 and 6.42, a numerical validation of the convergence result in Theorem 6.45 is provided.

The quasistatic Signorini frictionless contact problem for viscoelastic materials is studied in [94, 95]; there, the existence of the solution is obtained using a penalization method, based on the normal compliance contact condition. Results in the study of a quasistatic frictional contact problem involving viscoelastic materials can also be found in [132]; there, the contact is assumed to be bilateral, the variational formulation of the problem is derived and the unique weak solvability of the model is proved by using arguments of evolutionary variational inequalities and fixed point. Dynamic frictional contact problems with viscoelastic materials are analyzed in [30, 67, 68, 69, 80], using various functional methods.

Chapter 7 was written based on our original research, too. Thus, the existence and uniqueness results in the study of the electro-elastic problem in Section 7.1, together with analysis of the dual variational formulation of this problem, is published here for the first time. Similar results are obtained in [9, 141] in the study of a bilateral contact problem with Tresca friction law, and are completed with numerical simulations. The analysis of a model of static contact for electro-elastic materials with subdifferential boundary conditions can be found in [105]; there, the solvability of the problem is proved using arguments of hemivariational inequalities. The mathematical model in Section 7.2 is considered in [84], where a version of Theorem 7.9 is proved. This theorem represents an extension of a result previously obtained in [137], in the case when the foundation is assumed to be an insulator. And, finally, Section 7.3 follows our recent paper [18].

Dynamic contact problems with viscoelastic, viscoplastic and piezoelectic materials were considered by many authors and solved using various functional methods. For instance, a dynamic frictional contact problem with normal damped response is studied in [42] using a passage to the limit procedure. Other dynamic contact problems are solved using arguments of second order evolutionary hemivariational inequalities, see for instance [32, 88, 96, 98, 99, 100, 101, 102, 106] and the references therein. In particular, the results in [88] concern the study of a dynamic piezoelectric contact problem. A regularity result for the solution of a frictionless contact problem with normal compliance is provided in [81]; there, the material's behavior is described by a nonlinear viscoelastic constitutive law with short memory, the contact is modelled with normal compliance and it is proved that a regularity C^n of the normal compliance function implies a regularity C^{n+2} for the displacement field.

The variational analysis of various contact problems can also be found in the monographs [41, 55, 75, 119, 133, 138, 139]. For instance, existence and uniqueness results of the weak solution to various viscoelastic and viscoplastic contact problems is proved in [55]; there, fully discrete numerical schemes are described and implemented, and the results of the corresponding numerical simulations are presented. The monograph [133] contains a survey of results on quasistatic contact problems with elastic, viscoelastic and viscoplastic materials; there, various mathematical models are presented, existence and uniqueness results are stated and representative proofs are provided. The monograph [138] deals with the modelling, variational and numerical analysis of a special type of contact problem with viscoelastic and viscoplastic materials, in which the adhesion or damage phenomena are taken into account. Some results in this monograph are extended in [71], where a frictionless contact problem with normal compliance, unilateral constraint and adhesion is considered. And, finally, [139] presents existence, uniqueness, regularity and convergence results in the study of antiplane frictional contact problems. Literature concerning the study of antiplane frictional contact problems includes [103, 104, 107, 108]. There, the weak solvability of various models of antiplane problems is proved, using arguments of hemivariational inequalities.

Computational methods for problems in Contact Mechanics can be found in the works [82, 149, 151] and in the extensive lists of references therein. The state of the art in the field can also be found in the proceedings [92, 123, 150] and in the special issue [131], as well.

Symbols

Sets

$\mathbb{N}$: the set of positive integers

$\mathbb{R}$: the real line

$\mathbb{R}_+$: the set of nonnegative numbers

$\mathbb{R}^d$: the d-dimensional Euclidean space

$\mathbb{S}^d$: the space of second order symmetric tensors on $\mathbb{R}^d$

Ω: an open, bounded, connected set in $\mathbb{R}^d$ with a Lipschitz boundary Γ

$\overline{\Omega}$: the closure of Ω in $\mathbb{R}^d$, i.e. $\overline{\Omega} = \Omega \cup \Gamma$

Γ: the boundary of the domain Ω, that is decomposed as $\Gamma = \overline{\Gamma}_1 \cup \overline{\Gamma}_2 \cup \overline{\Gamma}_3$ with Γ_1, Γ_2 and Γ_3 being relatively open and mutually disjoint

meas (A): the $d-1$ Lebesgue measure of the measurable subset $A \subset \Gamma$

Γ_1: the part of the boundary where displacement condition is specified; meas $(\Gamma_1) > 0$ is assumed throughout the book

Γ_2: the part of the boundary where traction condition is specified

Γ_3: the part of the boundary for contact

Γ_a: the part of the boundary where the electric potential is specified; meas $(\Gamma_a) > 0$ is assumed throughout the book

Γ_b: the part of the boundary where the electric charges are specified

$[0, T]$: the time interval of interest, $T > 0$

$A \times B$: the cartesian product of the sets A and B

Spaces

X: an inner product space with inner product $(\cdot, \cdot)_X$ or a normed space with norm $\|\cdot\|_X$; sometimes X denotes a Hilbert space or a Banach space

X': the dual of X, i.e. the space of linear continuous functionals on X

$u \perp v$: two orthogonal elements in the space X, i.e. $(u, v)_X = 0$

$A^\perp$: the orthogonal complement of a subset $A \subset X$

0_X: the zero element of X

$X \times Y$: the product of the Hilbert spaces X and Y

$\mathcal{L}(X, Y)$: the space of linear continuous operators from X to a normed space Y

$\mathcal{L}(X) \equiv \mathcal{L}(X, X)$

$C([0, T]; X)$: the space of continuous functions defined on $[0, T]$ with values in X

$C^1([0, T]; X)$: the space of continuously differentiable functions defined on $[0, T]$ with values in X

$C([0, T]; K)$: the set of continuous functions defined on $[0, T]$ with values in $K \subset X$

$C^1([0,T];K)$: the set of continuously differentiable functions defined on $[0,T]$ with values in $K \subset X$

$C^m(\overline{\Omega})$: the space of functions whose derivatives up to and including order m are continuous up to the boundary Γ

$C_0^\infty(\Omega)$: the space of infinitely differentiable functions with compact support in Ω

$L^p(\Omega)$: the Lebesgue space of p-integrable functions on Ω, with the usual modification if $p = \infty$

$W^{m,p}(\Omega)$: the Sobolev space of functions whose weak derivatives of orders less than or equal to m are p-integrable on Ω

$H^m(\Omega) \equiv W^{m,2}(\Omega)$

$L^p(\Gamma)$: the Lebesgue space of p-integrable functions on Γ, with the usual modification if $p = \infty$

$L^p(\Gamma_i)$: the Lebesgue space of p-integrable functions on Γ_i ($i = 1, 2, 3$), with the usual modification if $p = \infty$

$L^2(\Omega)^d = \{\, \boldsymbol{v} = (v_i) \;:\; v_i \in L^2(\Omega),\ 1 \le i \le d \,\}$

$Q = \{\, \boldsymbol{\tau} = (\tau_{ij}) : \tau_{ij} = \tau_{ji} \in L^2(\Omega),\ 1 \le i, j \le d \,\}$

$Q_1 = \{\, \boldsymbol{\tau} \in Q : \operatorname{Div} \boldsymbol{\tau} \in L^2(\Omega)^d \,\}$

$V = \{\, \boldsymbol{v} \in H^1(\Omega)^d : \boldsymbol{v} = \boldsymbol{0}$ a.e. on $\Gamma_1 \,\}$

$V_1 = \{\, \boldsymbol{v} \in V : v_\nu = 0$ a.e. on $\Gamma_3 \,\}$

$W = \{\, \psi \in H^1(\Omega) : \psi = 0$ a.e. on $\Gamma_a \,\}$

$\mathcal{W}_1 = \{\, \boldsymbol{D} \in L^2(\Omega)^d : \operatorname{div} \boldsymbol{D} \in L^2(\Omega) \,\}$

Operators

A^{-1}: the inverse of the operator A

$\mathcal{P}_K$: the projection operator on a nonempty convex subset K

prox_j: the proximity operator of the function j

∇j: the gradient of the function j

γ: the trace operator

ε: the deformation operator

Div: the divergence operator for tensor-valued functions

div: the divergence operator for vector-valued functions

∇: the gradient operator

$\boldsymbol{I}_d$: the identity operator on $\mathbb{R}^d$

Other symbols

c: a generic positive constant

$|r|$: the absolute value of r

$r_+ = \max\{0, r\}$: the positive part of r

(x, y): the pair of components x and y

$\forall$: for all

$\exists$: there exist(s)

$\Longrightarrow$: implies

$\Longleftrightarrow$: is equivalent

δ_{ij}: the Kronecker delta

a.e.: almost everywhere

iff : if and only if

sup : least upper bound

inf : greatest lower bound

limsup : upper limit, or limit superior

liminf : lower limit, or limit inferior

l.s.c.: lower semicontinuous

$\dot{v}$: the time derivative of the function v

$\boldsymbol{u} \cdot \boldsymbol{v}$: the inner product of the vectors $\boldsymbol{u}$, $\boldsymbol{v} \in \mathbb{R}^d$

$\boldsymbol{\sigma} \cdot \boldsymbol{\tau}$: the inner product of the tensors $\boldsymbol{\sigma}$, $\boldsymbol{\tau} \in \mathbb{S}^d$

$\|\cdot\|$: the Euclidean norm on the spaces $\mathbb{R}^d$ and $\mathbb{S}^d$

$\boldsymbol{\nu}$: the unit outward normal vector to Γ

u_ν: the normal component of the vector $\boldsymbol{u}$, i.e. $u_\nu = \boldsymbol{u} \cdot \boldsymbol{\nu}$

$\boldsymbol{u}_\tau$: the tangential part of the vector $\boldsymbol{u}$, i.e. $\boldsymbol{u}_\tau = \boldsymbol{u} - u_\nu \boldsymbol{\nu}$

σ_ν: the normal component of the tensor $\boldsymbol{\sigma}$, i.e. $\sigma_\nu = \boldsymbol{\sigma}\boldsymbol{\nu} \cdot \boldsymbol{\nu}$

$\boldsymbol{\sigma}_\tau$: the tangential part of the tensor $\boldsymbol{\sigma}$, i.e. $\boldsymbol{\sigma}_\tau = \boldsymbol{\sigma}\boldsymbol{\nu} - \sigma_\nu \boldsymbol{\nu}$

References

[1] R. A. Adams, *Sobolev Spaces*, New York, Academic Press, 1975.

[2] R. P. Agarwal, D. O'Regan and D. R. Sahu, *Fixed Point Theory for Lipschitzian-type Mappings with Applications*, New York, Springer, 2009.

[3] S. S. Antman, *Nonlinear Problems of Elasticity*, New York, Springer-Verlag, 1995.

[4] K. Atkinson and W. Han, *Theoretical Numerical Analysis: a Functional Analysis Framework*, Texts in Applied Mathematics **39**, New York, Springer, 2001.

[5] C. Baiocchi and A. Capelo, *Variational and Quasivariational Inequalities: Applications to Free-Boundary Problems*, Chichester, John Wiley, 1984.

[6] C. Baiocchi, V. Comincioli, L. Guerri and G. Volpi, Free boundary problems in the theory of fluid flow through porous media: a numerical approach, *Calcolo* **10** (1973), 1–85.

[7] M. Barboteu, J. R. Fernández and T. Raffat, Numerical analysis of a dynamic piezoelectric contact problem arising in viscoelasticity, *Comput. Methods Appl. Mech. Engrg.* **197** (2008), 3724–3732.

[8] M. Barboteu, A. Matei and M. Sofonea, Analysis of quasistatic viscoplastic contact problems with normal compliance, submitted to *Quart. J. Mech. Appl. Math.*

[9] M. Barboteu, Y. Ouaffik and M. Sofonea, Analysis and simulation of a piezoelectric contact problem, in L. Beznea, V. Brinzanescu, R. Purice, *et al.*, eds., *Proceedings of the Sixth Congress of Romanian Mathematicians, Vol. I*, Bucharest, Editura Academiei Române, 2007, 485–492.

[10] M. Barboteu and M. Sofonea, Solvability of a dynamic contact problem between a piezoelectric body and a conductive foundation, *Ap. Math. Comp.* **215** (2009), 2978–2991.

[11] M. Barboteu and M. Sofonea, Modelling and analysis of the unilateral contact of a piezoelectric body with a conductive support, *J. Math. Anal. App.* **358** (2009), 110–124.

[12] V. Barbu, *Nonlinear Semigroups and Differential Equations in Banach Spaces*, Leyden, Editura Academiei, Bucharest-Noordhoff, 1976.

[13] V. Barbu, *Optimal Control of Variational Inequalities*, Boston, Pitman, 1984.

[14] V. Barbu and T. Precupanu, *Convexity and Optimization in Banach Spaces*, Dordrecht, D. Reidel Publishing Company, 1986.

[15] R. C. Batra and J. S. Yang, Saint-Venant's principle in linear piezoelectricity, *J. Elasticity* **38** (1995), 209–218.

[16] H. H. Bauschke and P. L. Combettes, *Convex Analysis and Monotone Operator Theory in Hilbert Spaces*, CMS Books in Mathematics, New York, Springer, 2011.

[17] P. Bisegna, F. Lebon and F. Maceri, The unilateral frictional contact of a piezoelectric body with a rigid support, in J. A. C. Martins and M. D. P. Monteiro Marques, eds., *Contact Mechanics*, Dordrecht, Kluwer, 2002, 347–354.

[18] M. Boureanu, A. Matei and M. Sofonea, Analysis of a contact problem for electro-elastic-visco-plastic materials, *Comm. Pure Appl. Anal.* **11** (2012), 1185–1203.

[19] H. Brézis, Equations et inéquations non linéaires dans les espaces vectoriels en dualité, *Ann. Inst. Fourier* **18** (1968), 115–175.

[20] H. Brézis, Problèmes unilatéraux, *J. Math. Pures et Appl.* **51** (1972), 1–168.

[21] H. Brézis, *Opérateurs maximaux monotones et semi-groupes de contractions dans les espaces de Hilbert*, Mathematics Studies, Amsterdam, North Holland, 1973.

[22] H. Brézis, *Analyse fonctionnelle – Théorie et applications*, Paris, Masson, 1987.

[23] M. Campillo and I. R. Ionescu, Initiation of antiplane shear instability under slip dependent friction, *J. Geophys. Res.* **102 B9** (1997), 363–371.

[24] T. Cazenave and A. Haraux, *Introduction aux problèmes d'évolution semi-linéaires*, Paris, Ellipses, 1990.

[25] P. G. Ciarlet, *Mathematical Elasticity, Volume I: Three Dimensional Elasticity*, Studies in Mathematics and its Applications **20**, Amsterdam, North-Holland, 1988.

[26] C. Ciulcu, D. Motreanu and M. Sofonea, Analysis of an elastic contact problem with slip dependent coefficient of friction, *Math. Ineq. Appl.* **4** (2001), 465–479.

[27] M. Cocu, Existence of solutions of Signorini problems with friction, *Int. J. Engng. Sci.* **22** (1984), 567–581.

[28] M. Cocu, E. Pratt and M. Raous, Existence d'une solution du problème quasistatique de contact unilatéral avec frottement non local, *C. R. Acad. Sci. Paris,* **320**, Série I (1995), 1413–1417.

[29] M. Cocu, E. Pratt and M. Raous, Formulation and approximation of quasistatic frictional contact, *Int. J. Engng. Sci.* **34** (1996), 783–798.

[30] M. Cocu and J. M. Ricaud, Analysis of a class of implicit evolution inequalities associated to dynamic contact problems with friction, *Int. J. Engng. Sci.* **328** (2000), 1534–1549.

[31] N. Cristescu and I. Suliciu, *Viscoplasticity*, Bucharest, Martinus Nijhoff Publishers, Editura Tehnică, 1982.

[32] Z. Denkowski, S. Migórski and A. Ochal, Existence and uniqueness to a dynamic bilateral frictional contact problem in viscoelasticity, *Acta Appl. Math.* **94** (2006), 251–276.

[33] Z. Denkowski, S. Migórski and N. S. Papageorgiu, *An Introduction to Nonlinear Analysis: Theory*, Boston, Kluwer Academic/Plenum Publishers, 2003.

[34] Z. Denkowski, S. Migórski and N. S. Papageorgiu, *An Introduction to Nonlinear Analysis: Applications*, Boston, Kluwer Academic/Plenum Publishers, 2003.

[35] I. Doghri, *Mechanics of Deformable Solids*, Berlin, Springer, 2000.

[36] S. Drabla and M. Sofonea, Analysis of a Signorini problem with friction, *IMA J. Appl. Math.* **62** (1999), 1–18.

[37] S. Drabla, M. Sofonea and B. Teniou, Analysis of some frictionless contact problems for elastic bodies, *Annales Polonici Mathematici* **LXIX** (1998), 75–88.

[38] A. D. Drozdov, *Finite Elasticity and Viscoelasticity – a Course in the Nonlinear Mechanics of Solids*, Singapore, World Scientific, 1996.

[39] G. Duvaut, Loi de frottement non locale, *J. Méc. Thé. Appl.* Special issue (1982), 73–78.

[40] G. Duvaut and J.-L. Lions, *Inequalities in Mechanics and Physics*, Berlin, Springer-Verlag, 1976.

[41] C. Eck, J. Jarušek and M. Krbeč, *Unilateral Contact Problems: Variational Methods and Existence Theorems*, Pure and Applied Mathematics **270**, New York, Chapman/CRC Press, 2005.

[42] C. Eck, J. Jarušek and M. Sofonea, A dynamic elastic-visco-plastic unilateral contact problem with normal damped response and Coulomb friction, *European J. Appl. Math.* **21** (2010), 229–251.

[43] I. Ekeland and R. Temam, *Convex Analysis and Variational Problems*, Amsterdam, North-Holland, 1976.

[44] L. C. Evans, *Partial Differential Equations*, Providence, AMS Press, 1999.

[45] G. Fichera, Problemi elastostatici con vincoli unilaterali. II. Problema di Signorini con ambique condizioni al contorno, *Mem. Accad. Naz. Lincei, S. VIII, Vol. VII, Sez. I*, **5** (1964) 91–140.

[46] A. Friedman, *Variational Principles and Free-boundary Problems*, New York, John Wiley, 1982.

[47] P. Germain and P. Muller, *Introduction à la mécanique des milieux continus*, Paris, Masson, 1980.

[48] R. Glowinski, *Numerical Methods for Nonlinear Variational Problems*, New York, Springer-Verlag, 1984.

[49] R. Glowinski, J.-L. Lions and R. Trémolières, *Numerical Analysis of Variational Inequalities*, Amsterdam, North-Holland, 1981.

[50] A. Guran, F. Pfeiffer and K. Popp, eds., *Dynamics with Friction: Modeling, Analysis and Experiment, Part I*, Singapore, World Scientific, 1996.

[51] M. E. Gurtin, *An Introduction to Continuum Mechanics*, New York, Academic Press, 1981.

[52] W. Han and B. D. Reddy, *Plasticity: Mathematical Theory and Numerical Analysis*, New York, Springer-Verlag, 1999.

[53] W. Han and M. Sofonea, Evolutionary variational inequalities arising in viscoelastic contact problems, *SIAM J. Num. Anal.* **38** (2000), 556–579.

[54] W. Han and M. Sofonea, Time-dependent variational inequalities for viscoelastic contact problems, *J. Comp. Appl. Math.* **136** (2001), 369–387.

[55] W. Han and M. Sofonea, *Quasistatic Contact Problems in Viscoelasticity and Viscoplasticity*, Studies in Advanced Mathematics **30**, Providence, RI, American Mathematical Society; Somerville, MA, International Press, 2002.

[56] W. Han, M. Sofonea and K. Kazmi, Analysis and numerical solution of a frictionless contact problem for electro-elastic-visco-plastic materials, *Comp. Meth. Appl. Mech. Eng.* **196** (2007), 3915–3926.

[57] J. Haslinger, I. Hlaváček and J. Nečas, Numerical methods for unilateral problems in solid mechanics, *Handbook of Numerical Analysis, Vol. IV*, P. G. Ciarlet and J.-L. Lions, eds., Amsterdam, North-Holland, 1996, 313–485.

[58] J.-B. Hiriart-Urruty and C. Lemaréchal, *Convex Analysis and Minimization Algorithms, I, II*, Berlin, Springer-Verlag, 1993.

[59] I. Hlaváček, J. Haslinger, J. Nečas and J. Lovíšek, *Solution of Variational Inequalities in Mechanics*, New York, Springer-Verlag, 1988.

[60] T. Ikeda, *Fundamentals of Piezoelectricity*, Oxford, Oxford University Press, 1990.

[61] I. R. Ionescu, C. Dascălu and M. Campillo, Slip-weakening friction on a periodic system of faults: spectral analysis, *Zeitschrift für Angewandte Mathematik und Physik (ZAMP)* **53** (2002), 980–995.

[62] I. R. Ionescu and Q.-L. Nguyen, Dynamic contact problems with slip dependent friction in viscoelasticity, *Int. J. Appl. Math. Comput. Sci.* **12** (2002), 71–80.

[63] I. R. Ionescu, Q.-L. Nguyen and S. Wolf, Slip displacement dependent friction in dynamic elasticity, *Nonlinear Analysis* **53** (2003), 375–390.

[64] I. R. Ionescu and J.-C. Paumier, Friction dynamique avec coefficient dépendant de la vitesse de glissement, *C. R. Acad. Sci. Paris* **316**, Série I (1993), 121–125.

[65] I. R. Ionescu and J.-C. Paumier, On the contact problem with slip rate dependent friction in elastodynamics, *European J. Mech., A – Solids* **13** (1994), 556–568.

[66] I. R. Ionescu and M. Sofonea, *Functional and Numerical Methods in Viscoplasticity*, Oxford, Oxford University Press, 1993.

[67] J. Jarušek, Contact problem with given time-dependent friction force in linear viscoelasticity, *Comment. Math. Univ. Carolinae* **31** (1990), 257–262.

[68] J. Jarušek, Dynamic contact problems with given friction for viscoelastic bodies, *Czechoslovak Mathematical Journal* **46** (1996), 475–487.

[69] J. Jarušek and C. Eck, Dynamic contact problems with small Coulomb friction for viscoelastic bodies. Existence of solutions, *Math. Models Meth. Appl. Sci.* **9** (1999), 11–34.

[70] J. Jarušek and M. Sofonea, On the solvability of dynamic elastic-visco-plastic contact problems, *Zeitschrift für Angewandte Mathematik und Mechanik* (*ZAMM*) **88** (2008), 3–22.

[71] J. Jarušek and M. Sofonea, On the solvability of dynamic elastic-visco-plastic contact problems with adhesion, *Annals of AOSR, Series on Mathematics and its Applications* **1** (2009), 191–214.

[72] K. L. Johnson, *Contact Mechanics*, Cambridge, Cambridge University Press, 1987.

[73] A. M. Khludnev and J. Sokolowski, *Modelling and Control in Solid Mechanics*, Basel, Birkhäuser-Verlag, 1997.

[74] N. Kikuchi and J. T. Oden, Theory of variational inequalities with applications to problems of flow through porous media, *Int. J. Engng. Sci.* **18** (1980), 1173–1284.

[75] N. Kikuchi and J. T. Oden, *Contact Problems in Elasticity: a Study of Variational Inequalities and Finite Element Methods*, Philadelphia, SIAM, 1988.

[76] D. Kinderlehrer and G. Stampacchia, *An Introduction to Variational Inequalities and their Applications*, Classics in Applied Mathematics **31**, Philadelphia, SIAM, 2000.

[77] A. Klarbring, A. Mikelič and M. Shillor, Frictional contact problems with normal compliance, *Int. J. Engng. Sci.* **26** (1988), 811–832.

[78] A. Klarbring, A. Mikelič and M. Shillor, On friction problems with normal compliance, *Nonlinear Analysis* **13** (1989), 935–955.

[79] A. J. Kurdila and M. Zabarankin, *Convex Functional Analysis*, Basel, Birkhäuser, 2005.

[80] K. L. Kuttler and M. Shillor, Set-valued pseudomonotone maps and degenerate evolution inclusions, *Comm. Contemp. Math.* **1** (1999), 87–123.

[81] K. L. Kuttler and M. Shillor, Regularity of solutions to a dynamic frictionless contact problem with normal compliance, *Nonlinear Analysis* **59** (2004), 1063–1075.

[82] T. A. Laursen, *Computational Contact and Impact Mechanics*, Berlin, Springer, 2002.

[83] J. Lemaître and J.-L. Chaboche, *Mechanics of Solids Materials*, Cambridge, Cambridge University Press, 1990.

[84] Z. Lerguet, M. Shillor and M. Sofonea, A frictional contact problem for an electro-viscoelastic body, *Electronic Journal of Differential Equations* **170** (2007), 1–16.

[85] J.-L. Lions, *Quelques méthodes de résolution des problèmes aux limites non linéaires*, Paris, Gauthiers-Villars, 1969.

[86] J.-L. Lions and E. Magenes, *Problèmes aux limites non-homogènes I*, Paris, Dunod, 1968.

[87] J.-L. Lions and G. Stampacchia, Variational inequalities *Comm. Pure Appl. Math.* **20**, 493 (1967).

[88] Z. Liu and S. Migórski, Noncoercive damping in dynamic hemivariational inequality with application to problem of piezoelectricity, *Discrete and Continuous Dynamical Systems, Series B* **9** (2008), 129–143.

[89] Z. Liu, S. Migórski and A. Ochal, Homogenization of boundary hemivariational inequalities in linear elasticity, *J. Math. Anal. Appl.* **340** (2008), 1347–1361.

[90] F. Maceri and P. Bisegna, The unilateral frictionless contact of a piezoelectric body with a rigid support, *Math. Comp. Modelling* **28** (1998), 19–28.

[91] L. E. Malvern, *Introduction to the Mechanics of a Continuum Medium*, New Jersey, Princeton-Hall, Inc., 1969.

[92] J. A. C. Martins and M. D. P. Monteiro Marques, eds., *Contact Mechanics*, Dordrecht, Kluwer, 2002.

[93] J. A. C. Martins and J. T. Oden, Existence and uniqueness results for dynamic contact problems with nonlinear normal and friction interface laws, *Nonlinear Analysis TMA* **11** (1987), 407–428.

[94] A. Matei, *Modélisation Mathématique en Mécanique du Contact*, Ph. D. Thesis, Université de Perpignan, Perpignan, 2002.

[95] A. Matei, V. V. Motreanu and M. Sofonea, A quasistatic antiplane contact problem with slip dependent friction, *Advances in Nonlinear Variational Inequalities* **4** (2001), 1–21.

[96] S. Migórski, Dynamic hemivariational inequality modeling viscoelastic contact problem with normal damped response and friction, *Appl. Anal.* **84** (2005), 669–699.

[97] S. Migórski, Hemivariational inequality for a frictional contact problem in elasto-piezoelectricity, *Discrete and Continuous Dynamical Systems, Series B* **6** (2006), 1339–1356.

[98] S. Migórski, Evolution hemivariational inequality for a class of dynamic viscoelastic nonmonotone frictional contact problems, *Comput. Math. Appl.* **52** (2006), 677–698.

[99] S. Migórski and A. Ochal, Hemivariational inequality for viscoelastic contact problem with slip-dependent friction, *Nonlinear Analysis* **61** (2005), 135–161.

[100] S. Migórski and A. Ochal, A unified approach to dynamic contact problems in viscoelasticity, *J. Elasticity* **83** (2006), 247–275.

[101] S. Migórski and A. Ochal, Existence of solutions for second order evolution inclusions with application to mechanical contact problems, *Optimization* **55** (2006), 101–120.

[102] S. Migórski, A. Ochal and M. Sofonea, Integrodifferential hemivariational inequalities with applications to viscoelastic frictional contact, *Math. Models Methods Appl. Sci.* **18** (2008), 271–290.

[103] S. Migórski, A. Ochal and M. Sofonea, Solvability of dynamic antiplane frictional contact problems for viscoelastic cylinders, *Nonlinear Analysis* **70** (2009), 3738–3748.

[104] S. Migórski, A. Ochal and M. Sofonea, Modeling and analysis of an antiplane piezoelectric contact problem, *Math. Models Methods Appl. Sci.* **19** (2009), 1295–1324.

[105] S. Migórski, A. Ochal and M. Sofonea, Variational analysis of static frictional contact problems for electro-elastic materials, *Math. Nachr.* **283** (2010), 1314–1335.

[106] S. Migórski, A. Ochal and M. Sofonea, A dynamic frictional contact problem for piezoelectric materials, *J. Math. Anal. Appl.* **361** (2010), 161–176.

[107] S. Migórski, A. Ochal and M. Sofonea, Weak solvability of antiplane frictional contact problems for elastic cylinders, *Nonlinear Analysis: Real World Applications* **11** (2010), 172–183.

[108] S. Migórski, A. Ochal and M. Sofonea, Analysis of a dynamic contact problem for electro-viscoelastic cylinders, *Nonlinear Analysis* **73** (2010), 1221–1238.

[109] S. Migórski, A. Ochal and M. Sofonea, Analysis of a frictional contact problem for viscoelastic materials with long memory, *Discrete and Continuous Dynamical Systems, Series B* **15** (2011), 687–705.

[110] S. Migórski, A. Ochal and M. Sofonea, History-dependent subdifferential inclusions and hemivariational inequalities in contact mechanics, *Nonlinear Analysis: Real World Applications* **12** (2011), 3384–3396.

[111] S. Migórski, A. Ochal and M. Sofonea, *Nonlinear Inclusions and Hemivariational Inequalities. Models and Analysis of Contact Problems*, Advances in Mechanics and Mathematics **26**, New York, Springer, 2012.

[112] J. J. Moreau, Proximité et dualité dans un espace hilbertien, *Bulletin de la Société Mathématique de France* **93** (1965), 273–283.

[113] D. Motreanu and M. Sofonea, Evolutionary variational inequalities arising in quasistatic frictional contact problems for elastic materials, *Abstract and Applied Analysis* **4** (1999), 255–279.

[114] D. Motreanu and M. Sofonea, Quasivariational inequalities and applications in frictional contact problems with normal compliance, *Adv. Math. Sci. Appl.* **10** (2000), 103–118.

[115] J. Nečas, *Les méthodes directes en théorie des équations elliptiques*, Praha, Academia, 1967.

[116] J. Nečas and I. Hlaváček, *Mathematical Theory of Elastic and Elastico-Plastic Bodies: an Introduction*, Amsterdam, Oxford, New York, Elsevier Scientific Publishing Company, 1981.

[117] Q. S. Nguyen, *Stability and Nonlinear Solid Mechanics*, Chichester, John Wiley & Sons, Ltd., 2000.

[118] J. T. Oden and J. A. C. Martins, Models and computational methods for dynamic friction phenomena, *Computer Methods in Applied Mechanics and Engineering* **52** (1985), 527–634.

[119] P. D. Panagiotopoulos, *Inequality Problems in Mechanics and Applications*, Boston, Birkhäuser, 1985.

[120] P. D. Panagiotopoulos, *Hemivariational Inequalities, Applications in Mechanics and Engineering*, Berlin, Springer-Verlag, 1993.

[121] V. Z. Patron and B. A. Kudryavtsev, *Electromagnetoelasticity, Piezoelectrics and Electrically Conductive Solids*, London, Gordon & Breach, 1988.

[122] E. Rabinowicz, *Friction and Wear of Materials*, second edition, New York, Wiley, 1995.

[123] M. Raous, M. Jean and J. J. Moreau, eds., *Contact Mechanics*, New York, Plenum Press, 1995.

[124] B. D. Reddy, *Introductory Functional Analysis with Applications to Boundary Value Problems and Finite Elements*, New York, Springer, 1998.

[125] M. Rochdi, M. Shillor and M. Sofonea, Quasistatic viscoelastic contact with normal compliance and friction, *J. Elasticity* **51** (1998), 105–126.

[126] M. Rochdi, M. Shillor and M. Sofonea, A quasistatic contact problem with directional friction and damped response, *Appl. Anal.* **68** (1998), 409-422.

[127] A. D. Rodríguez–Aros, M. Sofonea and J. M. Viaño, A class of evolutionary variational inequalities with Volterra-type integral term, *Math. Models Methods Appl. Sci.* **14** (2004), 555–577.

[128] A. D. Rodríguez–Aros, M. Sofonea and J. M. Viaño, Numerical approximation of a viscoelastic frictional contact problem, *C. R. Acad. Sci. Paris, Sér. II Méc.* **334** (2006), 279–284.

[129] A. D. Rodríguez–Aros, M. Sofonea and J. M. Viaño, Numerical analysis of a frictional contact problem for viscoelastic materials with long-term memory, *Numer. Math.* **198** (2007), 327–358.

[130] I. Roşca, Functional framework for linear variational equations, in L. Dragoş, ed., *Curent Topics in Continuum Mechanics*, Bucharest, Editura Academiei Române, 2002, 177–258.

[131] M. Shillor, ed., Recent advances in contact mechanics, Special issue of *Math. Comput. Modelling* **28** (4–8) (1998).

[132] M. Shillor and M. Sofonea, A quasistatic viscoelastic contact problem with friction, *Int. J. Engng. Sci.* **38** (2000), 1517–1533.

[133] M. Shillor, M. Sofonea and J. J. Telega, *Models and Analysis of Quasistatic Contact*, Lecture Notes in Physics **655**, Berlin, Springer, 2004.

[134] A. Signorini, Sopra alcune questioni di elastostatica, *Atti della Società Italiana per il Progresso delle Scienze*, 1933.

[135] M. Sofonea, C. Avramescu and A. Matei, A fixed point result with applications in the study of viscoplastic frictionless contact problems, *Comm. Pure Appl. Anal.* **7** (2008), 645–658.

[136] M. Sofonea and El H. Essoufi, A piezoelectric contact problem with slip dependent coefficient of friction, *Math. Model. Anal.* **9** (2004), 229–242.

[137] M. Sofonea and El H. Essoufi, Quasistatic frictional contact of a viscoelastic piezoelectric body, *Adv. Math. Sci. Appl.* **14** (2004), 613–631.

[138] M. Sofonea, W. Han and M. Shillor, *Analysis and Approximation of Contact Problems with Adhesion or Damage*, Pure and Applied Mathematics **276**, New York, Chapman-Hall/CRC Press, 2006.

[139] M. Sofonea and A. Matei, *Variational Inequalities with Applications. A Study of Antiplane Frictional Contact Problems*, Advances in Mechanics and Mathematics **18**, New York, Springer, 2009.

[140] M. Sofonea and A. Matei, History-dependent quasivariational inequalities arising in Contact Mechanics, *European J. Appl. Math.* **22** (2011), 471–491.

[141] M. Sofonea and Y. Ouafik, A piezoelectric contact problem with normal compliance, *Applicaciones Mathematicae* **32** (2005), 425–442.

[142] M. Sofonea, A. D. Rodríguez–Aros and J. M. Viaño, A class of integro-differential variational inequalities with applications to viscoelastic contact, *Math. Comput. Modelling* **41** (2005), 1355–1369.

[143] L. Solymar and L. B. Au, *Solutions Manual for Lectures on the Electrical Properties of Materials*, fifth edition, Oxford, Oxford University Press, 1993.

[144] N. Strömberg, *Thermomechanical Modelling of Tribological Systems*, Ph.D. Thesis **497**, Linköping University, Sweden, 1997.

[145] N. Strömberg, L. Johansson and A. Klarbring, Generalized standard model for contact friction and wear, in M. Raous, M. Jean and J. J. Moreau, eds., *Contact Mechanics*, New York, Plenum Press, 1995.

[146] N. Strömberg, L. Johansson and A. Klarbring, Derivation and analysis of a generalized standard model for contact friction and wear, *Int. J. Solids Structures* **33** (1996), 1817–1836.

[147] J. J. Telega, Topics on unilateral contact problems of elasticity and inelasticity, in J. J. Moreau and P. D. Panagiotopoulos, eds., *Nonsmooth Mechanics and Applications*, Wien, Springer-Verlag, 1988, 340–461.

[148] R. Temam and A. Miranville, *Mathematical Modeling in Continuum Mechanics*, Cambridge, Cambridge University Press, 2001.

[149] P. Wriggers, *Computational Contact Mechanics*, Chichester, Wiley, 2002.

[150] P. Wriggers and U. Nackenhorst, eds., *Analysis and Simulation of Contact Problems*, Lecture Notes in Applied and Computational Mechanics **27**, Berlin, Springer, 2006.

[151] P. Wriggers and P. D. Panagiotopoulos, eds., *New Developments in Contact Problems*, Wien, New York, Springer-Verlag, 1999.

[152] J. Yang, ed., *Special Topics in the Theory of Piezoelectricity*, New York, Springer, 2009.

[153] J. S. Yang, R. C. Batra and X. Q. Liang, The cylindrical bending vibration of a laminated elastic plate due to piezoelectric actuators, *Smart Mater. Struct.* **3** (1994), 485–493.

[154] J. Yang and J. S. Yang, *An Introduction to the Theory of Piezoelectricity*, New York, Springer, 2005.

[155] E. Zeidler, *Nonlinear Functional Analysis and its Applications. I: Fixed-point Theorems*, New York, Springer-Verlag, 1986.

[156] E. Zeidler, *Nonlinear Functional Analysis and its Applications. III: Variational Methods and Optimization*, New York, Springer-Verlag, 1986.

[157] E. Zeidler, *Nonlinear Functional Analysis and its Applications. IV: Applications to Mathematical Physics*, New York, Springer-Verlag, 1988.

[158] E. Zeidler, *Nonlinear Functional Analysis and its Applications,* II/A*: Linear Monotone Operators*, New York, Springer-Verlag, 1990.

[159] E. Zeidler, *Nonlinear Functional Analysis and its Applications,* II/B*: Nonlinear Monotone Operators*, New York, Springer-Verlag, 1990.

[160] E. Zeidler, *Applied Functional Analysis: Main Principles and Their Applications*, New York, Springer-Verlag, 1995.

Index

Banach
- fixed point theorem, 8
- space, 4

bilateral contact, 105, 112
bilinear form, 7
- continuous, 8
- positive, 8
- symmetric, 8
- X-elliptic, 8

boundary condition
- contact, 105
- displacement, 104, 112
- traction, 104, 112

Cauchy sequence, 4
Cauchy–Schwarz inequality, 5
classical formulation, 124
classical solution, 124
coefficient
- damping, 113
- deformability, 128
- electric conductivity, 121
- friction, 109
- Lamé, 93, 98
- relaxation, 98
- stiffness, 105, 128
- viscosity, 96

coefficient of friction, 109
compact subset
- relatively, 10
- relatively sequentially, 10

complementary energy, 138, 139
constitutive law, 92
- elastic, 92
- hardening, 100
- Hencky, 94
- Kelvin–Voigt, 95
- Perzyna, 99
- viscoelastic, 95–97
- viscoplastic, 98

contact
- bilateral, 105, 112
- condition, 105
- frictional, 108
- frictionless, 107
- law, 105
- viscous, 113

contact condition, 105
- bilateral, 105, 112
- electrical, 119, 120

contact condition (*cont.*)
  normal compliance, 105, 107, 112
  normal damped response, 113
  Signorini, 106, 112
contact law, 105
contact problem
  classical formulation, 124
  classical solution, 124
  variational formulation, 124
  weak solution, 124
contraction, 8
convergence
  strong, 4
  weak, 6
convex
  function, 24
  set, 7
  von Mises, 100
Coulomb friction law, 108
  classical version, 109
  modified version, 109
  quasistatic version, 113
  regularization, 110
  static version, 108

deformability coefficient, 128
deformation operator, 85
derivative, 58
deviatoric part, 94
direct sum, 15
displacement boundary condition, 104, 112
divergence, 88, 90
divergence operator, 88, 90
dual space, 6
dual variational formulation, 133, 156, 223

Eberlein–Smulyan theorem, 18
elastic constitutive law, 92
elasticity
  operator, 92, 94, 97, 98, 117, 118
  tensor, 93, 95, 98, 117
electric
  displacement, 116
  potential, 116
  conductivity coefficient, 121
  field, 117
  permittivity tensor, 117
electrical contact condition, 119, 120
electro-elastic material, 115
electro-viscoelastic material, 115
electro-viscoplastic material, 115
elliptic quasivariational inequality, 49
elliptic variational inequality
  first kind, 34
  penalization, 35
  regularization, 43
  second kind, 40
ellipticity of tensors, 93, 96, 117
energy function, 138, 160
  complementary, 138, 160
equation of equilibrium, 104, 112
evolutionary
  problem, 75
  quasivariational inequality, 76
  variational inequality, 76

fixed point, 8
fixed point theorem
  Banach, 8
  Schauder, 11
friction bound, 108, 113, 120
friction force, 107
friction law, 107
  classical Coulomb, 109
  modified Coulomb, 109
  quasistatic Coulomb, 113, 120
  quasistatic Tresca, 114
  static Coulomb, 108, 120
  static Tresca, 109

frictional
  condition, 107
  contact, 108
frictionless
  condition, 107, 113
  contact, 107
Friedrichs–Poincaré inequality, 89
function
  coercive, 29
  complementary energy, 138, 160
  continuously differentiable, 58
  convex, 24
  derivative, 58
  differentiable, 58
  energy, 138, 160
  Gâteaux differentiable, 26
  lower semicontinuous (l.s.c.), 24
  strictly convex, 24
  von Mises, 100
  weakly lower semicontinuous (weakly l.s.c.), 24
functional
  bounded, 6
  continuous, 6
  kernel, 17

gap, 91
Gâteaux differentiable function, 26
gradient operator, 26, 89
Green's formula, 89, 90
Gronwall inequality, 60

Hencky material, 94
Hilbert space, 5
history-dependent
  nonlinear equation, 65
  operator, 65, 69
  quasivariational inequality, 69

inequality
  Cauchy–Schwarz, 5
  Friedrichs–Poincaré, 89
  Gronwall, 60
  Korn, 87
inner product, 5
  norm, 5
  space, 5

Kelvin–Voigt constitutive law, 95
Korn inequality, 87
Kronecker symbol, 93

Lamé coefficients, 93, 98
limit
  strong, 4
  weak, 7
linear functional, 17
  bounded, 6
  continuous, 6
  kernel, 17
linear operator, 6
  bounded, 6
  continuous, 6
  operator norm, 6
lower semicontinuous (l.s.c), 24

material
  electro-elastic, 115
  electro-viscoelastic, 115
  electro-viscoplastic, 115
  hardening, 100
  piezoelectric, 115
Mazur theorem, 7

nonlinear equation
  history-dependent, 65
  Volterra integral term, 61
  Volterra-type, 61
norm, 3, 5
  equivalent, 4
  inner product, 5
normal compliance, 105, 107, 112
  with unilateral constraint, 106, 113

normal compliance function,
    107, 128, 162, 186
normal component, 86, 89
normal damped response, 113
normed space, 4
    complete, 4
    dual, 6

operator
    bounded, 6
    compact, 10
    continuous, 10, 20
    contraction, 8
    deformation, 85
    divergence, 88, 90
    elasticity, 92, 95, 97, 117,
        118
    fixed point, 8
    gradient, 26, 89
    hemicontinuous, 20
    history-dependent, 65, 69
    linear, 6
    Lipschitz continuous, 20
    monotone, 19
    nonexpansive, 20
    power, 9
    projection, 13
    proximity, 41
    pseudomonotone, 22
    relaxation, 97
    strictly monotone, 19
    strongly monotone, 20
    trace, 86
    viscosity, 95, 118
    Volterra, 64, 69
operator norm, 6
orthogonal
    complement, 15
    decomposition, 15
    elements, 15

parallelogram
    identity, 5
    law, 5
penalization parameter, 128
Perzyna law, 99
piezoelectric
    material, 115
    tensor, 117
primal variational formulation,
    132, 156, 223
process
    quasistatic, 112, 116
    static, 103, 116
projection, 13
    lemma, 11
    operator, 13
proximal element, 41
proximity operator, 41
pseudomonotone operator, 22

quasistatic process, 112, 116
quasivariational inequality
    elliptic, 49
    evolutionary, 76
    history-dependent, 69

relaxation
    coefficients, 98
    operator, 97
    tensor, 98
Riesz representation theorem, 17

Schauder fixed point theorem, 11
seminorm, 27
sequence
    bounded, 4
    Cauchy, 4
    strong limit, 4
    strongly convergent, 4
    weak limit, 7
    weakly convergent, 6
Signorini contact condition, 106,
    112
slip, 108
slip rate, 113, 120
slip zone, 109, 114
Sobolev space, 84
solution
    classical, 124

weak, 124, 127, 137, 147, 148, 160, 167, 176, 178, 181, 183, 223, 227, 236, 241
space, 5
Banach, 4
complete, 4
continuous differentiable functions, 58
continuous functions, 58
dual, 6
Hilbert, 5
inner product, 5
normed, 4
static process, 103, 116
stick zone, 109, 114
stiffness coefficient, 105, 128
strain tensor
deviatoric part, 94
linearized, 85
small, 85
trace, 93
strong convergence, 4
strong limit, 4
subset
bounded, 10
closed, 7
convex, 7
relatively compact, 10
relatively sequentially compact, 10
weakly closed, 7
symmetry of tensors, 93, 96, 117

tangential part, 86, 89
tensor
elasticity, 93, 95, 98
electric permittivity, 117
ellipticity, 93, 96, 117
piezoelectric, 117
relaxation, 97, 98
strain, 85
symmetry, 93, 96, 117
trace, 93
viscosity, 95, 98
theorem
Banach fixed point, 8
Eberlein–Smulyan, 18
Mazur, 7
Riesz representation, 17
Schauder fixed point, 11
Weierstrass, 29
trace, 86
function, 86
operator, 86
tensor, 93
traction boundary condition, 104, 112
Tresca friction law
quasistatic version, 114
static version, 109

variational formulation, 124
dual, 133, 156, 224
primal, 132, 156, 223
variational inequality
elliptic, 34, 40
evolutionary, 76
time-dependent, 68
velocity field, 111
viscoelastic constitutive law, 96
Kelvin–Voigt, 95
long memory, 97
short memory, 95
viscoplastic constitutive law, 98
viscosity
coefficients, 96
operator, 95, 118
tensor, 95, 98
term, 75, 187
viscous contact, 113
Volterra
integral term, 61
nonlinear equation, 61
operator, 64, 69
von Mises
convex, 100
function, 100

weak convergence, 6
weak divergence, 88, 90
weak limit, 7
weak solution, 124, 127, 137, 147, 148, 160, 167, 176, 178, 181, 183, 223, 227, 236, 241
weakly lower semicontinuous (weakly l.s.c.), 24
Weierstrass theorem, 29

yield limit, 100

rinted in the United States
by Baker & Taylor Publisher Services

Printed in the United States
by Baker & Taylor Publisher Services